# Lecture Notes in Computer Scie

T0237999

Commenced Publication in 1973
Founding and Former Series Editors:
Gerhard Goos, Juris Hartmanis, and Jan van Leeuwei

Peter F. Patel-Schneider   Yue Pan
Pascal Hitzler   Peter Mika
Lei Zhang   Jeff Z. Pan
Ian Horrocks   Birte Glimm (Eds.)

# The Semantic Web – ISWC 2010

9th International Semantic Web Conference, ISWC 2010
Shanghai, China, November 7-11, 2010
Revised Selected Papers, Part II

 Springer

Volume Editors

Peter F. Patel-Schneider
Bell Labs Research, Murray Hill, NJ 07974, USA
E-mail: pfps@research.bell-labs.com

Yue Pan
IBM Research Labs, Beijing 100193, China
E-mail: panyue@cn.ibm.com

Pascal Hitzler
Wright State University, Dayton, OH 45435, USA
E-mail: pascal.hitzler@wright.edu

Peter Mika
Yahoo! Research, 08018 Barcelona, Spain
E-mail: pmika@yahoo-inc.com

Lei Zhang
IBM Research Labs, Shanghai 201203, China
E-mail: lzhangl@cn.ibm.com

Jeff Z. Pan
The University of Aberdeen, Aberdeen, AB24 3UE, UK
E-mail: jeff.z.pan@abdn.ac.uk

Ian Horrocks
University of Oxford, Oxford, OX1 3QD, UK
E-mail: ian.horrocks@comlab.ox.ac.uk

Birte Glimm
University of Oxford, Oxford, OX1 3QD, UK
E-mail: birte.glimm@comlab.ox.ac.uk

The cover photo was taken by Nicolas Rollier (flickr user nrollier).

Library of Congress Control Number: 2010940710

CR Subject Classification (1998): C.2, H.3, D.2, H.4, H.5, H.2.8

LNCS Sublibrary: SL 3 – Information Systems and Application, incl. Internet/Web and HCI

| | |
|---|---|
| ISSN | 0302-9743 |
| ISBN-10 | 3-642-17748-4 Springer Berlin Heidelberg New York |
| ISBN-13 | 978-3-642-17748-4 Springer Berlin Heidelberg New York |

springer.com

© Springer-Verlag Berlin Heidelberg 2010
Printed in Germany

Typesetting: Camera-ready by author, data conversion by Scientific Publishing Services, Chennai, India
Printed on acid-free paper        06/3180

# Preface

The International Semantic Web Conferences (ISWC) constitute the major international venue where the latest research results and technical innovations on all aspects of the Semantic Web are presented. ISWC brings together researchers, practitioners, and users from the areas of artificial intelligence, databases, social networks, distributed computing, Web engineering, information systems, natural language processing, soft computing, and human–computer interaction to discuss the major challenges and proposed solutions, the success stories and failures, as well the visions that can advance research and drive innovation in the Semantic Web.

This volume contains the main proceedings of ISWC 2010, including papers accepted in the Research and Semantic-Web-in-Use Tracks of the conference, as well as long papers accepted in the Doctoral Consortium, and information on the invited talks.

This year the Research Track received 350 abstracts and 228 full papers from around the world. The Program Committee for the track was recruited from researchers in the field, and had world-wide membership. Each submitted paper received at least three reviews as well as a meta-review. The reviewers participated in many spirited discussions concerning their reviews. Authors had the opportunity to submit a rebuttal, leading to further discussions among the reviewers and sometimes to additional reviews. Final decisions were made during a meeting between the Track Chairs and senior Program Committee members. There were 51 papers accepted in the track, a 22% acceptance rate.

The Semantic-Web-in-Use Track, targeted at deployed applications with significant research content, received 66 submissions, and had the same reviewing process as the Research Track, except without the rebuttal phase. There were 18 papers accepted in this track, a 27% acceptance rate.

For the sixth consecutive year, ISWC also had a Doctoral Consortium Track for PhD students within the Semantic Web community, giving them the opportunity not only to present their work but also to discuss in detail their research topics and plans, and to receive extensive feedback from leading scientists in the field, from both academia and industry. Out of 24 submissions, 6 were accepted as long papers, and a further 7 were accepted for short presentations. Each student was assigned a mentor who led the discussions following the presentation of the work, and provided detailed feedback and comments, focusing on the PhD proposal itself and presentation style, as well as on the actual work presented.

The ISWC program also included four invited talks given by leading figures from both the academic and business world. This year talks were given by Li Xiaoming of Peking University, China; mc schraefel of the University of Southampton, UK; Austin Haugen of Facebook; and Evan Sandhaus of the New York Times.

The ISWC conference included the Semantic Web Challenge, as in the past. In the challenge, organized this year by Christian Bizer and Diana Maynard, practitioners and scientists are encouraged to showcase useful and leading-edge applications of Semantic Web technology, either on Semantic Web data in general or on a particular data set containing 3.2 billion triples. ISWC also included a large tutorial and workshop program, organized by Philippe Cudré-Mauroux and Bijan Parsia, with 13 workshops and 8 tutorials spread over two days. ISWC again included a Poster and Demo session, organized by Axel Polleres and Huajun Chen, for presentation of late-breaking work and work in progress, and a series of industry talks.

A conference as complex as ISWC requires the services of a multitude of people. First and foremost, we thank all the members of the Program Committees for the Research Track, the Semantic-Web-In-Use Track, and the Doctorial Consortium. They took considerable time, during summer vacation season for most of them, to read, review, respond to rebuttals, discuss, and re-discuss the submissions. We also thank the people involved in the other portions of the conference, particularly Birte Glimm, the Proceedings Chair; Lin Clark and Yuan Tian, the webmasters; Axel Polleres and Huajun Chen, the Posters and Demos Chairs, and their Program Committee; Yong Yu, the Local Arrangements Chair, Haofen Wang, who managed most aspects of the local arrangements, and Dingyi Han, Gui-Rong Xue and Lei Zhang, the Local Arrangements Committee; Sebastian Rudolph, the Publicity Chair; Jie Bao, the Metadata Chair; Anand Ranganathan and Kendall Clark, the Sponsor Chairs; and Jeff Heflin, the Fellowship Chair.

September 2010

Yue Pan and Peter F. Patel-Schneider
Program Chairs, Research Track Chairs

Pascal Hitzler, Peter Mika, and Lei Zhang
Semantic-Web-In-Use and Industry Track Chairs

Jeff Z. Pan
Doctoral Consortium Chair

Ian Horrocks
Conference Chair

# Conference Organization

## Organizing Committee

### Conference Chair

Ian Horrocks                    University of Oxford, UK

### Program Chairs, Research Track Chairs

Yue Pan                         IBM Research Labs, China
Peter F. Patel-Schneider        Bell Labs, USA

### Semantic-Web-In-Use and Industry Chairs

Pascal Hitzler                  Wright State University, USA
Peter Mika                      Yahoo! Research, Spain
Lei Zhang                       IBM Research Labs, China

### Posters and Demos Chairs

Axel Polleres                   National University of Ireland, Ireland
Huajun Chen                     Shanghai Jiao Tong University, China

### Doctoral Consortium Chair

Jeff Z. Pan                     The University of Aberdeen, UK

### Workshops and Tutorials Chairs

Philippe Cudré-Mauroux          Massachusetts Institute of Technology, USA
Bijan Parsia                    University of Manchester, UK

### Semantic Web Challenge Chairs

Chris Bizer                     Freie Universität Berlin, Germany
Diana Maynard                   University of Sheffield, UK

### Metadata Chair

Jie Bao                         Rensselaer Polytechnic Institute, USA

### Local Organization Chair

Yong Yu                         Shanghai Jiao Tong University, China

### Local Organization Committee

Dingyi Han                      Shanghai Jiao Tong University, China
Gui-Rong Xue                    Shanghai Jiao Tong University, China
Haofen Wang                     Shanghai Jiao Tong University, China
Lei Zhang                       IBM Research Labs, China

**Publicity Chair**

Sebastian Rudolph          Karlsruher Institut für Technologie, Germany

**Webmasters**

Lin Clark                  National University of Ireland, Ireland
Yuan Tian                  Shanghai Jiao Tong University, China

**Proceedings Chair**

Birte Glimm                University of Oxford, UK

**Sponsor Chairs**

Anand Ranganathan          IBM T.J. Watson Research Center, USA
Kendall Clark              Clark & Parsia, LLC, USA

**Fellowship Chair**

Jeff Heflin                Lehigh University, USA

## Senior Program Committee — Research

Hassan Ait-Kaci                    Jeff Heflin
Abraham Bernstein                  Aditya Kalyanpur
Paul Buitelaar                     David Karger
Ciro Cattuto                       Juanzi Li
Vinay Chaudhri                     Li Ma
Bob DuCharme                       Natasha Noy
Michel Dumontier                   Jacco van Ossenbruggen
Tim Finin                          Yuzhong Qu
Asunción Gómez-Pérez               Evren Sirin
Claudio Gutierrez

## Program Committee — Research

Sudhir Agarwal                     Mark Burstein
Harith Alani                       Diego Calvanese
Paul André                         Enhong Chen
Melliyal Annamalai                 Key-Sun Choi
Kemafor Anyanwu                    Philipp Cimiano
Knarig Arabshian                   Lin Clark
Marcelo Arenas                     Oscar Corcho
Jie Bao                            Melanie Courtot
Michael Benedikt                   Isabel Cruz
Chris Bizer                        Claudia d'Amato
Eva Blomqvist                      Mathieu d'Aquin
Kalina Bontcheva                   David De Roure

Mike Dean
Stefan Decker
Ian Dickinson
Xiaoyong Du
Thomas Eiter
Robert H.P. Engels
Achille Fokoue
Enrico Franconi
Zhiqiang Gao
Nikesh Garera
Yolanda Gil
Stefan Gradmann
Michael Gruninger
Volker Haarslev
Harry Halpin
Siegfried Handschuh
Tom Heath
Nicola Henze
Martin Hepp
Nathalie Hernandez
Stijn Heymans
Kaoru Hiramatsu
Rinke Hoekstra
Andreas Hotho
Wei Hu
Zhisheng Huang
Jane Hunter
David Huynh
Eero Hyvönen
Zhi Jin
Lalana Kagal
Anastasios Kementsietsidis
Vladimir Kolovski
Markus Krötzsch
Ora Lassila
Georg Lausen
Faith Lawrence
Shengping Liu
Pankaj Mehra
Jing Mei
Riichiro Mizoguchi
Knud Moeller
Paola Monachesi
William Murray

Wolfgang Nejdl
Yuan Ni
Alexandre Passant
Chintan Patel
Alun Preece
Guilin Qi
Anand Ranganathan
Riccardo Rosati
Sebastian Rudolph
Uli Sattler
Ansgar Scherp
Daniel Schwabe
Yi-Dong Shen
Michael Sintek
Sergej Sizov
Kavitha Srinivas
Steffen Staab
Giorgos Stamou
Robert Stevens
Umberto Straccia
Heiner Stuckenschmidt
Mari Carmen Suárez-Figueroa
V.S. Subrahmanian
Xingzhi Sun
York Sure
Jie Tang
Christopher Thomas
Lieven Trappeniers
Tania Tudorache
Anni-Yasmin Turhan
Octavian Udrea
Michael Uschold
Haixun Wang
Haofen Wang
Fang Wei
Max Wilson
Katy Wolstencroft
Zhe Wu
Bin Xu
Peter Yeh
Yong Yu
Lei Zhang
Ming Zhang
Hai Zhuge

# Program Committee — Semantic-Web-In-Use and Industry

Harith Alani
Sören Auer
Mathieu d'Aquin
Dave Beckett
Chris Bizer
Boyan Brodaric
Vinay Chaudri
Huajun Chen
Gong Cheng
Kendall Clark
John Davies
Leigh Dodds
Michel Dumontier
Aldo Gangemi
Paul Gearon
Mark Greaves
Stephan Grimm
Peter Haase
Michael Hausenblas
Manfred Hauswirth
Ivan Herman
Rinke Hoekstra
David Huynh
Eero Hyvönen

Renato Iannella
Krzysztof Janowicz
Atanas Kiryakov
Markus Krötzsch
Mark Musen
Knud Möller
Chimezie Ogbuji
Daniel Olmedilla
Eric Prud'hommeaux
Yuzhong Qu
Yves Raimond
Marta Sabou
Satya S. Sahoo
Andy Seaborne
Susie Stephens
Hideaki Takeda
Jie Tang
Jamie Taylor
Andraz Tori
Holger Wache
Haofen Wang
Jan Wielemaker
David Wood
Guo-Qiang Zhang

# Program Committee — Doctoral Consortium

Abraham Bernstein
Meghyn Bienvenu
Huajun Chen
Ying Ding
Jianfeng Du
Jérôme Euzenat
Giorgos Flouris
Zhiqiang Gao
Marko Grobelnik
Siegfried Handschuh
Andreas Harth
Stijn Heymans
Wei Hu
Zhisheng Huang
Roman Kontchakov

Diana Maynard
Enrico Motta
Lyndon Nixon
Guilin Qi
Manuel Salvadores
Guus Schreiber
Pavel Shvaiko
Yi-Dong Shen
Amit Sheth
Elena Simperl
Giorgos Stamou
Giorgos Stoilos
Heiner Stuckenschmidt
Vojtech Svatek
Anni-Yasmin Turhan

Denny Vrandecic
Holger Wache
Haofen Wang

Shenghui Wang
Ming Zhang
Yuting Zhao

## External Reviewers

Nor Azlinayati Abdul Manaf
Alessandro Adamou
Mark van Assem
Cosmin Basca
Sujoy Basu
Elena Botoeva
Jos de Bruijn
Carlos Buil-Aranda
Catherina Burghart
Jean Paul Calbimonte
Xiong Chenyan
DongHyun Choi
Alexandros Chortaras
Maria Copeland
Enrico Daga
Brian Davis
Renaud Delbru
Alexander DeLeon
Zhongli Ding
Laura Dragan
Fang Du
Liang Du
Alistair Duke
George Eadon
Jinan El-Hachem
Sean Falconer
Jun Fang
Nicola Fanizzi
Sébastien Ferré
Björn Forcher
Andrés García-Silva
Birte Glimm
Gunnar Aastrand Grimnes
Tudor Groza
Christian Hachenberg
Olaf Hartig
Norman Heino
Daniel Hienert
Aidan Hogan

Thomas Hornung
Matthew Horridge
Julia Hoxha
Gearoid Hynes
Robert Isele
Max Jakob
Martin Junghans
Aditya Kalyanpur
Kamal Kc
Malte Kiesel
Jörg-Uwe Kietz
Eun-Kyung Kim
Yoshinobu Kitamura
Pavel Klinov
Kouji Kozaki
Beate Krause
Thomas Krennwallner
Markus Krötzsch
Maurizio Lenzerini
Paea LePendu
Xuan Li
Yuan-Fang Li
Feiyu Lin
Maxim Lukichev
Sen Luo
Yue Ma
Frederick Maier
Theofilos Mailis
Michael Martin
Philipp Mayr
Anees ul Mehdi
Michael Meier
Pablo Mendes
Eleni Mikroyannidi
Fleur Mougin
Zhi Nie
Mathias Niepert
Nadejda Nikitina
Andriy Nikolov

Vit Novacek
Andrea Nuzzolese
Jasmin Opitz
Magdalena Ortiz
Raul Palma
Rafael Peñaloza
Jorge Pérez
Danh Le Phuoc
Axel Polleres
Freddy Priyatna
Jörg Pührer
Guilin Qi
Timothy Redmond
Yuan Ren
Achim Rettinger
Vinny Reynolds
Ismael Rivera
Mariano Rodriguez-Muro
Dmitry Ryashchentsev
Anne Schlicht
Florian Schmedding
Michael Schmidt
Thomas Schneider
mc schraefel
Floarea Serban
Wei Shen
Rob Shearer
Fuming Shih
Andrey Simanovsky
Mantas Simkus
Evren Sirin
Sebastian Speiser
Giorgos Stoilos
Cosmin Stroe
Mari Carmen Suárez-Figueroa

Kewu Sun
Xiaoping Sun
Martin Szomszor
Christer Thörn
VinhTuan Thai
Christopher Thomas
Despoina Trivela
Eleni Tsalapati
Dmitry Tsarkov
Alexander Ulanov
Natalia Vassilieva
Tasos Venetis
Kunal Verma
Boris Villazón-Terrazas
Denny Vrandecic
Bo Wang
Xiaoyuan Wang
Zhe Wang
Zhichun Wang
Jens Wissmann
Gang Wu
Kejia Wu
Linhao Xu
Yixin Yan
Fangkai Yang
Amapali Zaveri
Benjamin Zapilko
Maciej Zaremba
Lei Zhang
Xiao Zhang
Dmitriy Zheleznyakov
Hai-Tao Zheng
Qian Zhong
Ming Zuo

# Sponsors

| Platinum Sponsors | Gold Sponsors | Silver Sponsors |
| --- | --- | --- |
| AI Journal | fluid Operations AG | IBM |
| Elsevier | LarKC | EMC$^2$ |
| OntoText | SaltLux | W3C |
| | Yahoo! | Amiando |

# Table of Contents – Part II

## Semantic-Web-In-Use Track

## Doctoral Consortium

# Invited Talks

# Table of Contents – Part I

## Research Track

# I18n of Semantic Web Applications

Sören Auer[1], Matthias Weidl[1], Jens Lehmann[1],
Amrapali J. Zaveri[1], and Key-Sun Choi[2]

[1] Universität Leipzig, Institut für Informatik, Johannisgasse 26,
D-04103 Leipzig, Germany
lastname@informatik.uni-leipzig.de
http://aksw.org
[2] KAIST, Semantic Web Research Center
335 Gwahangno Yuseong, Daejeon 305-701, Korea
kschoi@cs.kaist.ac.kr
http://www.kaist.edu

**Abstract.** Recently, the use of semantic technologies has gained quite
some traction. With increased use of these technologies, their maturation not only in terms of performance, robustness but also with regard
to support of non-latin-based languages and regional differences is of
paramount importance. In this paper, we provide a comprehensive review of the current state of the internationalization (I18n) of Semantic
Web technologies. Since resource identifiers play a crucial role for the
Semantic Web, the internatinalization of resource identifiers is of high
importance. It turns out that the prevalent resource identification mechanism on the Semantic Web, i.e. URIs, are not sufficient for an efficient
internationalization of knowledge bases. Fortunately, with IRIs a standard for international resource identifiers is available, but its support
needs much more penetration and homogenization in various semantic
web technology stacks. In addition, we review various RDF serializations
with regard to their support for internationalized knowledge bases. The
paper also contains an in-depth review of popular semantic web tools
and APIs with regard to their support for internationalization.

## 1 Introduction

Recently, the use of semantic technologies has gained quite some traction.
With the growing use of these technologies, their maturation not only in terms
of performance, robustness but also with regard to support of non-latin-based
languages and regional differences is of paramount importance. Internationalization and localization are means of adapting computer software to different
languages and regional differences. *Internationalization* is the process of designing a software application so that it can be adapted to various languages and
regions without engineering changes. *Localization* is the process of adapting internationalized software for a specific region or language by adding locale-specific
components and translating text. For the localization of Semantic Web applications, existing software methodologies (such as GNU gettext for translation or

P.F. Patel-Schneider et al.(Eds.): ISWC 2010, Part II, LNCS 6497, pp. 1–16, 2010.

different locales for region-specific data formating) can be applied. Also, with the *datatype* and *language tags* for RDF literals, there is good support for localization of knowledge bases. With regard to internationalization of Semantic Web technologies the situation, however, is much more challenging as we experienced during the process of internationalizing the DBpedia extraction framework [4] for creating a Korean version of DBpedia.

We noticed in particular, that Asian languages and resources pose a special challenge for Semantic Web and Linked Data applications, tools and technologies. The (non-standard) generation of URIs (for Asian language resources) can have a substantial impact also with regard to classification, interlinking, fusing and information quality assessment. The importance of tackling a proper internationalization of the Semantic Web technology stack is stressed by the fact that Asia has compared to Europe and the USA the largest number of Internet users[1] and many Asian languages are based on fundamentally different linguistic paradigms and scripts. Hence, for the success of individual tools and the Web of Data as a whole it is crucial (a) to incorporate and outreach to user communities beyond the western world and (b) to consider the varying scripting paradigms in order to achieve a wider applicability of the Semantic Web research and development results.

In this paper we want to contribute to a successful internationalization of Semantic Web technologies by summarizing our findings, providing a review of the current state of the internationalization (I18n) of Semantic Web technologies and outlining some best-practices for Semantic Web tool and application developers as well as knowledge engineers.

We are starting to look at the situation with one of the uttermost important building blocks of the Semantic Web – resource identifiers. It turns out that the currently prevalent resource identification mechanism on the Semantic Web, i.e. *URIs*, are not sufficient for an effective internationalization of knowledge bases. Fortunately, with IRIs a standard for international resource identifiers is available, but its support needs much more penetration and homogenization in various semantic web technology stacks. Hence, one goal of this paper is to sensitize the Semantic Web community for the use of IRIs instead of URIs. We review various RDF serializations with regard to their support for internationalized ontologies and knowledge bases. Surprisingly, also here, the currently prevalent serialization technology - RDF/XML - is not adequate for serializing internationalized knowledge bases. The paper also contains an in-depth review of popular Semantic Web tools and APIs with regard to support for I18n.

The paper is structured as follows: We describe the internationalization issues with URIs and possible solutions in the Sections 2. We describe problems with regard to internationalization of the RDF/XML serialization in Section 3. We survey the other available RDF serialization techniques for their compatibility with internationalization in Section 4. We also provide a comprehensive evaluation of internationalization support in popular Semantic Web tools and APIs in Section 5 and conclude in Section 6.

---

[1] http://www.internetworldstats.com/stats3.htm

# 2  What's Wrong with URIs?

Resource identifiers are one of the main building blocks of the Semantic Web. The concept of Universal Resource Identifiers (URIs) is the prevalent mechanism for identifying resources on the Semantic Web. URIs only use US-ASCII characters for names of the resources. However, from the standpoint of an internationalization, URIs are not suitable since characters from non-latin alphabets or special characters have to be encoded in a cumbersome way. The W3C suggests to use percent-encoding in such cases, where a special character is encoded using its two digit hexadecimal value prefixed with the percent character "%". For example, "%20" is the percent-encoding for the US-ASCII space character.

There exist different character encodings for different languages or language families, such as *ISO 8859-1* for Western Europe, *KS X 1001* for the Korean language or *Big-5* for traditional Chinese characters. If these encodings would be used for URIs, conflicts would arise when merging knowledge bases using different encodings for their URIs. For this purpose, the W3C suggested the use of UTF-8 for encoding in URIs[2] as it supports almost every language currently used in the world; it is widely supported, needs one to four bytes to encode the characters and also preserves US-ASCII characters[3]. First the characters not allowed in URIs are encoded according to UTF-8 and then each byte of the sequence is percent-encoded. The Korean word 베를린 (transcription of Berlin), for example, encoded in UTF-8 and percent-encoded looks as follows:

1. byte (베): EB B2 A0
2. byte (를): EB A5 BC
3. byte (린): EB A6 B0

http://ko.wikipedia.org/wiki/%EB%B2%A0%EB%A5%BC%EB%A6%B0

Even though the use of percent-encoding solves the problem of representing special characters, the URIs are (as the above example demonstrates) not easily readable by humans. In software applications it adds additional overhead, since URIs have to be encoded and decoded. It also has to be considered that different parts of the URI (such as the server name, the path and the local name) have different sets of allowed characters or that have to be encoded differently.

*IRIs for the Semantic Web.* To eliminate the disadvantages of URIs, the idea to use UTF-8 without percent-encoding in resource identifiers was raised by Francois Yergeau in 1996[4]. The W3C introduced IRIs (Internationalized Resource Identifier) in 2001 [3]. With the use of IRIs and UTF-8 it is possible to use all characters of the Unicode standard, which covers 107,000 characters and 90 different scripts[5]. With this technology users can easily use, read, alter, and create

---

[2] http://www.w3.org/TR/REC-html40/appendix/notes.html
[3] http://unicode.org/faq/utf_bom.html
[4] http://www.w3.org/International/O-URL-and-ident.html
[5] http://www.unicode.org/standard/principles.html

IRIs in their native language. The above mentioned URI for Berlin in Korean as an IRI would look as follows:

http://ko.wikipedia.org/wiki/베를린

Since non-US-ASCII characters do not have to be encoded, the "%" character does not have to be used in most cases. XML also does support IRIs and UTF-8 but does not allow certain characters in XML tags like %, (, ), or & and some others. But since IRIs can contain all these characters, it can cause problems when serializing RDF in XML as we discuss in detail in the next section. Since most of the knowledge bases on the Semantic Web are currently represented using URIs, they have to be converted to IRIs. The challenge is to figure out whether a percent-encoded sequence was created from a legacy encoding or from UTF-8. But there exists a high chance of heuristic identification of UTF-8 [2] and rare coincidences with byte sequences from legacy encodings as is pointed out in [3]. Furthermore, URIs and IRIs are identical as long as no special characters are used [3].

The main security problem with IRIs is spoofing, because some characters are visually almost indistinguishable. This problem is an extension of those for URIs. However, because UTF-8 contains far more characters the chance for spoofing may increase. One example is the similarity of the Latin "A", the Greek "Alpha", and the Cyrillic "A" [5].

## 3   What's wrong with RDF/XML?

RDF/XML and the RDF embedding in XHTML (i.e. RDFa) are the only RDF serializations officially recommended by the W3C. In essence, RDF/XML is an XML dialect for representing data adhering to the RDF data model. In XML documents, it is possible to use different languages simultaneously even when they use different alphabets. To specify the language for content in the document the *xml:lang* attribute is used. XML markup, however, such as tag and attribute names, is not affected by this attribute. RDF resources are in RDF/XML represented as XML markup, i.e. XML tags and attributes of XML tags. XML tag names, for example, are limited to the following characters [7]:

```
Name ::= NameStartChar (NameChar)*
NameStartChar ::= ":" | [A-Z] | "_" | [a-z] | [#xC0-#xD6] |
[#xD8-#xF6] | [#xF8-#x2FF] | [#x370-#x37D] | [#x37F-#x1FFF] |
[#x200C-#x200D] | [#x2070-#x218F] | [#x2C00-#x2FEF] |
[#x3001-#xD7FF] | [#xF900-#xFDCF] | [#xFDF0-#xFFFD] |
[#x10000-#xEFFFF]
NameChar ::= NameStartChar | "-" | "." | [0-9] | #xB7 |
[#x0300-#x036F] | [#x203F-#x2040]
```

Consequently, some IRIs, which use certain characters of the UTF8 encoding, cannot be used with RDF/XML and there are no simple workarounds or solutions to this dilemma. A solution to this problem would either require a change of the XML standard to allow UTF-8 encoded XML tag names, or a substantial change or extension of RDF/XML, which allows to represent IRIs as XML content.

As we described in Section 2, non US-ASCII characters can be represented in URIs by using the percent-encoding. The "%" character in UTF-8 is defined as #x0025 and not allowed for XML tag names (cf. XML tag name definition above). Thus, a RDF graph with non US-ASCII characters cannot be serialized in RDF/XML. The possible workarounds to this problem are:

1. Use of a different character or sequence of characters instead of the "%" character.
2. Add an underscore to the end of an URI with encoded characters.
3. Use of a different RDF serialization, possibly with IRIs instead of URIs.

*Encoding of the % character.* The solution used by DBpedia for the English DBpedia edition was to replace the "%" character by "_percent_". With this solution the default Wikipedia encoding (i.e. percent-encoding) could be maintained. But this solution produced errors during the DBpedia extraction process for languages with many special characters like Korean, for example. This is due to the fact that after replacing the "%" character with e.g. "_percent_" URIs get very long and although the URI length is not a constraint, some tools (such as the Internet Explorer[6]) have trouble processing very long URIs. This solution is also problematic since RFC3986 states: "URI producers should ... limit these names (URIs) ... to no more than 255 characters in length" [6]. If we would limit URIs to 255 characters, Korean URIs could only encode 7 Korean characters. To increase this number, it would be possible to use a shorter replacement, for example, "_p_" instead of "_percent_" for the "%" character. Thus, a Korean URI could reach up to 17 characters. Such a solution may be sufficient for some applications but renders URIs with non-Latin characters unreadable by humans. Also exchanging data using these URIs with other applications not being aware of the non-standard % encoding (or making already otherwise use of the encoding sequence) will render the encoding irreversible.

*Underscore workaround.* During the search for different solutions it was discovered that adding an underscore at the end of every URI with special characters makes it possible that this URI can be serialized in XML with certain tools e.g. Jena [7] even though this did not adhere to the XML standard as mentioned above. This approach was used for the Korean DBpedia, because it enabled to fully maintain the Wikipedia encoding. Furthermore, this workaround did not require dramatic changes, since only the underscore at the end of an URI had to be considered.

---

[6] http://support.microsoft.com/kb/208427
[7] http://lists.w3.org/Archives/Public/semantic-web/2009Nov/0116.html

*Use of a different RDF serialization.* The last mentioned possibility is to use a different RDF serialization with IRIs instead of URIs. This solution is discussed more in detail in the following Section and different serializations are evaluated wrt. their support of IRIs.

Furthermore, the idea was raised to use XML entities for special characters instead of percent-encoding. First, the special characters have to be encoded according to UTF-8. The encoded character can be represented in two formats: &#nnnn; for the decimal form and &#xhhhh; for the hexadecimal form. Unfortunately, the & and the # character are both reserved characters in URIs and, thus, XML entities cannot be used for replacing percent-encoding.

## 4   Looking at Other RDF Serializations

In addition to RDF/XML, there exist a number of other RDF serialization formats, which partially go beyond the RDF data model in their expressivity and differently accentuate the balance between human readability and simple machine processability. In this section, we assess these different RDF serializations with regard to their support for internationalized resource identifiers. In our comparison, we include JSON, N-Triples, Notation 3, Turtle and RDFa. In order to demonstrate the support for international resource identifiers in the different formats, we showcase the RDF triples from Listing 1 serialized in each of the different formats. These example triples are an exerpt from the Korean DBpedia describing the Korean Advanced Institute of Science and Technology (KAIST).

```
1   http://ko.dbpedia.org/resource/KAIST
2   http://ko.dbpedia.org/property/ 이름
3   "KAIST 한국과학기술원"
4
5   http://ko.dbpedia.org/resource/KAIST
6   http://ko.dbpedia.org/property/ 설립
7   "1971"
8
9   http://ko.dbpedia.org/resource/KAIST
10  http://ko.dbpedia.org/property/ 종류
11  http://ko.dbpedia.org/resource/ 국립대학
```

**Listing 1.** RDF triples of KAIST.

*JSON* JSON (JavaScript Object Notation) was developed for easy data interchange between human beings and applications as well. JSON, allthough carrying JavaScript in its name and being a subset of JavaScript, meanwhile became a language independent format which can be used for exchanging all kinds of data structures and is widely supported in different programming languages. Compared to XML, JSON does require less overhead wrt. parsing and serializing. There is a non-standardized specification[8] for RDF serialization in JSON.

--------

[8] http://n2.talis.com/wiki/RDF_JSON_Specification

Text in JSON and, thus, also RDF resource identifiers are encoded in Unicode and hence can contain IRIs. Also, there is no problem to use the "%" character for URIs in JSON. The only characters that must be escaped are quotation marks, reverse solidus, and the control characters (U+0000 through U+001F)[9]. Thus, JSON can be considered to be an excellent solution for the serialization of internationalized RDF.

```
1   {
2     "http://ko.dbpedia.org/resource/KAIST" : {
3       "http://ko.dbpedia.org/property/ 이름" : [ { "type" : " literal ",
4          "lang" : "ko", "value" : "KAIST 한국과학기술원" } ],
5       "http://ko.dbpedia.org/property/ 설립" : [ { "type" : " literal ",
6          "value" : "1971" } ],
7       "http://ko.dbpedia.org/property/ 종류" : [ { "type" : "uri", "lang" :
8          "ko", "value" : "http://ko.dbpedia.org/resource/ 국립대학" } ]
9     }
10  }
```

**Listing 2.** RDF triples in RDF/JSON.

*N-Triples.* This serialization format was developed specificially for RDF graphs. The goal was to create a serialization format which is very simple. N-Triples are easy to parse and generate by software. They are a subset of *Notation 3* and *Turtle* but lack, for example, shortcuts such as CURIEs. This makes them less readable and more difficult to create manually. Another disadvantage is that N-triples use only the 7-bit US-ASCII character encoding instead of UTF-8. Thus, it does not support IRIs but can handle the "%" character.

```
1  <http://ko.dbpedia.org/resource/KAIST> <http://ko.dbpedia.org/property/%EC
      %9D%B4%EB%A6%84> "KAIST\n\uD55C\uAD6D\uACFC\uD559\uAE30\
      uC220\uC6D0"@ko .
2  <http://ko.dbpedia.org/resource/KAIST> <http://ko.dbpedia.org/property/%EC
      %84%A4%EB%A6%BD> "1971"^^<http://www.w3.org/2001/XMLSchema#
      gYear> .
3  <http://ko.dbpedia.org/resource/KAIST> <http://ko.dbpedia.org/property/%EC%
      A2%85%EB%A5%98> <http://ko.dbpedia.org/resource/%EA%B5%AD%EB%
      A6%BD%EB%8C%80%ED%95%99> .
```

**Listing 3.** RDF triples in N-Triple.

*Notation 3.* N3 (Notation 3) was devised by Tim Berners-Lee and developed just for the purpose of serializing RDF. The main aim was to create a very human-readable serialization. That's why a RDF model serialized in N3 is much more compact than the same model in RDF/XML but still allows a great deal of expressiveness. The encoding for N3 files is UTF-8. Thus, the use of IRIs does not pose a problem. The "%" character can be used at any place that is allowed in RDF.

---

[9] http://www.ietf.org/rfc/rfc4627.txt

```
1   @base <http://ko.dbpedia.org/resource/> .
2   @prefix koprop: <http://ko.dbpedia.org/property/> .
3   @prefix xsd: <http://www.w3.org/2001/XMLSchema#> .
4
5   <KAIST>    koprop: 이름     "KAIST 한국과학기술원"@ko ;
6              koprop: 설립     "1971"^^xsd:gYear ;
7              koprop: 종류     < 국립대학 > .
```

**Listing 4.** RDF triples in Notation 3 (or Turtle).

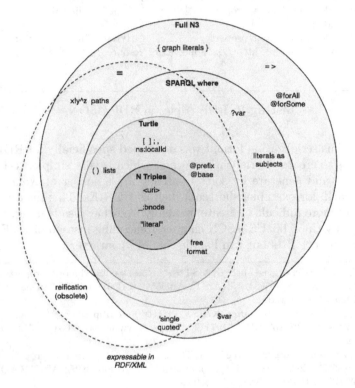

**Fig. 1.** N3 subsets [1]

*Turtle.* Turtle (Terse RDF Triple Language) is a subset of, and compatible with, Notation 3 and a superset of the minimal N-Triples format (cf. Figure 1). The goal was to use the essential parts of Notation 3 for the serialization of RDF models and omit everything else. Turtle became part of the SPARQL query language for expressing graph patterns. Turtle, just like Notation 3, is human-readable, can handle the "%" character in URIs as well as IRIs due to its UTF-8 encoding. Our example in Turtle format is exactly the same as in Notation 3, since Notation 3 specific syntax is required.

**Table 1.** Overview of RDF serialization techniques

| Technique | Percent-encoding | UTF-8/IRI support | Expressivity | Readability | Overhead |
|-----------|------------------|-------------------|--------------|-------------|----------|
| RDF/XML | n | r | + | ◯ | ◯ |
| JSON | s | s | + | + | + |
| N-Triples | s | n | - | + | + |
| Notation 3 | s | s | + | ++ | ++ |
| Turtle | s | s | + | ++ | ++ |
| RDFa | s | s | + | I | ◯ |

++: Very good +: Good ◯ : Moderate -: Poor.
s:supported r: supported with some restrictions n: not supported.

*RDFa.* RDFa[10] (RDF in Attributes) was developed for embedding RDF into XHTML pages. Since it is an extension to the XML based XHTML, UTF-8 and UTF-16 are used for encoding. The "%" character for URIs in triples can be used because RDFa tags are not used for a part of a RDF statement. Thus IRIs are usable, too. Because RDFa is embedded in XHTML, the overhead is bigger compared to other serialization technologies and also reduces the readability.

```
1   <!DOCTYPE html PUBLIC "-//W3C//DTD XHTML+RDFa 1.0//EN" "http://
        www.w3.org/MarkUp/DTD/xhtml-rdfa-1.dtd">
2   <html xmlns="http://www.w3.org/1999/xhtml"
3         xmlns:kores="http://ko.dbpedia.org/resource"
4         xmlns:koprop="http://ko.dbpedia.org/property">
5     <head>
6       <title></title>
7       <meta http-equiv="Content-Type" content="text/html; charset=utf-8"/>
8     </head>
9     <body>
10      <div about="kores:KAIST">
11        <span property="koprop: 이름">KAIST 한국과학기술원</span>
12        <span property="koprop: 설립">1971</span>
13        <span property="koprop: 종류"
14           resource="[kores: 국립대학국립대학]"></span>
15      </div>
16    </body>
17  </html>
```

**Listing 5.** RDF triples in RDFa.

---

[10] http://www.w3.org/2001/sw/wiki/RDFa

# 5   Tool Evaluation

In this section, we review some popular Semantic Web applications and APIs with regard to their support for internationalization. For testing purposes we use the Korean DBpedia edition in two different versions: firstly percent-encoded and secondly with IRIs.

## 5.1   OntoWiki and Erfurt API

OntoWiki[11] is a tool for agile, distributed knowledge engineering scenarios. It follows a Wiki like approach for browsing and authoring of RDF knowledge bases. It offers different views on the stored information and an inline editing mode for RDF data. Social collaboration aspects are added as well. It can be used with relational databases and triple stores and is based on the Erfurt API, an API for developing Semantic Web applications, which is written in PHP. Besides RDF/XML, OntoWiki also supports Turtle, RDF/JSON, and Notation 3 for exporting RDF data.

We tested OntoWiki 0.9 with a MySQL database with UTF-8 encoding and collation for storing the RDF data. OntoWiki supports percent-encoded URIs as well as IRIs. It was possible to load both Korean DBpedia editions in OntoWiki. Unfortunately, when accessing triples with IRIs, the error message "Illegal mix of collations" appeared (but this seems to be fixed in version 0.9.5 alpha). Exporting percent-encoded triples did not cause problems. The serialization of RDF/XML with percent-encoded properties was performed, although this is not allowed as outlined above. When exporting triples with IRIs, the properties only contained question marks, independently of the serialization format used.

## 5.2   Protégé

Protégé[12] is a Java desktop application for developing ontologies and knowledge bases. Protégé 3 supports OWL 1. The first version of OWL does not support IRIs and accordingly Protégé 3 does not support IRIs, either. Protégé 4 supports OWL 2 and IRIs. Protégé serialized IRIs without errors but IRIs were encoded using HTML entities:

```
http://ko.dbpedia.org/Ontology1276166885490.owl#CA_&#50724;
&#49324;&#49688;&#45208;/
```

Working with percent-encoded URIs in Protégé does not pose a problem. But when serializing such URIs in RDF/XML, which should not work, Protégé often serialized the triples without any error. The supported serialization formats of Protégé include Notation 3 and Turtle.

---

[11] http://ontowiki.net/
[12] http://protege.stanford.edu/

## 5.3   Virtuoso Universal Server

Virtuoso Universal Server is a middleware and database engine which contains a triple store to save and query RDF graphs. There exists an open source and a commercial edition of Virtuoso, which vary with regard to support and some functionality (e.g. clustering). Virtuoso handles RDF graphs with percent-encoding in URIs very well. When loading the Korean DBpedia data set, which was serialized in N-Triples format, no problems have been discovered and the data could be easily accessed using the integrated SPARQL endpoint. When uploading data sets with IRIs, however, Virtuoso did not accept all triples. 92 triples of the dataset, consisting of 5.21 million triples, were not processed correctly. The reason is that some triples contain the "$>$" character, which in Notation 3 format represents the end of the subject, predicate, or object, respectively. If this character occurs, the Virtuoso parser probably assumes that the end of the N-Triple representation of the triple has been reached and expects the "." character. We will analyze this issue in more detail, since it leads to a part of the IRI specification, which is problematic in our opinion. The following is an example triple from the Korean DBpedia extraction:

```
<http://ko.dbpedia.org/resource/더 _ 로드>
<http://ko.dbpedia.org/property/wikilink>
<http://ko.dbpedia.org/resource/< 로드> >   .
```

In Listing 6 we look at the ABNF of the IRI specification (RFC 3987) [5] in order to determine whether this kind of IRI is allowed and should be accepted by Virtuoso.

```
1  IRI = scheme ":" ihier−part [ "?" iquery ] [ "#" ifragment ]
2  ihier−part = "//" iauthority ipath−abempty / ipath−absolute
3                / ipath−rootless / ipath−empty
4  iauthority = [ iuserinfo "@" ] ihost [ ":" port ]
5  ipath−abempty = *( "/" isegment )
6  isegment = *ipchar
7  ipchar = iunreserved / pct−encoded / sub−delims / ":" / "@"
```

**Listing 6.** Excerpt of the IRI specification ABNF.

The observed issue is related to the *ihier-part*. The *iauthority* part starts with a double slash ("//") and is terminated by another slash ("/"), a question ("?") mark or number sign ("#"), or the end of the IRI. In RDF *iuserinfo* and *port* are not needed and *ihost specifies* the server address, in this case http://ko.dbpedia.org. The second part of the *ihier-part* is *ipath-abempty* which is formed by a slash ("/") followed by *isegment*: The first time *ipath-abempty* is used for "resource" and the second time for "/< 로드 >". The first part ("resource") only uses US-ASCII characters and, thus, is correct. *ipchar* are formed as follows: *iunreserved* contains all unreserved characters from URIs (see Table 2) as well as *ucschars*:

**Table 2.** Unreserved characters in URIs [6]

| A | B | C | D | E | F | G | H | I | J | K | L | M | N | O | P | Q | R | S | T | U | V | W | X | Y | Z |
|---|---|---|---|---|---|---|---|---|---|---|---|---|---|---|---|---|---|---|---|---|---|---|---|---|---|
| a | b | c | d | e | f | g | h | i | j | k | l | m | n | o | p | q | r | s | t | u | v | w | x | y | z |
| 0 | 1 | 2 | 3 | 4 | 5 | 6 | 7 | 8 | 9 | - | _ | . | ~ |   |   |   |   |   |   |   |   |   |   |   |   |

```
ucschar = %xA0-D7FF / %xF900-FDCF / %xFDF0-FFEF
        / %x10000-1FFFD / %x20000-2FFFD / %x30000-3FFFD
        / %x40000-4FFFD / %x50000-5FFFD / %x60000-6FFFD
        / %x70000-7FFFD / %x80000-8FFFD / %x90000-9FFFD
        / %xA0000-AFFFD / %xB0000-BFFFD / %xC0000-CFFFD
        / %xD0000-DFFFD / %xE1000-EFFFD
```

*pct-encoded* refers to percent encoding ("%" HEXDIG HEXDIG) and *sub-delims* are a part of the reserved characters of URIs (see Table 3).

The *isegment* part of the object of our example IRI uses Korean characters (로 드), which are allowed in IRIs (these are contained in *ucschar*). The characters "<" and ">" that caused the problems (003C and 003E) are not included in the unreserved characters but are also not mentioned in the reserved characters for IRIs. Furthermore, RFC 3987 [5] states "Systems accepting IRIs MAY also deal with the printable characters in US-ASCII that are not allowed in URIs, namely "<", ">", ' " ', space, "{", "}", "|", "\", "^", and ' " ' ". In our opinion, this part of the specification is problematic, because it allows different tools and APIs to handle such IRIs differently, since it is not defined exactly how to handle these characters. This conflicts with the goals of Semantic Web technologies, which aim at a clear and unambiguous transfer of knowledge between different systems.

**Table 3.** sub-delim characters in IRIs [5]

Due to these reasons and because "<" and ">" are used for the start and end of the subject, predicate, or object in Notation 3, these characters should be percent-encoded to avoid misunderstandings. In this case Virtuoso accepted all triples. Thus, Virtuoso appears to be a very good solution for storing RDF graphs with URIs as well as IRIs.

### 5.4 Jena

Jena[13] is an open-source Java framework for developing Semantic Web applications. It includes a RDF API, an OWL API, a SPARQL query engine, an inference engine, and it can read and write RDF in RDF/XML, Notation 3 and N-Triples.

---

[13] http://jena.sourceforge.net/

| http://ko.dbpedia.org/resource/ 도동_분기점/고속도로_나 들목3 | http://ko.dbpedia.org/property/형태 | IC X.svg |
|---|---|---|
| http://ko.dbpedia.org/resource/ 도리데_역/좌표/coor_dms | http://ko.dbpedia.org/property/wikiPageUsesTemplate | http://ko.dbpedia.org/resource/Template:coor_dms |
| http://ko.dbpedia.org/resource/ 도리데_역/좌표/coor_dms | http://ko.dbpedia.org/property/coorDmsProperty | E |
| http://ko.dbpedia.org/resource/ 도리데_역/좌표/coor_dms | http://ko.dbpedia.org/property/coorDmsProperty | N |
| http://ko.dbpedia.org/resource/ 도리데_역/좌표/coor_dms | http://ko.dbpedia.org/property/coorDmsProperty | 3 |
| http://ko.dbpedia.org/resource/ 도리데_역/좌표/coor_dms | http://ko.dbpedia.org/property/coorDmsProperty | 35 |
| http://ko.dbpedia.org/resource/ 도리데_역/좌표/coor_dms | http://ko.dbpedia.org/property/coorDmsProperty | 47.2164 |
| http://ko.dbpedia.org/resource/ 도리데_역/좌표/coor_dms | http://ko.dbpedia.org/property/coorDmsProperty | 47.2428 |
| http://ko.dbpedia.org/resource/ 도리데_역/좌표/coor_dms | http://ko.dbpedia.org/property/coorDmsProperty | 53 |
| http://ko.dbpedia.org/resource/ 도리데_역/좌표/coor_dms | http://ko.dbpedia.org/property/coorDmsProperty | 140 |
| http://ko.dbpedia.org/resource/ 도리데_역/인접정차역1 | http://ko.dbpedia.org/property/wikiPageUsesTemplate | http://ko.dbpedia.org/resource/Template:인접정차역 |

**Fig. 2.** Virtuoso with the Korean DBpedia data set and IRIs

When testing the Korean DBpedia with the Jena framework and the output format RDF/XML, Jena stopped and threw an `InvalidPropertyURIException`. This behaviour was expected, because percent-encoded URIs cannot be serialized in XML. When adding an underscore to properties with such URIs, Jena magically could serialize the triples. In this workaround the namespace is (due to a "misimplementation" of the URI segmentation algorithm in Jena) extended with a part of the property name. Only the underscore is written in the start-tag and end-tag. Thus, the tags do not contain a percent character and the XML file becomes valid. An example of Jena's RDF/XML serialization of triples with the underscore appended to URIs is shown in Listing 7.

```
1  <?xml version="1.0"?>
2  <rdf:RDF
3      xmlns:rdf="http://www.w3.org/1999/02/22-rdf-syntax-ns#"
4      xmlns:j.0="http://ko.dbpedia.org/property/%EC%A0%9C%EB%AA%A9">
5    <rdf:Description rdf:about="http://ko.dbpedia.org/resource/Rock_U_%281
          st_Mini_Album%29">
6      <j.0:_ xml:lang="ko">Good Day</j.0:_>
7    </rdf:Description>
8  </rdf:RDF>
```

**Listing 7.** Korean DBpedia in RDF/XML by Jena framework.

The Jena framework handles IRIs generally well. However, there exist some triples which were not accepted, such as the following:

`<http://ko.dbpedia.org/resource/2006 년 _ 태풍/태풍 _ 정보 _(소)1>`

```
<http://ko.dbpedia.org/property/최대강풍반경 (kma)>
"550"^^<http://www.w3.org/2001/XMLSchema#integer> .
```

This triple was not accepted, because of the "(" and ")" in the property part. During the test with Jena no triples with such brackets have been accepted. This could be caused by a code migration problem, because usually brackets are not allowed at this position in URIs. However, as discussed in Section 5.3, in IRIs additional characters can be used at this position. These characters, which are reserved in a URI but allowed in an IRI, are summarised in Table 3.

## 5.5  Sesame

Sesame[14] is an open-source triple store implemented in Java. It also supports RDFS inference.

Sesame did fine in the most tests. There were no errors during the tests with percent-encoded triples. When loading the Korean DBpedia data set with IRIs we observed an anomaly when processing IRIs containing the bracket "{" character. This is not a reserved character in IRIs but also not part of the unreserved characters. This issue has been discussed in Section 5.3 which points out that the specification does not define exactly how to handle these characters.

**Table 4.** Reserved characters in URIs [6]

| ! | * | ' | ( | ) | ; | : | @ | & | = | + | $ | , | / | ? | # | [ | ] |
|---|---|---|---|---|---|---|---|---|---|---|---|---|---|---|---|---|---|

## 5.6  OWL API

The OWL API is an open-source API written in Java for creating, manipulating and serializing OWL ontologies. It includes a parser and writer for RDF/XML, OWL/XML, and Turtle[15]. Since version 3.0, the OWL API uses the OWL 2 specification. This is a requirement for using IRIs in OWL.

Percent-encoded triples were used for the first test. As expected, the OWL API did not accept these triples resulting in the error message "Illegal Element Name (Element Is Not A QName)". When adding an underscore at problematic triples, the OWL API could serialize the triples. It uses the same strategy as Jena, i.e. usage of an extra namespace for such properties.

OWL API also supports Turtle as output format. When using Turtle, serializing of percent-encoded triples works well. The OWL API failed to load RDF triples with IRIs when non-ASCII character occurred in a resource or property (with exeption "org.xml.sax.SAXParseException: Element or attribute do not match, QName production: QName::=(NCName':')?NCName."). It seems that OWL API was not able to find an appropriate QName even though such a

---

[14] http://www.openrdf.org/
[15] http://owlapi.sourceforge.net/index.html

QName does exist and only allowed characters have been used. However, the OWL API was able to serialize triples into XML with IRIs. It uses non-ASCII characters in tags as intended by the XML specification. Unfortunately, XML entities where used in XML tag attributes, as the following example shows:

```
1  <?xml version="1.0"?>
2  ...
3  <owl:NamedIndividual rdf:about="http://ko.dbpedia.org/#M2_
      (&#52380;&#52404;)">
4     <rdf:type rdf:resource="http://ko.dbpedia.org/#&#51060;&#47492;"/>
5     <ko: 별자리 rdf:resource="http://ko.dbpedia.org
             /#&#47932;&#48337;&#51088;&#47532;"/>
6  </owl:NamedIndividual>
7  ...
```

**Listing 8.** Korean DBpedia in RDF/XML with IRIs by the OWL API.

**Table 5.** Overview of I18n support in popular tools and APIs

| Tools | Percent-encoding | Underscore workaround | UTF-8/IRI support | Problematic Characters with IRIs | Output formats |
|-------|-----------------|----------------------|-------------------|----------------------------------|----------------|
| OntoWiki 0.9 | + | - | ◯ | | 1, 2, 3, 5 |
| Protégé 4.1 | + | - | ◯ | | 1, 2, 3 |
| Jena 2.6.2 | + | + | ◯ | ( | 1, 2, 4 |
| Virtuoso 6 | + | - | + | > | 1, 4, 5, 6 |
| Sesame 2.3.1 | + | + | + | { | 1, 2, 3, 4, 6 |
| OWL API 3 | + | + | ◯ | | 1, 3 |

+: Available; ◯ : Available with some restrictions; -: Not available.
1: RDF/XML 2: Notation 3: Turtle 4: N-Triples 5: RDF/JSON 6: HTML.

# 6   Conclusions

With the maturing of Semantic Web technologies proper support for internationalization is a crucial issue. This particularly involves the internationalization of resource identifiers, RDF serializations and corresponding tool support. As it was, for example, noted by Richard Cyganiak "the relationships between URIs, IRIs, RDF, XML, UTF-8 etc are incredibly complex"[16]. With this work we aimed at shedding some light on this intricate matter and providing some guidelines for knowledge engineers and tool developers alike. It turned out, that there are serious issues with the two most prevalent building blocks of the Semantic Web,

---

[16] http://lists.w3.org/Archives/Public/semantic-web/2009Nov/0122.html

namely URIs and the RDF/XML serialization. With IRIs, some workarounds for RDF/XML or the use of alternative serialization techniques these issues, however, can be circumnavigated. Unfortunately, the semantic web tools landscape is very diverse with regard to support for IRIs, RDF/XML workarounds and alternative serializations. As a result of our evaluation of some prominent tools, it turns out, that an internationalization is possible, when one limits oneself to IRIs, which do not contain a relatively small set of characters commonly causing problems with certain tools or RDF serializations. For the future, it would be desirable, if all RDF serializations could cope with IRIs in a natural way and the tool support would be more homogeneous. However, besides some engineering effort, this will also require some modifications to better align the different standards and specifications.

## Acknowledgments

We would like to thank the members of Semantic Web Research Center at KAIST as well as the anonymous reviewers for their helpful comments on earlier versions on this document. This work was partially supported by a grant from the European Union's 7th Framework Programme provided for the project LOD2 (GA no. 257943).

## References

1. Berners-Lee, T.: Notation 3 (1998),
   http://www.w3.org/DesignIssues/Notation3.html
2. Dürst, M.J.: The properties and promises of UTF-8. In: Unicode Consortium (ed.) 11th International Unicode Conference, The Unicode Consortium (1997),
   http://www.ifi.unizh.ch/mml/mduerst/papers/PDF/IUC11-UTF-8.pdf
3. Dürst, M.J.: Internationalized resource identifiers: From specification to testing. In: Proc. of the 19th Internationalization and Unicode Conference, San Jose, California (September 2001)
4. Lehmann, J., Bizer, C., Kobilarov, G., Auer, S., Becker, C., Cyganiak, R., Hellmann, S.: Dbpedia -a crystallization point for the web of data. Journal of Web Semantics (3), 154–165 (2009)
5. Duerst, M., Suignard, M.: Internationalized Resource Identifiers (IRIs). Network Working Group (2005), ftp://ftp.rfc-editor.org/in-notes/rfc3987.txt
6. Berners-Lee, T., Fielding, R., Masinter, L.: Uniform Resource Identifier (URI): Generic Syntax. Network Working Group (2005),
   ftp://ftp.rfc-editor.org/in-notes/rfc3986.txt
7. Bray, T., Paoli, J., Sperberg-McQueen, C.M., Maler, E., Yergeau, F.: Extensible markup language (XML) 1.0, 5th edn. (2008), http://www.w3.org/TR/REC-xml/

# Social Dynamics in Conferences: Analyses of Data from the Live Social Semantics Application

Alain Barrat[1,2], Ciro Cattuto[2], Martin Szomszor[3], Wouter Van den Broeck[2], and Harith Alani[4]

[1] Centre de Physique Théorique (CNRS UMR 6207), Marseille, France
alain.barrat@cpt.univ-mrs.fr
[2] Complex Networks and Systems Group
Institute for Scientific Interchange (ISI) Foundation, Turin, Italy
ciro.cattuto@isi.it, wouter.vandenbroeck@isi.it
[3] City eHealth Research Centre, City University, UK
[4] Knowledge Media institute, Open University, UK
h.alani@open.ac.uk

**Abstract.** Popularity and spread of online social networking in recent years has given a great momentum to the study of dynamics and patterns of social interactions. However, these studies have often been confined to the online world, neglecting its interdependencies with the offline world. This is mainly due to the lack of real data that spans across this divide. The Live Social Semantics application is a novel platform that dissolves this divide, by collecting and integrating data about people from (a) their online social networks and tagging activities from popular social networking sites, (b) their publications and co-authorship networks from semantic repositories, and (c) their real-world face-to-face contacts with other attendees collected via a network of wearable active sensors. This paper investigates the data collected by this application during its deployment at three major conferences, where it was used by more than 400 people. Our analyses show the robustness of the patterns of contacts at various conferences, and the influence of various personal properties (e.g. seniority, conference attendance) on social networking patterns.

## 1 Introduction

Participation in online social networking has been growing at an unprecedented speed, with sites such as Facebook logging more than 400 million active users in only a few years since its birth. This new online phenomena is arming today's researchers in many disciplines with very rich and rapidly evolving social environment which is proving invaluable for the study and analyses of social dynamics, collective behaviour, community formation, etc.

Nevertheless, in spite of the surge in investigations of online social networks, these studies have largely overlooked the association of these networks with each other, and with the offline, real-world networks. Social networks, in all their shapes and forms, often reflect each other in a variety of ways. The lack of comparative analyses of such heterogenous networks is mainly due to the shortage of data that spans this online-offline

P.F. Patel-Schneider et al.(Eds.): ISWC 2010, Part II, LNCS 6497, pp. 17–33, 2010.
© Springer-Verlag Berlin Heidelberg 2010

divide. Additionally, to better inform the analyses of real-world face-to-face (F2F) contact networks, researchers need to take into account the already-existing social relationships between users. Existing relationships can have a high impact on the shape, dynamics, and strengths of interaction between the subjects. Such multi-relation analyses (called multiplexity in social networks [6,21]) remains underinvestigated [13,17].

We have designed, developed, and deployed an application that bridges the divides between offline and online, and between real-time and historical social networks and relationships. This is achieved by integrating various heterogeneous and distributed networks. More specifically, Live Social Semantics (LSS) collects and integrates data about people from (a) their online social networks and tagging activities from popular social networking sites such as Facebook, Twitter, Flickr, and Last.fm (b) their publications and co-authorship networks from semantic repositories of publications, such as data.semanticweb.org and rkbexplorer.org, and (c) their real-world face-to-face (F2F) proximity, considered as a proxy for a social interaction, recorded with a network of wearable sensors (sociopatterns.org). To the best of our knowledge, this is the first time that an application is deployed that is capable of gathering and integrating this type of data.

## 1.1 Main Contribution

The main contribution of this paper is the investigation of the data collected by the LSS application during its deployment at three major conferences (section 4), where it was used by more than 400 people. We analyse the data for contacts patterns and the impact of parameters such as seniority and role on these patterns. We also compare the networks from online social networking sites with those generated from real F2F contacts, as well as with co-authorship networks. More specifically, we investigate the following:

- **Face-to-face interactions in scientific conferences** (section 5.1): We start our analyses by looking for common statistical characteristics in the F2F interaction networks we collected from three scientific conferences. We focus our attention to F2F contacts frequency and duration and how they compare across all three conferences.
- **Networking behaviour of frequent users** (sections 5.2). Here we focus our analysis on users who participated in two LSS deployments (section 4). We measure the networking behaviour of these users quantitatively and qualitatively, and across conferences, in comparison to the behavior of other one-time users of LSS.
- **Scientific seniority of users** (section 5.3). This analysis aims to study the impact of seniority on social activity. Seniority is approximated from (a) number or publications, (b) h-index, and (c) organisational roles at the conference where LSS was deployed. In this analysis we search for correlations between seniority of users and the seniority of their F2F contacts, as well as the general strength of their social network. We also compare scientific seniority of users to the number of their Twitter followers.
- **Comparison of F2F contact network with Twitter and Facebook** (section 5.4): We compare the size of F2F network of users to the size of their Facebook and/or

Twitter social network. The idea is to see if there is a clear correlation between the two parameters; i.e whether people with strong online social presence are also very active in F2F networking and vice versa.

– **Social networking with online and offline friends** (section 5.5): We analyse F2F contact networks while taking into account any co-authorship relationships between users, which we obtain from data.semanticweb.org, and any online social relationship, taken from Facebook or Twitter.

The purpose of our analyses is to provide novel insights into the comparability of online and offline networks, and to better understand the impact of specific drivers and parameters on the social contact behaviour of individuals and groups in scientific communities and gatherings. Such knowledge can feed into the design of better tools for supporting networking at conferences and at similar events. It can also be used for the identification of future scientific leaders and event organisers.

In the following section we describe some work related to monitoring live social interactions and to online network analysis. In section 3 we briefly describe the application and its main components, and then summarise the outcome of its deployments so far in section 4. Section 5 details the analyses we applied to our data collection and the main results obtained. Discussion and future work related to LSS and to our results is given in section 6, followed by conclusions in section 7.

## 2   Related Work

Using sensor devices for detecting contacts at conferences is not a novel idea by itself. IBM used RFIDs to track sessions and meal attendance at a conference [20]. Bluetooth-enabled mobiles were also used to track networking of conference attendees [9] and for sensing organisational aspects [5]. Networks from blutoothed mobiles were also studied for characterising some statistical properties of human mobility and contact [16]. Wu and colleagues used what they call "sociometric badges" to investigate impact of F2F interactions on productivity [22]. These badges used radio frequency to detect physical proximity, infra red to detect F2F body alignments, and voice sensors to detect conversations. All these works focus on only one type of network which is based on proximity of users, irrespective of whether these users interacted with each other (e.g. had a F2F contact) or were already closely linked in other social contexts. Nishimura and colleagues used passive RFIDs to monitor and support conference communities [15].

Recently, the SocioPatterns project (http://sociopatterns.org) developed an RFID platform that is scalable and attains reliable detection of F2F interactions [2,3]. They used this platform to investigate patterns of human contacts at various social gatherings [10]. The LSS application presented here leverage that platform to mine real-time social contacts. To the best of our knowledge, the Live Social Semantics application is the first where real-world F2F contacts are mashed up in real time with semantic data from on-line social networking systems.

Social scientists identify several parameters that influence and motivate social and communication networks, such as physical and digital proximity, social support and community belonging, and homophily; similarity of individuals [14]. Such parameters were the focus of many works on characterising social networks (e.g. [8,12,11]). However, such works are often limited to online social networks.

The novelty of the analyses we present in this paper resides in the integration of heterogeneous data sources for the analysis of social networks.

## 3   Live Social Semantics Application

The Live Social Semantics (LSS) [1,19] is an innovative application that tracks and supports social networking between conference attendees. The application integrates data and technologies from the Semantic Web, online social networks, and a F2F contact sensing platform. It helps researchers to find like-minded and influential researchers, to identify and meet people in their community of practice, and to capture and later retrace their real-world networking activities at conferences.

LSS integrates (a) the available wealth of linked semantic data, (b) the rich social data from existing major social networking systems, and (c) a physical-presence awareness infrastructure based on active radio-frequency identification (RFID).

Figure 1 shows the main components of LSS. At the center of this architecture is a 4Store[1] triplestore for storing, integrating, and accessing the heterogenous data collected by LSS from various distributed resources. LSS gathers tagging and social networking information on registered users from selected sites (component 2 in Figure 1). This tagging data is then used by component 3 for building semantic user profiles, which applies a series of services (component 4) for filtering tags [18], disambiguating tags [7], and associating them with semantics from DBpedia [19].

In [19], we focused on describing and evaluating the generation of semantic profiles of interest from the tags shared by LSS users on Delicious and Flickr. The evaluation demonstrated the relative high accuracy of 85% achieved by our fully automatic tag-to-URI association algorithm which maps every tag to a DBpedia URI.

Information on user's publications and co-authorship networks are collected from data.semanticweb.org and rkbexplorer.com. Co-authorship networks represent another type of social networking that LSS integrates with the networks it collects from online social networking sites and from F2F contact networks.

Real-world F2F interactions of conference attendees are mined using RFID hardware and software infrastructure developed by the SocioPatterns project [2,3]. The RFID platform is represented by component 6 in the architecture, and it is responsible for collecting and processing readings from active wearable RFIDs carried by the conference attendees who participated in using the LSS application. This information is periodically uploaded to the triple store via RDF/HTTP and integrated with the other data layers. Details of using RFID in LSS can be found in [4].

## 4   Data from LSS Deployments

Live Social Semantics was deployed at three conferences. Below are some statistics on participation in each of these deployments.

---

[1] http://4store.org/

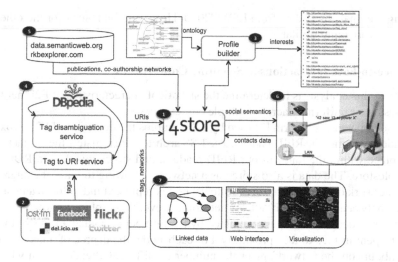

**Fig. 1.** General architecture of LSS application. A triple store is the central point of integration for all LSS components.

- **ESWC 2009:** The first deployment of LSS was at the European Semantic Web Conference (ESWC) in Crete, 1-4 June 2009. This conference was attended by 305 people, out of which 187 participated in LSS. Out of the 187 who collected an RFID badge, 139 of them also created accounts on our LSS application website. LSS participants in this conference were allowed to declare their Facebook, Delicious, Flickr, and LastFm accounts. Results of this deployment are fully described in [1].
- **HT 2009:** HyperText (HT), Turin, June 29-July 1, 2009: Attended by around 150 people. 113 of them collected an RFID, and 97 registered with LSS. Full description of these results can be found in [19].
- **ESWC 2010:** Extended Semantic Web Conference (ESWC) in Crete, May 31-June 3, 2010. There were around 315 attendees at this conference. 175 people collected an RFID, and 132 of them registered on the LSS site.

## 5    Data Analyses and Results

Understanding the correlations between the characteristics of users who are linked in a social network is a long-standing problem in social sciences, ecology and epidemiology: a typical pattern, referred to as "assortative mixing", describes the tendency of nodes of a network (here, the users), to link to other nodes with similar properties.

In this section we describe a variety of analyses that we applied to the data we gathered from LSS deployments. In this analyses we take several parameters into account, such as degree and strength of F2F networks, size of online social networks, co-authorship relations, conference chairing roles, and scientific seniority of users. These analyses are aimed at identifying patterns, or testing and verifying various conceptions, on how people connect socially at conferences.

We start by showing the high similarity of the social networks we obtained from all three deployments. Therefore, to save space, we sometimes only report the results of applying our analysis to data from ESWC2010. However, the results for the other two conferences are quite similar.

### 5.1    Face-to-Face Interactions in Scientific Conferences

The aim of this analysis is to determine the statistical characteristics of F2F networks, and assess their uniformity across multiple conferences.

The Sociopatterns platform [3] used by LSS enables the detection of F2F proximity of attendees wearing the RFID badges. The LSS architecture registers the contact events taking place within the range of our RFID readers, and stores this data in RDF in the LSS triplestore. The data is also stored as a network, allowing to build the aggregated contact network of the conference as follows: nodes represent individuals, and an edge is drawn between two nodes if at least one contact event took place between the corresponding attendees. Each edge is weighted by the number of contact events or the total duration spent in face to face proximity. For each node, the degree of a node (number of neighbours on the network) gives the number of different attendees with whom the user has been in contact, and the "strength" (sum of the weights of the links) is defined by the total time this person spent in F2F interaction with other attendees.

Tables 1 and 2 give the main characteristics of the observed behavior of the participants in the three LSS deployments. The data show a very high level of uniformity across the three conferences.

More in detail, Table 1 shows that most contacts are very short, but that some very long contacts are also measured. In fact, the distributions of contact durations are broad,

**Table 1.** Some characteristics of the contact events between LSS participants during three conferences. The F2F contact pattern is very similar for all three conferences.

| Contact characteristics | ESWC 2009 | HT 2009 | ESWC 2010 |
|---|---|---|---|
| Number of contact events | 16258 | 9875 | 14671 |
| Average contact length (s) | 46 | 42 | 42 |
| Fraction of contacts ≤ 1 mn | 0.87 | 0.89 | 0.88 |
| Fraction of contacts ≤ 2 mn | 0.94 | 0.96 | 0.95 |
| Fraction of contacts ≤ 5 mn | 0.99 | 0.99 | 0.99 |
| Fraction of contacts ≤ 10 mn | 0.998 | 0.998 | 0.998 |

**Table 2.** Some characteristics of the aggregated network of contacts between participants. The degree of a user is the number of other users with whom s/he has had at least one contact. The weight of an edge between two users is given by the total time they have spent in F2F interaction.

| Network characteristics | ESWC 2009 | HT 2009 | ESWC 2010 |
|---|---|---|---|
| Number of users | 175 | 113 | 158 |
| Average degree | 54 | 39 | 55 |
| Average strength (s) | 8590 | 7374 | 7807 |
| Average edge weight (s) | 159 | 189 | 141 |
| Fraction of weights ≤ 1 mn | 0.7 | 0.67 | 0.74 |
| Fraction of weights ≤ 5 mn | 0.9 | 0.89 | 0.93 |
| Fraction of weights ≤ 10 mn | 0.95 | 0.94 | 0.96 |

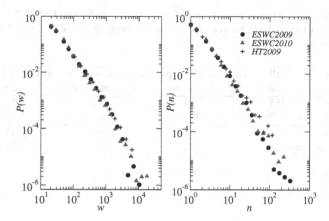

**Fig. 2.** Probability distribution of (left) the total time spent in F2F interaction, and (right) the number of contact events, between two participants to the LSS deployments. The X-axis is the total time (left) and the number (right) of contact events, and the Y-axis gives the probability to observe such a value.

as also observed in other settings [3,10]. Figure 2 shows that the distributions of total time spent in F2F interaction by two attendees, and of the number of contact events between two attendees, are also broad, and are very similar in all three conferences.

Interestingly, the **observed general behaviours across conferences are remarkably similar**, both qualitatively and quantitatively, from the point of view of the contact durations and for what regards cumulated contacts between participants: The average contact durations and total time spent F2F by two individuals are very close,[2] and in fact, the whole statistical distributions can be superimposed, as shown in Figure 2.

## 5.2   Face-to-Face Networking Behaviour of Frequent Attendees

The successive deployment of the LSS architecture at ESWC in 2009 and 2010 enables not only the comparison of the overall attendees behavior, as shown in the previous paragraphs, but also to focus on the persons who attended both deployments. These common participants turn out to be 33. It is thus interesting to investigate their characteristics, in order to understand if these participants are in some aspects different from the others.

Table 3 compares the main characteristics of the contacts between returning participants with the overall average characteristics. It highlights how the attendees who participated in LSS in both ESWC 2009 and ESWC 2010 conferences were much more active, in terms of F2F interactions, than those who used LSS only once.

We observe that the average number of distinct participants with whom returning attendees have interacted is larger. The total time spent in F2F proximity with other attendees (strength) is close to twice the interaction time averaged over all participants.

---

[2] Note that since the distributions are broad, the precise value of the averages is rather sensitive to rare events in the distribution tail.

**Table 3.** Average characteristics in each year of the participants to both ESWC 2009 and ESWC 2010, and of the contact patterns between these returning participants, as compared to the average over all participants

| Characteristics | all participants, 2009 | all participants, 2010 | common partici-pants, 2009 | common partici-pants, 2010 |
|---|---|---|---|---|
| Average degree | 55 | 54 | 73 | 62 |
| Average strength | 8590 | 7807 | 16426 | 13216 |
| Average weight | 159 | 141 | 416 | 404 |
| Average contact duration in seconds | 46 | 42 | 52 | 57 |
| Average number of contact events per edge | 3.44 | 3.37 | 8 | 7 |

This feature can be investigated in more details by measuring the average weight of a link between a returning attendee and any of his/her neighbours, or also between two returning attendees. We find that returning attendees have a larger average interaction time (212 seconds, against a global average of 141 seconds) and interact more frequently with their neighbours (4.3 contact event per edge, against 3.44 overall).

When focusing the analyses to only those interactions that took place *between* the returning attendees, Table 3 shows an even stronger effect, with an average total duration of interaction (link weight) of about 400 seconds. Interestingly, this strong difference in total interaction time comes mostly from a much larger number of contact events, while the average duration of a single contact event is only slightly larger for returning attendees. **Overall, returning attendees interact more frequently and longer than average, especially among each other**.

**Stability of F2F interactions across conferences.** In section 5.1 we showed that the general statistical patterns of F2F networking are very similar across all three conferences where LSS was deployed. Then in section 5.2 we showed that frequent users have stronger F2F networks. Another interesting question is whether these common users show similar social-networking behaviour from one year to the next.

To this end, we study the correlation between the properties of individuals and of their links in the interaction network in 2009 and 2010. More precisely, we plot in Figure 3 for each individual the number of neighbours in 2010 versus the number of neighbours in 2009 (top). We also plot the total time spent in F2F interaction in 2010, versus the same quantity measured at ESWC2009 (middle plot). For the links observed in both 2009 and 2010, we also plot the weight in 2010 versus the weight in 2009 (bottom plot). The plots show a clear correlation pattern. More quantitatively, the Pearson correlation coefficients [3] are 0.37 for the degrees, 0.76 for the total time spent in interaction, and 0.75 for the link weights. What this implies is that **although people interacted with a**

---

[3] The Pearson correlation coefficient between two variables is defined as the covariance of the two variables divided by the product of their standard deviations. It measures the correlation (linear dependence) between these two variables, and is comprised between $-1$ when the variables are perfectly anti-correlated and $+1$ when they are perfectly correlated.

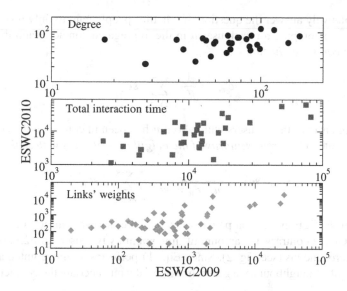

**Fig. 3.** Correlation between the characteristics of a returning attendee in 2009 and 2010. The X-axis gives the value of the charateristics of an attendee measured in 2009, and the Y-axis gives the value of the same characteristics measured in 2010. Black circles: degree, giving the number of other participants with whom an attendee has interacted. Red squares: total time spent in interaction by an attendee. Green diamonds: total time spent in F2F interaction by a pair of returning attendees in 2010, versus the same quantity measured in 2009.

**different set of people during these two conferences, the time they spent in these interactions was very similar**.

Since people's social behaviour seems to remain rather stable from one conference to the next, we can assume that they will show homogenous behaviour at other similar conferences, and that the typical changes in conference programs and events have little impact on the behaviour of attendees.

### 5.3   Scientific Seniority and F2F Network Patterns

One interesting parameter to investigate in conference F2F networks is the scientific seniority of people. This section investigates this parameter and its influence on F2F networking dynamics.

We consider two different ways to quantify the scientific seniority $se_u$ of a user $u$: (i) the number of papers authored by an individual at semantic web related conferences,[4] and (ii) the h-index.[5] While the publication and citation patterns change from one community to the next, we are here dealing with scientists coming from the same community, so that these quantities are reasonable indicators of how senior a person is.

---

[4] Number of papers is obtained from data.semanticweb.org and is therefore limited to the conferences metadata available in this repository. However, these numbers give a good approximation of seniority for the attendees of the conferences in question.

[5] From scholarometer http://scholarometer.indiana.edu/

To quantitatively answer the question of whether people tend to only mix with their peers or not, we compute for each user $u$ in the aggregated contact graph the *average seniority of nearest neighbours*

$$se_{nn}^u = \frac{1}{k_u} \sum_{v \in \mathcal{V}(u)} se_v \tag{1}$$

where the sum is over the $k_u$ users with whom $u$ has been in contact at the conference. We also compute the *average seniority of the neighbours of users with seniority se*

$$se_{nn}(se) = \frac{1}{|u/se_u = se|} \sum_{u/se_u=se} se_{nn}^u . \tag{2}$$

The study of the F2F interaction patterns has however shown that not all contacts are equivalent. On the contrary, the amount of time spent by two users in F2F proximity is strongly heterogeneous (see Fig. 2). Since Eq. (1) performs an unweighted average of the seniority of all neighbours, we generalize it to take into account the contact diversity:

$$se_{nn,w}^u = \frac{1}{s_u} \sum_{v \in \mathcal{V}(u)} w_{uv} se_v \tag{3}$$

where $w_{uv}$ is the total time spent in F2F proximity by $u$ and $v$, and $s_u = \sum_{t \in \mathcal{V}(u)} w_{ut}$ is the total time spent by $u$ in F2F proximity with other users. We also consider for each user $u$ the *strongest* link, and define $se_{nn,max}^u$ as the seniority of the user $v$ with whom the corresponding contacts took place.

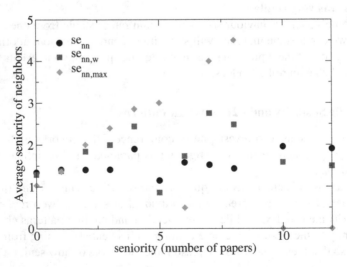

**Fig. 4.** Average seniority of the neighbours of a user, versus the user's own seniority. Black circles correspond to an unweighted average over all neighbours, Eqs. (1) and (2). Red squares show the weighted average Eq. (3), and green diamonds show the seniority of the neighbour with whom the strongest link is observed.

**Table 4.** Some characteristics of the ESWC 2010 chairs, and of the links between chairs, compared with the overall averages

| Characteristics | all participants, 2009 | chairs 2009 | all participants, 2010 | chairs, 2010 |
|---|---|---|---|---|
| average degree | 55 | 77.7 | 54 | 77.6 |
| average strength | 8590 | 19590 | 7807 | 22520 |
| average weight | 159 | 500 | 141 | 674 |
| average number of events per edge | 3.44 | 8 | 3.37 | 12 |

Figure 4 displays the average seniority of the neighbours of users with seniority $se$, Eq. (2), as a function of the seniority $se$, measured as the number of papers authored by an individual. No clear pattern is observed if the unweighted average over all neighbours in the aggregated network, Eq. (1), is considered. The picture is different when the time spent in F2F interaction is considered in order to compute an average in which each neighbour is weighted by the time spent with him/her, as in Eq. (3). An assortative trend is then observed, which is even stronger if considering for each individual only the neighbour with whom the most time has been spent. Such procedures allow to filter out short encounters which are then given small importance, or completely ignored.

Our analyses unveils a clear assortative mixing behavior, in which **people tend to mix with others with similar seniority levels**. In other words, more senior individuals tend to spend more time with other senior individuals, and junior people are more likely to mix with their peers. Similar results are obtained when the h-index is considered as a measure of seniority. It is important to note that relying only on unweighted contacts, i.e., the only knowledge of who has met whom, would not have allowed to reveal this assortative mixing, and that information on the temporal aspects is crucial in this respect.

**Conference chairing and F2F networks.** Another indicator of seniority is whether a person has taken a chairing role at the conference or not. Conference chairing and organisational roles were retrieved for all LSS users at ESWC2010 from data.semanticweb.org and used in this analysis.

Table 4 explores this particular aspect of the relationship between "seniority" and social activity at the conference. Track chairs are indeed typically more senior. We observe that the **chairs interact with more distinct people (larger average degree), and spend more time in F2F interaction (almost three times as much as a random participant)**. Moreover, the contact events between chairs tend to be longer, and edges between chairs in the aggregated network correspond to many contact events. The density of the subnetwork of chairs is also very large (80% of all possible encounters are observed, against 35% for the possible encounters between all participants).

**Scientific seniority and twitter followers.** A comparison between seniority and the number of followers on Twitter (i.e. number of Twitter users who follow the person in question) is given in Figure 5 for ESWC2010. It is interesting to see that of all LSS users, the **most senior scientists are not the mostly followed on Twitter**. Also, a number of less senior people in terms of h-index were followed by many Twitter

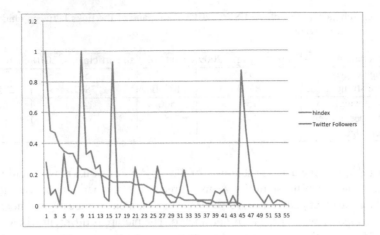

**Fig. 5.** Comparison between people's scientific seniority and the number of people following them on Twitter. Y-axis is the h-index normalized by the maximum h-index among participants, or the number of Twitter followers, as well normalized by the maximum number of followers observed, and X-axis is the 55 people who gave LSS their Twitter accounts during ESWC2010, ranked in decreasing order of h-index.

users. It is worth noting that the first two peaks on the Twitter line in the figure belong to researchers with high visibility and who have taken on chairing responsibilities in other conference events (sessions, tracks, workshops, etc.). The third peak belong to a developer in a semantic web company and not to a researcher (hence the zero h-index). In future work we will consider other parameters, such as user's Twitter activity levels and time since Twitter account was created.

What these results show is that the number of Twitter followers is not necessarily a good indication of pure seniority in the context of scientific communities, but rather it is a reflection of popularity of individuals and of the work they do (more in section 6).

### 5.4   Face-to-Face Interactions and Size of Online Social Networks

Figure 6 shows the average amount of F2F networking of participants during ESWC 2010, alongside the size of their online social networks from Facebook and/or Twitter (followers and followees). The figure does not show any strong correlation between these two parameters. In other words, **people who were active in F2F contacts do not necessarily have the largest online social networks, and vice versa**. Note that these online networks include people who were not present at the conference, or who were present but did not participate in LSS. These online networks also include people from outside the research community (e.g. family, friends, or even spam). The figure also shows a large discrepancy for some of those with low degree of F2F contacts. In a closer look at the data, we found that these readings belong to people who were not researchers in the semantic web field, but were nevertheless present at the conference. It is therefore reasonable to expect these people not to know many of the attendees, which limited their social interactions at the conference.

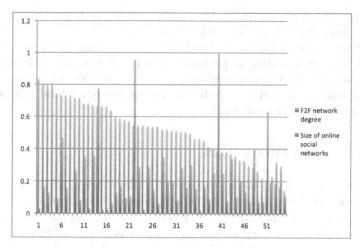

**Fig. 6.** X-axis shows the LSS users who declared their Facebook and/or Twitter accounts during ESWC2010 deployment. The Y-axis shows the total size of their online social networks, normalized by the maximal size observed, and the degree in the F2F interaction network of ESWC2010, divided by the maximum degree. There is no strong correlation between the amount of F2F contact activity and the size of online social networks. In other words, it appears that people who have a large number of friends on Twitter and/or Facebook are not necessarily the most socially active in the offline world.

## 5.5   Social Networking with Online and Offline "Friends"

A social relationship between two individuals can be defined from different points of view. They can be friends, colleagues, co-workers, and the relationship can exist in real life and/or in online social networking sites. The concept of multiplexity refers to the extent to which multiple ties coexist between the same persons. The LSS platform offers an interesting way to crosslink data concerning on the one hand real life interactions taking place on short times and on the other hand professional or online social links. We focus here on the ESWC 2010 deployment, although the other deployments give very similar results.

Among the participants to the ESWC 2010 deployment, 26 links of co-authorship are found, together with 194 links of Facebook friendship, and 112 pairs in which at least one individual follows the other on Twitter. Table 5 gives the average contact characteristics for pairs of LSS users who share a social relationship either at a professional

**Table 5.** F2F contact characteristics between (i) all LSS users, (ii) LSS users who are coauthors, (iii) LSS users who are friends on Facebook, and (iv) pairs of users who are linked on Twitter

| Characteristics | all participants | coauthors | FB | Twitter |
|---|---|---|---|---|
| average contact duration (s) | 42 | 75 | 63 | 72 |
| average edge weight (s) | 141 | 4470 | 830 | 1010 |
| average number of events per edge | 3.37 | 60 | 13 | 14 |

level or online. The average duration of a contact is much larger than for a random pair of attendees, but remains of the same order of magnitude.

The total time spent in F2F interaction is instead much larger, due to the fact that **individuals sharing an online or professional social link meet much more often than other individuals**. Moreover, while the two different online social networks give very similar results, the **average number of encounters -and total time spent in interaction- is highest for co-authors**.

# 6    Discussion and Future Work

The analyses we report in this paper is based on data from three conferences. When we closely compared the list of users of HT2009 with those of ESWC2009 and ESWC2010, it was clear that there was negligible overlap between these users lists. For this reason, some of our analyses that required common users was limited to ESWC data only. However, in our analyses we showed that behaviour in F2F networking of groups and of individuals is very similarly from one conference to the other (sections 5.1 and 5.2).

Our data is naturally sparse, since not all conference attendees participated in LSS, and not all users of one LSS deployment attended, or participated in other LSS deployments. Also, for some deployments, many F2F contacts were taking place outside the perimeters of our RFID readers (e.g. at the bar, during meals), and therefore could not be logged by LSS. However, we believe that the patterns we identified were strong enough in spite of this data sparsity. To overcome this problem, SocioPatterns.org is currently developing RFID with on-board memory, thus enabling F2F contacts to be logged regardless of distance to RFID readers.

As we report in this paper, there are often many parameters and types of relationships that influence social networks and their analyses. In this paper we focused our analyses on a number of such parameters, and our results are based on the network data we gathered from LSS deployments. Such data however, might contain some bias, caused by that data's inherited limitation to only those users who registered to use LSS, and to only those conferences where LSS was deployed. Other parameters can be taken into account in future deployments and analyses, such as users' age, affiliations, and group or project membership (e.g. from rkbexplorer.org). Deploying LSS at non-computer science conferences will also help to widen the scope of our analyses.

Chronology of social relationships could be taken into account when analyzing social networks, to investigate the influence of existing relationships between users on the dynamics of their networks. Some temporal relationships can be obtained from further LSS deployments over longer periods of time, where F2F, and online social networks can be monitored more frequently, and compared with each other over time.

Additionally, we currently do not consider when an online social networking account was set up, or whether the owner is an active user of these accounts or not. Such information can strengthen the analyses of these networks. We are currently building models and tools for generating rich user profiles that can acquire and represent user's activities in various social networking systems. Such profiles can then be analysed to identify usage and social behaviour, influence, trends, and interests.

With respect to estimating scientific seniority of LSS users, we relied on the number of their papers, their h-index, and their chairing roles at the conferences where LSS was deployed (section 5.3). Other features could be taken into account, such as their chairing roles in other previous conferences, or their overall number of publications. However, based on our knowledge of who's who in the semantic web community, we find that the seniority results from our approaches were very realistic approximations.

Linked Data resources such as data.semanticweb.org and rkbexplorer.org proved invaluable for this work. We used these resources for obtaining information on publications, co-authorships, chairing roles at various conferences, etc. Such initiatives should be supported and extended further, emphasizing quality as well as quantity of the data they store.

Work on LSS so far has concentrated on building the platform and website. Future work will focus on providing users with added-value services that use the collected data and analyses results to, for example, recommend new contacts to add to online social networks, to meet or collaborate with certain users F2F, attend specific talks, etc.

# 7  Conclusions

Data from LSS gave us the opportunity to analyse and compare various types of online as well as offline social networks for conference attendees, and to better understand their characteristics, dynamics, and dependencies. Below we summarize the main finding from our analyses:

- Statistical properties of the F2F social contact patterns were very similar across all three conferences. (section 5.1).
- Frequent conference attendees (i.e. used LSS in more than one conference) were more socially active in F2F networking than others, with %22 more F2F interactions and %50 more interaction time than other users (section 5.2).
- Time spent on F2F networking by frequent conference attendees remained stable, even though the list of people they networked with change (section 5.2).
- Conference attendees tend to networks with others of similar levels of scientific seniority. We also show that conference chairs meet more people and spend 3 times as much time in F2F networking than other users (section 5.3).
- People who have the highest number of Twitter follower are not necessarily the most senior in terms of their h-index, although they have high visibility, popularity, and experience (section 5.3).
- No visible correlation is found between size of online social networks of users in Facebook and Twitter and the number of people they met face to face (section 5.4).
- People's F2F contacts with their Facebook friends and Twitter mutual followers were respectively %50 and %71 longer, and %286 and %315 more frequent than with others. They have also spent %79 more time in F2F contacts with the people they co-authored papers with, and they met them %1680 more times than they met non co-authors (section 5.5).

# References

1. Alani, H., Szomszor, M., Cattuto, C., den Broeck, W.V., Correndo, G., Barrat, A.: Live social semantics. In: Bernstein, A., Karger, D.R., Heath, T., Feigenbaum, L., Maynard, D., Motta, E., Thirunarayan, K. (eds.) ISWC 2009. LNCS, vol. 5823, pp. 698–714. Springer, Heidelberg (2009)
2. Barrat, A., Cattuto, C., Colizza, V., Pinton, J.-F., den Broeck, W.V., Vespignani, A.: High resolution dynamical mapping of social interactions with active RFID (2008), http://arxiv.org/abs/0811.4170
3. Cattuto, C., den Broeck, W.V., Barrat, A., Colizza, V., Pinton, J.-F., Vespignani, A.: Dynamics of person-to-person interactions from distributed RFID sensor networks. In: PLoS ONE (2010)
4. den Broeck, W.V., Cattuto, C., Barrat, A., Szomszor, M., Correndo, G., Alani, H.: The live social semantics application: a platform for integrating face-to-face proximity with on-line social networking. In: Workshop on Communication, Collaboration and Social Networking in Pervasive Computing Environments (PerCol 2010), IEEE Int. Conf. on Pervasive Computing and Communications (PerCom), Mannheim, Germany (2010)
5. Eagle, N., (Sandy) Pentland, A.: Reality mining: sensing complex social systems. Personal Ubiquitous Comput. 10(4), 255–268 (2006)
6. Fischer, C.S.: To dwell among friends. University Chicago Press, Chicago (1982)
7. Garca-Silva, A., Szomszor, M., Alani, H., Corcho, O.: Preliminary results in tag disambiguation using dbpedia. In: Knowledge Capture (K-Cap 2009) - Workshop on Collective Knowledge Capturing and Representation - CKCaR 2009, CA, USA (2009)
8. Golder, S., Wilkinson, D.M., Huberman, B.A.: Rhythms of social interaction: Messaging within a massive online network. In: Communities and Technologies 2007: Proceedings of the Third Communities and Technologies Conference, Michigan State University (2007)
9. Hui, P., Chaintreau, A., Scott, J., Gass, R., Crowcroft, J., Diot, C.: Pocket switched networks and human mobility in conference environments. In: WDTN 2005: Proc. 2005 ACM SIG-COMM Workshop on Delay-Tolerant Networking. ACM, New York (2005)
10. Isella, L., Stehlé, J., Barrat, A., Cattuto, C., Pinton, J.-F., den Broeck, W.V.: What's in a crowd? analysis of face-to-face behavioral networks. arXiv:1006.1260 (2010)
11. Kumar, R., Novak, J., Tomkins, A.: Structure and evolution of online social networks. In: Proc. 12th ACM SIGKDD Int. Conf. on Knowledge Discovery and Data Mining, Phil. (2006)
12. Marlow, C., Naaman, M., boyd, d., Davis, M.: Ht 2006, tagging paper, taxonomy, flickr, academic article, to read. In: Proc. 17th ACM Conference on Hypertext and Hypermedia (HT). ACM Press, New York (2006)
13. McPherson, M., Smith-Lovin, L., Cook, J.: Birds of a feather: Homophily in social networks. Annual Review of Sociology 27, 415–444 (2001)
14. Monge, P.R., Contractor, N.S.: Theories of Communication Networks. Oxford University Press, Oxford (2003)
15. Nishimura, T., Matsuo, Y., Hamasaki, M., Fujimura, N., Ishida, K., Hope, T., Nakamura, Y.: Ubiquitous community support system for ubicomp 2005. demo at ubicomp 2005, tokyo (2005)
16. Scherrer, A., Borgnat, P., Fleury, E., Guillaume, J.-L., Robardet, C.: Description and simulation of dynamic mobility networks. Comput. Netw. 52(15), 2842–2858 (2008)
17. Szell, M., Lambiotte, R., Thurner, S.: Trade, conflict and sentiments: multi-relational organization of large-scale social networks (2010), arXiv:1003.5137

18. Szomszor, M., Cantador, I., Alani, H.: Correlating user profiles from multiple folksonomies. In: Proc. Int. Conf. Hypertext (HT 2008), Pittsburgh, PA, USA (2008)
19. Szomszor, M., Cattuto, C., den Broeck, W.V., Barrat, A., Alani, H.: Semantics, sensors, and the social web: The live social semantics experiments. In: Aroyo, L., Antoniou, G., Hyvönen, E., ten Teije, A., Stuckenschmidt, H., Cabral, L., Tudorache, T. (eds.) ESWC 2010. LNCS, vol. 6088. Springer, Heidelberg (2010)
20. Thibodeau, P.: IBM uses RFID to track conference attendees (2007),
    `http://pcworld.about.com/od/businesscenter/`
    `IBM-uses-RFID-to-track-confere.htm`
21. Wasserman, A., Faust, K.: Social Network Analysis: Methods and applications. Cambridge University Press, Cambridge (1994)
22. Wu, L., Waber, B., Aral, S., Brynjolfsson, E., Pentland, S.: Mining face-to-face interaction networks using sociometric badges: Evidence predicting productivity in it configuration. In: The 2008 Winter Conference on Business Intelligence, University of Utah (2008)

# Using Semantic Web Technologies for Clinical Trial Recruitment

Paolo Besana[1], Marc Cuggia[1], Oussama Zekri[2], Annabel Bourde[1],
and Anita Burgun[1]

[1] Université de Rennes 1
[2] Centre Eugéne Marquis

**Abstract.** Clinical trials are fundamental for medical science: they provide the evaluation for new treatments and new diagnostic approaches. One of the most difficult parts of clinical trials is the recruitment of patients: many trials fail due to lack of participants. Recruitment is done by matching the eligibility criteria of trials to patient conditions. This is usually done manually, but both the large number of active trials and the lack of time available for matching keep the recruitment ratio low.

In this paper we present a method, entirely based on standard semantic web technologies and tool, that allows the automatic recruitment of a patient to the available clinical trials. We use a domain specific ontology to represent data from patients' health records and we use SWRL to verify the eligibility of patients to clinical trials.

## 1 Introduction

Clinical trials are the gold standard for testing therapies or new diagnostic techniques that may improve clinical care. Patients are enrolled to clinical trials if they match the eligibility criteria that define the trials. The recruitment process is a particular point of weakness for clinical trials. The development of information technology in medicine, and particularly in hospitals, offers a good opportunity to support and improve the recruitment process.

The work presented in this paper aims at suggesting the clinical trials to which a patient could be enrolled. It fits into the standard procedure, mandated by the French national oncology guideline, of evaluating cancer patients in multidisciplinary meetings. Doctors from different disciplines meet periodically to discuss and decide the treatment of patients. Clinical trials are about treatments, and the decision of enrolling a patient is taken during these meetings.

The system, by finding clinical trials in which a patient may be enrolled, takes a different perspective to clinical trial recruitment, usually oriented towards finding patients for a clinical trial. It is centred on patients, and it is possibly more acceptable by doctors, as it does not interrupt their workflow. Our project analysis and evaluation is based on the clinical trials and patients' data discussed in the multidisciplinary meetings in the Centre Hospitalier Universitaire of Rennes (France), between September and December 2009.

P.F. Patel-Schneider et al.(Eds.): ISWC 2010, Part II, LNCS 6497, pp. 34–49, 2010.

Matching trials to patients requires the formalisation of patient conditions and of eligibility criteria in such a way that their correspondences can be computed and found. Patients' data and criteria have usually a different level of abstraction. Data are specific and precise (for example, Prostate Ductal Adenocarcinoma, Cribriform Pattern), while criteria need to include cases that are different within determined boundaries (for example, Invasive Prostate Carcinoma).

The fundamental hypothesis we make in the project is that mapping terms from patients' records and eligibility criteria to a formal ontology both minimises the risk of ambiguity and allows automated reasoning. The medical domain, and oncology in particular, has a wealth of well-established ontologies that can be used, which were originally developed as terminology services for uniquely identifying diseases, symptoms, and therapies. Additionally, ontologies written in OWL have clearly defined expressivity and computational properties and they can also exploit tools and applications both for authoring (such as Protï¿œgï¿œ) and for reasoning (such as Jena, Pellet, Fact++).

The work presented in this paper focuses on the use of OWL and SWRL for representing patients' data and eligibility criteria, and reason about them. The goal is to show that it is possible to identify a workable formalism within the boundaries of Description Logics that can be used for the whole process of recruitment, without the need to add external resources.

## 2  Clinical Trials

Clinical trials (CTs) are fundamental for evaluating therapies or new diagnosis techniques. They are the most common research studies designed to test the safety and/or the effectiveness of interventions. A CT may address issues such as prevention, screening, diagnosis, treatment, quality of life or genetics, and each trial is designed to answer specific scientific questions. CTs are based on statistical tests and population sampling, and because they rely on adequate sample sizes it is common for CTs to fail in their objectives because of the difficulty of meeting the necessary recruitment targets in an effective time and at reasonable cost.

The two most important issues that must be decided early in the design of a CT are the population of interest (which determines the eligibility criteria of the trial) and the sample size required to give sufficient statistical power for analysis. Reduced sample size reduces the power of the study, but relaxing eligibility criteria to allow a larger population of interest (and hence a larger pool from which to recruit) introduces a confounding element where factors that are not the prime focus of the study cannot be excluded.

Patient data, either acquired during the clinical care process or contained in Electronic Health Records (EHR), could be reused to automatically apply eligibility criteria

The features of the population of interest for a clinical trial are defined by the eligibility criteria of the trial. These characteristics determine the rules to be applied for building the sample of subjects. They may include age, gender, medical history and current health status. Eligibility criteria for treatment studies often require that patients have a particular type and are at a particular stage of their disease.

Enrolling participants with similar characteristics helps to ensure that the results of the trial will be due to what is under study and not other factors. A second function of eligibility criteria is to exclude patients who are likely to be put at risk by the study, minimizing the risk of a subject's condition worsening through participation.

## 3    Problem Description

The goal of the system presented in this paper is to select the clinical trials in which a patient might be enrolled, among those currently active in a hospital. The list of selected clinical trials are then evaluated by the doctors in the multidisciplinary meeting. In particular, it is important to remove trials that are either not relevant or with mismatch conditions, in order to provide the physicians with a list of focussed suggestions.

The suggestions are based on the available information at the moment of the meeting, which can be weeks before the trial actually starts. Criteria referring to the patient conditions at the moment of the trial cannot be considered in this stage and are discarded: the project focuses on a subset of the criteria, called pre-screening criteria. The system is applied to clinical trials concerning prostate cancer.

There are three main aspects to be considered: how to represent patients' data in a format that can be queried; how to represent the eligibility criteria; and how to match criteria to patients, dealing with the difference in abstraction discussed in Section 1. In this paper we present an approach that uses only OWL and SWRL to represent data and criteria, and to reason. It can be considered a low-level representation: directly computable, but not for human use. However, it is possible to find representations at a slightly higher level that can be directly converted into SWRL using re-write rules.

### 3.1    Patient Data

Results of exams are stored in patients' records. While historically kept in physical folders, they are beginning to be stored in an Electronic Health Records (EHR), often in natural language, or as scanned images. There is an ongoing effort to formalise their representation, in order to simplify search, and to allow interoperability between different systems.

In our project, the data is currently in free text, but an expert operator will convert it into a more formal model before running the matching process.

Patients' data are represented according to an Information Model, such as the HL7 Reference Information Model (RIM). An information model defines what is the information that needs to be collected and gives it a semantics. It is often a set of classes (also called templates) with the attributes and methods

*date of birth: 11 October 1935*
*Relevant elements:*
*diagnosis: prostate adenocarcinoma*
*June 2007: radical prostatectomy*
*Gleason 6 = 3+3, pT3a R0*
*Initial PSA (Prostate Specific Antigene) marker = 9.48*
*one month after surgery=undetectable*
*after 12 months=0.26 ng/ml*

**Fig. 1.** Example of patient record available at multidisciplinary meeting

(the class patient, for example). The HL7 RIM contains 5000 attributes to cover terminology requirements. Semantics is provided by a reference ontology.

## 3.2 Eligibility Criteria

The eligibility criteria consist of a set of inclusion criteria defining the characteristics mandatory in the population of interest and a set of exclusion criteria defining the characteristics to be avoided. Usually the negation of an exclusion criterion becomes an inclusion criterion and viceversa.

Eligibility criteria can be simple, stating the value of a single observable entity, such as the diagnosis (diagnosis of prostate adenocarcinoma), or can be qualified by other properties (diagnosis of prostate adenocarcinoma, confirmed by histology). The values can be at different level of abstraction: for example, the diagnosis can specify a particular type of cancer, or can be more generic and include different forms of cancer, possibly by specifying only the location (diagnosis of prostate neoplasm) or some features of the cancer (invasive cancer).

Criteria can define the acceptable value of some medical parameters (Prostate Specific Antigene PSA > 5ng/l), of some personal attribute of the patient (age > 18, age < 75) or can specify the staging of the disease (such as the Classification of Malignant Tumours, TMN, or the Karnofsky score and the Zubrod score, used to measure the general patient's well-being). There can be alternative values (patient in stage pT2, pT3 or pT4). The staging system used in the patients data may be different from the one used in the criteria. Conversion tables can address this issue.

The criteria can specify time constraints that refer to other events: PSA value > 9 ng/l 6 months after surgery, or PSA < 2 a month before inclusion.

A criterion can contain a conjunction of other criteria (diagnosis of X and grade pT2), or can be the disjunction of criteria (PSA > 9 or grade pT2).

Inclusion criteria are usually positive expression, while exclusion criteria are introduced by negation ("no ...", "absence of..."). Some composite criteria may contain both positive and negative statements (invasive cancers, excluding skin cancer).

Figure 2 shows an example of the eligibility criteria used for selecting patients in clinical trial for a therapy for prostate cancer run in the university hospital of Rennes in 2009.

---

**Inclusion:**

*1) Histologically proven cancer localised in the prostate*

*2) Absence of metastases*

*3) Cancer in intermediate prognostic group : - T2a <= T < T3a - or T1b/c with PSA >= 10ng/ml - or T1b/c with Gleason score >= 7*

*4) PSA < 30ng/ml with a normal calibration value of 4ng/ml*

*5) age < 77 years*

*6) life expectancy >= 10 years*

*7) OMS-WHO=ZUBROD <= 1 [Zubrod]*

**Exclusion:**

*1) history of invasive cancer unless it is older than 5 years*

*3) PSA >=30ng/ml in two successive measurement (even if the latest is lower than 30)*

*4) history of pelvic radiotherapy*

*5) history of radical anterior prostectomy due to cancer*

*6) previous hormonotherapy or castration*

---

**Fig. 2.** Eligibility criteria for a clinical trial for an adjuvant therapy for prostate cancer

## 4   State of the Art

A detailed and extensive overview of the formalisms used for representing eligibility criteria is given in [10]. In the paper the authors distinguish different types of expression languages for eligibility criteria. We summarise three categories, which include most of the projects.

**ad hoc expression** normally driven by use cases more than by theoretical basis. Ad hoc languages define a set of parameters that can take boolean, numeric or enumerated values. The languages provide comparison and logical operators. Some ad hoc languages are based on a rich information model, such as the HL7 RIM. In general they have a limited capability of using formal reasoning methods such as temporal constraints or predicate logic. However, ad hoc languages proved very popular, and are used in various projects.

**Arden Syntax** is a hybrid between a production rule system and a procedural formalism. It has been chosen as standard for HL7. It provides rich time functions and explicit links to clinical data embedded in curly brackets. It is more expressive than most ad hoc languages. It lacks declarative properties and defined semantics for temporal comparison and data abstraction. It is well supported.

**logic based languages** vary in expressivity. Systems overviewed include one based on SQL (based on relational algebra), one on Protege Constraint language (PAL) and one on Description Logic [9].

Because of its relation with the project presented in the paper, we will describe [9] in more detail. This work aims at demonstrating how it is possible to use ontologies for reasoning in health informatics. The authors use the problem of matching eligibility criteria to patients' condition as case study. They focus on the problems of knowledge engineering and of scalability. They analyse the mapping of the representation of patients' data used in a hospital, based on a local terminology, to a formal medical ontology - SNOMED CT (Systematized Nomenclature of Medicine – Clinical Terms)[1]. The problem of scalability is addressed using SHER, an OWL-DL reasoner developed by IBM for dealing with large ontologies. Patients' data are observations that connect instances of SNOMED CT classes. Eligibility criteria are class definitions that are used to query observations via subsumption.

In the Epoch project [7] the researchers have developed a framework for managing the overall process of clinical trials. They created a suite of OWL ontologies covering the different phases of the process, but they do not provide a detailed explanation about the representation of patients' data and clinical trials.

## 5    Method

As we have seen in Section 3, formalising and reasoning over eligibility criteria requires the possibility of reasoning over ontologies when criteria are expressed as generic conditions, or when only some attributes of the diseases are specified,. It also needs to be able to represent and reason over data types and over time. It should allow the composition of criteria both through disjunction and conjunction. An additional requirement is the traceability of the results: if a patient is selected or rejected for a clinical trial, it must be possible to identify the observations supporting the criteria.

As stated in Section 1, we make the fundamental hypothesis that mapping terms from patients' records and eligibility criteria to a formal ontology minimises the risk of ambiguity and allows automatic reasoning. The ontology plays both the role of reference terminology and of background knowledge for reasoning.

The system needs to find the clinical trials to which a patient might be enrolled. Before a multidisciplinary meeting, all the active Trials of a hospital are loaded into the system. Then each of the patients to be discussed in the meeting is loaded into the system, one at the time. The eligibility criteria are matched to patient's observations. The criteria are then aggregated and a list of clinical trials with satisfied inclusion criteria and without satisfied exclusion criteria is extracted.

The system exclusively uses OWL and SWRL to represent patient data and eligibility criteria: everything is loaded into an ontology, and all the operations take place within the ontology.

### 5.1    Choosing the Medical Ontology

The first step is to identify the ontology that can be used for these roles. For its role as reference terminology, an ontology needs to provide a good coverage

---

[1] http://www.ihtsdo.org/snomed-ct/

of the terms appearing in patients' records and in criteria: it should be possible to map terms in the text to entities already defined in the ontology, or to easily define new entities, using other entities and the compositional grammar of the ontology. Following the criteria presented in [4], we also evaluated availability (open source or licensed) and its format.

In order to evaluate coverage an expert in clinical trial selected 200 criteria from clinicaltrials.gov. We used MetaMap [2] to map the extracted criteria, written in free text, to concepts in the UMLS metathesaurus [3]. UMLS (Unified Medical Language System) connects terms from over 100 medical terminologies: each concept in UMLS is mapped to different terminologies. We checked, for each UMLS concept found by MetaMap, whether there was a corresponding term in the different terminologies. From the evaluation, it resulted that NCI Thesaurus (NCI-T) [8] provides the best coverage with 75%. NCI-T contains 75000 classes and it is developed specifically for oncology by the National Cancer Institute, US. NCI-T is followed by SNOMED CT, a large ontology (over a million classes) covering all medical domains, with a coverage of 65%.

Regarding availability, NCI-T is open source, while SNOMED CT has a very expensive licence. Both are written in a Description Logic: SNOMED is $\mathcal{ER}++$, NCI Thesaurus is $\mathcal{SH}(D)$. However, NCI-T is directly available in OWL1.1 [6], while SNOMED is distributed in database tables and requires conversion . NCI-T introduces a particular idiom, intended for human use, for properties. Some of its properties have the prefix may_have, while some of the others have the prefix excludes_. The may_have properties are used when a class of diseases have subclasses that may or may not present a particular feature. A may_have_[feature] property has a corresponding subproperty has_[feature] that is used in subclasses that have the feature. The excludes_[feature] property is used to specify that a particular disease does not present a feature or a symptom. These properties are intended for human use: from a Description Logics perspective these properties have no particular meaning, and it is not possible to use them to verify consistency.

At the end of the evaluation process, we opted for NCI-Thesaurus as background knowledge. Additionally, the use of a domain-specific ontology requires a system that should be minimally coupled with the ontology itself, to allow portability in other domains.

The support for data types in OWL1.1 is not particularly powerful, as it is not possible to specify ranges. Similarly the reasoning over time is hard in OWL. The use of SWRL (Semantic Web Rule Language) allows us to overcome these limitations: in SWRL it is possible to write rules that state ontological properties and contain built-in functions. In Protege 3.4, the available built-ins include numerical comparators (greaterThan, lessThan,...) and temporal comparators.

The core of the system is a simple ontology, that glues together the components: it imports NCI-T ontology and the ontologies that define SWRL built-ins and possibly the classes they require. This glue ontology defines the classes that are used to represent patient data and the eligibility criteria in clinical trials.

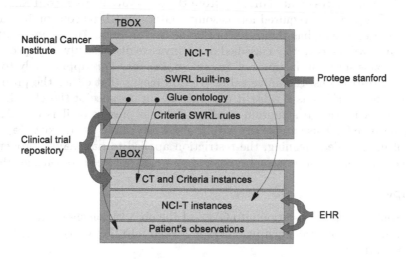

**Fig. 3.** General architecture of the system

Figure 3 shows the general architecture: TBox contains the glue ontology that imports NCI-T and the SWRL ontologies. The SWRL rules are definitions, and therefore are part of the TBox. The ABox contains the instances of NCI-T classes, that are linked by observations, and the instances of the criteria.

## 5.2   Representing Patient Data

Medical ontologies like SNOMED CT or NCI Thesaurus were originally developed as terminology services, and were not conceived for representing patients's data directly. The information about patients is captured by the Information Model. Some of the Information Models, such as the HL7 v3.0, use an ontology as reference for its terms. In order to allow portability, we use a thin information layer composed by observations that connect values taken from the real information model.

In our system, an observation is the tuple:

$$\langle observable\_entity, observable\_property, observed\_value \rangle$$

It is a reified relation connecting an *observable entity* (a measurement, a medical finding, the result of an exam), an *observable property* (value, numeric value, date, method, ...), and the *observed value* of the property (it can be an instance of a class, or a datatype). In Owl, it is represented as an instance of the class Observation. All entities and values, unless they are datatypes, need to be instances: OWL-DL does not support the relations between classes (it would make it into OWL-FULL which is undecidable). If the class definition contains as sufficient condition properties with existential restriction, then an instance for the restricted class is created. To avoid a cascading effect of having to create instances for all the properties of the instances created to fill properties, only the properties of instances created directly from the observation are filled. Filling the restricted properties is required for reasoning later: SWRL reasons on instances only. If a disease class has the restriction `has_lesion some Invasive_Lesion`, the instance will be correctly classified as an invasive disease only if its property `has_lesion` is set to an instance of `Invasive_Lesion`. This applies only to the existential restrictions: the universal restriction means that either the property has no value, or if it has a value it can only be an instance of the class in the restriction. A particular symptom or effect either is present or it is not. As we stated above, NCI-T use properties with prefixes `may_have_` and `exclude_`. Because of its intended meaning, the restriction applied to `may_have_` properties can be exclusively be of universal type, and so we do not need to consider them.

## Example

We present here the translation into OWL of the observation shown in Figure 1. First of all, a set of predefined instances is loaded into the ontology, used in all the observations:

```
valuep instance of ncit:Value;
numvaluep of ncit:NumericValue;
datep of ncit:Date; procp of ncit:Procedure:;
now of temporal:ValidInstant
```
*Date of Birth: 10 November 1935*

The observable entity is an instance of NCI class `ncit:Birthday`, the observable property is an instance of the class `ncit:Value` (instantiated as `valuep` in the previous step) and the observed value is represented as an instance of the class `validInstant` from the temporal ontology,

```
ncit:BirthDay:birth_day_1;
Temporal:validInstant:i1:[has_time:10November1935];
Observation:o1:[has_observable=Birth_Day_1,
    has_observable_property=valuep,has_value:i1];
```
*diagnosis: prostate adenocarcinoma*

The observation links the instances of the class `ncit:Diagnosis` and of the class `ncit:Prostate_Adenocarcinoma`

```
ncit:Diagnosis:diagnosis1;ncit:Prostate_Adenocarcinoma:ac1;
Observation:o2:[has_observable=diagnosis1,
    has_observable_property=valuep,has_value=ac1];
```
*june 2007: radical prostatectomy*

We need to observations: one to cover the value of Clinical procedure, the other to cover its date

```
ncit:Clinical_Procedure:cp1;  ncit:Radical_Prostatectomy:rp1;
Temporal:validInstant:i3:[has_time:1July2007];
Observation:o3:[has_observable=cp1,
    has_observable_property=valuep,has_value=rp1];
Observation:o4:[has_observable=cp1,
    has_observable_property=datep, has_value=i3];
```
*PSA after 12 months=0.26 ng/ml*

We need one observation for the numerical value of the PSA and one for the date of the exam

```
ncit:PSA_Assay:psa1;    Temporal:Instant:i6:[has_time:1July2008];
Observation:o7:[has_observable=psa2,
   has_observable_property=numvalp, has_numeric_value=0.26];
Observation:o8:[has_observable=psa2,
   has_observable_property=datep, has_ value=i6];
```

## 5.3  Representing Clinical Trials

Criteria are queries over the observations containing patients' data. Patients whose data match all the inclusion criteria are included in the clinical trial, unless they match an exclusion criteria.

SWRL [5] provides a high-level syntax for horn-like rules: a SWRL rule has the form of an implication with an antecedent (body) and a consequent (head): if the antecedent holds, the condition specified in the consequent holds. SWRL maintains the expressivity of OWL-DL, with a set of additional features such as built-in functions for data types. SWRL is monotonic: it is not possible to retract or change what is already assessed, but only to add something new. OWL and consequently SWRL rely on the Open World Assumption: the lack of some information is considered ignorance. On the contrary, the Closed World Assumption, used in most of the other programming languages and in databases, considers to be false that which is not known to be true . With the Open World Assumption, it is not possible to verify whether something is false unless it is explicitly stated as false. It is a realistic assumption in the medical domain: it may not be possible to collect all the information, both because in different phases different information is available, and because some exams are probabilistic (for example, the presence of metastases cannot be completely ruled out if some samples give negative results).

In Protégé 3.4 SWRL rules are computed using Jess, a proprietary library based on the RETE algorithm and developed by the Sandia Laboratories . The ontology and rules need to be converted to Jess (simple with Protégé), and then the engine is run. However, a limitation of Jess is its lack of support for inferred classifications in the ontology. Protege 4.1 uses only OWL reasoners for SWRL, and therefore it supports inferred classifications. We started the development before the availability of version 4.1,. However, our system can work on both versions of Protégé.

Because of SWRL monotonicity, we can only add new information: every satisfied rule adds the observations it matches to the support list of a criterion. Supported inclusion criteria are added as supporting arguments to a clinical trial, while supported exclusion criterion are added as arguments against the clinical trial. If a clinical trial has arguments against it, it is excluded from the list of possible trial for a patient.

In our system, criteria are instances of Inclusion or Exclusion Criteria classes, defined in the glue ontology. Each criterion has a set of rules: if one matches any of the observations, it adds the observation to the list of support of the criterion.

After running all the criteria rules, the criteria are aggregated either in favour or against the clinical trials

Criteria also can be annotated with a human-readable description, to facilitate the final report. The left side of the rule represents the condition, and must match one or more observations. The right hand side adds the matching observations to the list of supporting observations:

Observation(?o) ∧ C1 ∧ C2... → is_supported(Pid, ?o)

where Pid is the identifier of the criterion supported by the observation. Criteria that verify the value of an observable entity can be represented directly:

Observation(?o) ∧ has_observable(?o, Entity)∧
has_value(?o, ?V) ∧ ExpectedEntity(?V)
→ is_supported(Pid, ?o)

Similarly, for criteria about the numeric value of an observable entity:

Observation(?o) ∧ has_observable(?o, Entity)∧
has_numeric_value(?o, ?V) ∧ swrlb : comparator(?V, Num)
→ is_supported(Pid, ?o)

Where the swrlb:comparator is one of the numeric built-in functions (such as greaterThan, lessThan, greaterOrEqual, lessOrEqual) and Num is constant. When an observable can have more than a value, it is necessary to create a rule per value, all supporting the same criterion.

When the criteria have temporal requirements, the rule needs to extract the observation about the date relative to the observed entity (if available), and use the temporal built-in operators of SWRL. The glue ontology must therefore import the Temporal Ontology[2] that defines temporal operations, based on Allen's temporal logic [1], and some basic classes used in defining time intervals. Some of the criteria refer to dates that are implicit (like the current time, inserted before running, as we saw above).

When the criteria refer to partially defined concepts that do not have a corresponding entity in the background ontology the clean solution would be to define a new class with sufficient conditions and push it in the TBox before loading the rules. However, as said above, Jess, the engine used by Protege 3.4 for SWRL does not support inferred relations. Equivalently, the sufficient conditions are specified in SWRL.

The work presented in this paper focuses on the use of OWL and SWRL for representing patients' data, eligibility criteria and reasoning about them. The goal is to show that it is possible to identify a workable formalism within the boundaries of DL. While the representation is computable, and results can be obtained, it is awkward for human operators. Most of the criteria fit in a relatively limited set of patterns. The criteria can be inserted using these patterns, and then SWRL rules are generated using rewrite rules.

---

[2] http://swrl.stanford.edu/ontologies/built-ins/3.3/temporal.owl

## Example

We present as example the eligibility criteria shown in Figure 2. We first create an instance for the clinical trial:

ClinicalTrial:ct1

We then show the representation of a subset of the criteria.

*Cancer localised in the prostate*

In this case, we need to exploit the subsumption mechanism and the multi-hierarchy nature of NCI-T. The observable entity is of type ncit:Diagnosis, and the value needs to be an instance of both the class cancer and the class prostate disorder.

InclusionCriterion:ic1:[CT=ct1]

r1:
Observation(?o)has_observable(?o,?e) ∧ ncit : Diagnosis(?e)∧
has_observable_property(?o,valuep)∧
has_value(?o,?v) ∧ ncit : Neoplasm(?v)∧
ncit : Prostate_Disorder(?v) → supported_by(ic1,?o)

*Absence of metastases*

This criterion, presented as an inclusion criterion, needs to be treated as an exclusion criterion: the enrollment is excluded if there is a metastasis. We use the diagnosis to verify whether it is a metastatic disease. We could also use rules about the stage of cancer.

ExclusionCriterion:ec1:[CT=ct1]

r2:
Observation(?o) ∧ has_observable(?o,?e)∧has_observable_
property(?o,valuep) ∧ ncit : Disease_has_finding(?e,?f)∧
ncit : Metastatic_Lesion(?f) → supported_by(ec1,?o)

*Cancer in intermediate prognostic group : - T2a <= T < T3a - or T1b/c with PSA >= 10ng/ml - or T1b/c with Gleason score >= 7*

We split the criterion into 10 alternative rules (4 for the stages T2a-T3a, 2 for the T1b/c with the specified PSA value, and two for T1b/c with specified Gleason score), one of which has to be supported by an observation

InclusionCriterion:ic2a:[CT=ct1]

r3a:
Observation(?o) ∧ has_observable(?o,?e) ∧ ncit : Finding(?o,?e)∧
has_observable_property(?o,valuep) ∧ has_value(?o,?f)∧
ncit : pT2a_Stage_Finding(?f) → supported_by(ic2,?o)
...

r3d: ····ncit : pT3a_Stage_Finding(?f) → supported_by(ic2,?o)

The criterion is supported also if the finding is T1b/c with a value of PSA above 10ng/ml, or with a gleason score above 7

r3e: :
Observation(?o1) ∧ has_observable(?o1,?e) ∧ ncit : Finding(?e)∧
has_value(?o1,?f) ∧ ncit : pT1b_Stage_Finding(?f)
∧Observation(?o2) ∧ has_observable(?o2,?e2)∧
ncit : PSA_Assay(?e2) ∧ has_numeric_value(?o2,?v)∧
swrlb : greaterOrEqual(?v, 10) → supported_by(ic2,?o)

r3f:
···∧ncit : pT1c_Stage_Finding(?f)∧
··· → supported_by(ic2,?o)

The Gleason score follows the same principle

*NO history of invasive cancer unless it is older than 5 years*

In this case, the criterion would be false if there was an observation about an invasive cancer more recent than 5 years before screening. Therefore the exclusion criteria queries about such events, using the observable Personal_ Medical_ History, that can be used to report medical events in the past, and the Invasive_ Malignant_Neoplasm class from NCI-T. The Adenocarcinoma of the Prostate, seen above, is one of its subclasses. We need to verify two observations about the same observable (the medical history): one relative to its value, and the other about its date.

ExclusionCriteria:ec2: [CT=ct1]

r4:
Observation(?o1) ∧ has_observable(?o1, ?e) ∧ Personal_Medical_History(?e)∧
has_observable_property(?o1, valuep)∧has_value(?o1, ?v)∧
ncit : Invasive_Malignant_Neoplasm(?v) ∧ Observation(?o2)∧
has_observable(?o2, ?e) ∧ has_observable_property(?o2, datep)∧
has_value(?o, ?d) ∧ temporal : duration(?p, now, ?d) ∧ swrlb : lessThan(?p, 5)
→ supported_by(ec2, ?o)

where the built-in function `temporal:duration(?p, now, ?D, "Months")` instantiates ?p with the length of the interval beween the date ?d and the date now

## 5.4  Aggregating the Results

A clinical trial has a list of arguments in favour and one against enrollment of the patient. The list in favour is filled by inclusion criteria, while the list against by exclusion criteria.

Once the criteria rules are run, the instances of the criteria will have the property supported_by either filled with one or more observations, or empty. Inclusion criteria with the property filled are supported, and they in turn support the clinical trial. Exclusion criteria with the supported_by property filled are arguments against the enrolling of the patient to the clinical trial.

At this point, rules specific to clinical trials are run to aggregate the results of the criteria.All inclusion criteria of a clinical trial need to be verified: there must be, for each clinical trial, a rule stating that the conjunction of all inclusion criteria must have at least one supporting argument:

supported_by(cid1, ?a) ∧ supported_by(cid2, ?b) ∧ ...
→ is_supported(CT1, true)

Then a generic rule for the exclusion criteria is run:

ExclusionCriterion(?c) ∧ has_ct(?c, ?ct) ∧ supported_by(?c, ?a)
→ argument_against(?ct, ?c)

It is also possible to trace the criteria that were against the enrollment to a criteria, and for each criterion it is possible to trace the observation that supports it.

However, extracting the clinical trials that are supported without arguments against it requires to reason with the closed world assumption: according to the open world assumption, the empty list of arguments against enrollment is considered ignorance, and cannot be used to infer that the there are no arguments against enrollment. The last step needs to be performed outside the ontology.

**Table 1.** Example of results at the end of the execution

| criteria | description | CT | support |
|----------|-------------|-----|---------|
| ic1 | cancer localised in prostate | CT1 | o2 |
| ec1 | absence of metastases | CT1 | |
| ic3 | intermediate prognostic group | CT1 | o5 |

| CT | favour | against |
|-----|--------|---------|
| CT1 | ic1,ic3 | - |
| CT2 | ... | ... |

An external program obtains the list of all clinical trials and selects those that are both supported and have no arguments against.

Table 1 shows the state of the criteria and of the clinical trials at the end of the execution on the example data and criteria.

# 6  Evaluation and Discussion

The aim of this work is to show how it is possible to represent eligibility criteria of clinical trials using SWRL on top of a large domain specific ontology such as NCI Thesaurus. The first step is to assess how well eligibility criteria can be represented. An expert in clinical trials selected 97 criteria from clinicaltrials.gov, that are particularly representative pre-screening criteria. We started from a larger set, but some were removed as they required information which is not available at pre-screening time.

Out of the 97, 92 could be fully represented using SWRL. The problematic ones were caused by the lack of the corresponding entity in NCI. 11 needed disjunction, 20 contained numerical comparison, 14 required some form of temporal reasoning. About a third contained queries over more than one observation.

We also extracted four real clinical trials active during 2009 in the University Hospital of Rennes, and we selected 129 patients that were examined and assessed for the trials during the same year. The four trials have 67 different eligibility criteria, some appearing in more than a trial. Overall, 7 could not be represented: all contained terms that cannot be mapped to corresponding entities in NCI (life expectancy, hip replacement, cardio-vascular pathology, neuropathology, hypertension, under tutelage). It is easy to notice how terms are missing when they come from a domain that differs from oncology, the domain of NCI. The criteria are separated into inclusion or exclusion.

Some of the criteria can be directly translated into SWRL - we have seen some examples above. Others require more thought, especially these which involve temporal reasoning.

In our project time is not a stringent requirement: the matching of the patients to the available clinical trials is done offline, before the multidisciplinary meeting. The slowest step in the overall procedure is loading NCI-T ontology: on a dual core machine, with 8Gb of memory, takes over 100 seconds and 2Gb of memory. Importing data into the ontology is nearly instantaneous: we load one patient at the time, and only the clinical trials currently active in the hospital are loaded. The next bottleneck is the conversion of the ontology and of the SWRL rules into Jess, operation that takes on average 10 seconds. The actual running of the engine takes less than a third of a second (but as we explained above, the inferred relations are not considered by Jess). Compared to [9] we use a much smaller ontology (SNOMED CT is over a million classes, while NCI-T is only 75000). Loading the background ontology is performed once. The criteria and the patients observation need to be inserted for every patient and deleted at the end of the matching.

## 7     Future Work

We plan to proceed in three directions, one addressing the cause of failure in representing criteria, another one dealing with the new version of NCI-T that will soon released and finally a third studying portability to other domains.

The choice of NCI-T over SNOMED CT has a few advantages, among which the smaller size of the ontology. However, there is a trade-off between smaller size and possibility of representing all the criteria: the main cause of failure was shown to be the lack of the corresponding concept in NCI-T for terms in the criterion. We need to address the problem of entities not defined in NCI-T: we plan to study the feasibility of importing fragments of other medical ontologies to cover these gaps.

A new version of NCI-T is currently under development: it will be release in OWL2.0, and it will exploit the new features on datatypes available in OWL2.0. Once the new version will be released, we will move the system to OWL2.0.

Patients and trials studied in this project concerned prostate cancer only. We plan to assess the results applying the system to different types of cancers. Moving to a domain different from oncology requires either identifying another domain-specific ontology, or using SNOMED CT, possibly extracting a relevant portion of this large ontology.

## 8     Conclusion

Clinical trials are required for the evaluation of medical treatments. Their weakness lies in the difficulty of recruiting enough patients in order to make them statistically meanigful. In this paper we have presented an approach based on OWL and SWRL that addresses the problem of recruitment of patients.

The patients' data, extracted from the Electronic Health Record are converted into observations, that are reified relations linking observable entities, such as measurements, diagnoses, results of exams, to attributes, such as value, date or method. The eligibility criteria are SWRL rules that match observations.

The evaluation showed that it is possible to represent the great majority of criteria, and the difficulties raise when entity in the background ontology cannot be found for terms in the criteria. Compared to the work in [9], the approach based on SWRL allows the representation and reasoning over temporal constraints in the criteria. Compared to the Epoch framework [7], this work focusses on the representation of patient data and eligibility criteria using a domain specific ontology.

## Acknowledgments

We are grateful to Olivier Dameron and Fiona McNeill for their valuable comments. The project was funded by the French National Agency of research (ANR) under the TECSAN program 2008.

## References

1. Allen, J.F.: An interval-based representation of temporal knowledge. In: Proceedings of the 7th IJCAI, pp. 221–226 (1981)
2. Aronson, A.R.: Effective mapping of biomedical text to the UMLS Metathesaurus: the MetaMap program. In: Proceedings of the AMIA Symposium, p. 17. American Medical Informatics Association (2001)
3. Bodenreider, O.: The unified medical language system (UMLS): integrating biomedical terminology. Nucleic Acids Research 32(database issue), D267 (2004)
4. Bodenreider I, O., Burgun, A.: Towards desiderata for an ontology of diseases for the annotation of biological datasets. In: International Conference on Biomedical Ontology 2009 (July 2009)
5. Horrocks, I., Patel-Schneider, P.F., Boley, H., Tabet, S., Grosof, B., Dean, M.: SWRL: A Semantic Web Rule Language Combining OWL and RuleML. W3C Member Submission 21 May 2004 In: World Wide Web Consortium (2004)
6. Noy, N.F., de Coronado, S., Solbrig, H., Fragoso, G., Hartel, F.W., Musen, M.A.: Representing the NCI Thesaurus in OWL DL: Modeling tools help modeling languages. Applied Ontology 3(3), 173–190 (2008)
7. Shankar, R.D., Martins, S.B., O'Connor, M.J., Parrish, D.B., Das, A.K.: Epoch: an ontological framework to support clinical trials management. In: Proceedings of the International Workshop on Healthcare Information and Knowledge Management, p. 32. ACM, New York (2006)
8. Sioutos, N., Coronado, S., Haber, M.W., Hartel, F.W., Shaiu, W.L., Wright, L.W.: NCI Thesaurus: a semantic model integrating cancer-related clinical and molecular information. Journal of biomedical informatics 40(1), 30–43 (2007)
9. Srinivas, K., Patel, C., Cimino, J., Ma, L., Dolby, J., Fokoue, A., Kalyanpur, A., Kershenbaum, A., Schonberg, E.: Matching patient records to clinical trials using ontologies. In: Aberer, K., Choi, K.-S., Noy, N., Allemang, D., Lee, K.-I., Nixon, L.J.B., Golbeck, J., Mika, P., Maynard, D., Mizoguchi, R., Schreiber, G., Cudré-Mauroux, P. (eds.) ASWC 2007 and ISWC 2007. LNCS, vol. 4825, pp. 809–822. Springer, Heidelberg (2007)
10. Weng, C., Tu, S.W., Sim, I., Richesson, R.: Methodological Review: Formal representation of eligibility criteria: A literature review. Journal of Biomedical Informatics 43(3), 451–467 (2010)

# Experience of Using OWL Ontologies for Automated Inference of Routine Pre-operative Screening Tests

Matt-Mouley Bouamrane[1], Alan Rector[2], and Martin Hurrell[3]

[1] College of Medical, Veterinary and Life Sciences
Centre for Population and Health Sciences
University of Glasgow, Scotland, U.K.
Matt-Mouley.Bouamrane@glasgow.ac.uk
[2] School of Computer Science, Manchester University, U.K.
Rector@cs.man.ac.uk
[3] CIS Informatics, Glasgow, Scotland, U.K.
martin.hurrell@informatics.co.uk

**Abstract.** We describe our experience of designing and implementing a knowledge-based pre-operative assessment decision support system. We developed the system using semantic web technology, including modular ontologies developed in the OWL Web Ontology Language, the OWL Java Application Programming Interface and an automated logic reasoner. Using ontologies at the core of the system's architecture permits to efficiently manage a vast repository of pre-operative assessment domain knowledge, including classification of surgical procedures, classification of morbidities, and guidelines for routine pre-operative screening tests. Logical inference on the domain knowledge, according to individual patient's medical context (medical history combined with planned surgical procedure) enables to generate personalised patients' reports, consisting of a risk assessment and clinical recommendations, including relevant pre-operative screening tests.

## 1 Introduction

In the U.K., a typical patient pathway to surgery involves the following steps: referral from primary care to an outpatient clinic in a hospital, pre-operative assessment of the patient at the hospital, the actual surgery, discharge from hospital and a return to community-based care. Pre-operative assessment is a routine medical evaluation and screening process which takes place prior to surgery in order to assess a patient's fitness for surgery, while identifying potential risk factors. Pre-operative screening is designed as an opportunity for taking appropriate action which can be beneficial to the patient. Screening is an opportunity to alter the clinical management of a patient scheduled for elective surgery, as new information come to the attention of health professionals. This information may warrant additional specific precautions, such as making provisions for additional resources (e.g. requesting some specialist equipment or booking a bed

P.F. Patel-Schneider et al.(Eds.): ISWC 2010, Part II, LNCS 6497, pp. 50–65, 2010.

in an Intensive Care Unit), requesting a specific intervention (e.g. pre-operative treatment or intervention, referral to specialist consultant) or even cancelling the planned surgery. Thus, the period between pre-operative assessment and surgery is seen as an important opportunity to anticipate risks and optimise the patient fitness for surgery. At this stage, routine ordering of pre-operative screening tests has long been a common practice within hospitals. Pre-operative tests may be requested both for asymptomatic patients and patients with specific risk factors (e.g. patients with a history of cardio-vascular or respiratory co-morbidities).

We describe our experience of developing a knowledge-based decision support system designed to assist health professionals in secondary care during the pre-operative assessment of patient prior to elective surgery. We review related work on medical decision support systems and background information on the pre-operative assessment process in section 2. We discuss design features of the system, including technology and specificity of the domain knowledge in section 3. We describe the iterative development of the system in section 4. We discuss our experience of implementing pre-operative test guidelines in section 5 and conclude with some final remarks and future work.

## 2   Pre-operative Assessment and Decision Support Systems

### 2.1   Medical Computer Decision Support Systems

The majority of errors within health delivery systems are not necessarily due to human errors, but are rather often consequences of broader systemic flaws in the organisation of processes and services [1]. The potential benefits of integrating computer-based Clinical Decision Support Systems (CDSS) [2,3,4,5,6,7,8,9,10] within work practices include the ability to:

- (i) influence clinicians behaviour and reduce variability of outcomes across various health professionals and increase the standardisation of processes towards evidence-based guidelines.
- (ii) combine and synthesise complex related pieces of information.
- (iii) facilitate access to clinical information and reporting of results through greater accessibility of data and improved display of information (e.g. using graphs, charts...)
- (iv) support the generation of patient-specific (medical history) and context specific (e.g. according to morbidity, surgical intervention, local hospital rules, etc.) prompts and reminders.
- (v) reduce medication adverse events with computer assisted order entry through a reduction of misread manual writing, notifications of adverse interaction, allergies, etc.
- (vi) identify patterns within the patient data which must be acted upon (e.g. abnormal or inconsistent findings, alerts, ordering of tests and further investigations, referral to specialist consultant...)

- (vii) doing all of the above while preserving health professionals' independence and ability to tailor patient care according to individual circumstances, specific needs, availability of resources or other constraints.

If successfully embedded within routine work practices CDSSs can become important process standardisation and error preventing tools. While CDSSs have generally proved reliable whenever rules and guidelines are clearly applicable, their record on emulating medical diagnosis is rather less obvious. This is due to the inherent difficulty and complexity in designing explicit conceptual models of the medical diagnosis thought process, except perhaps in the most straightforward cases, which would limit the usefulness of such systems.

CDSSs have other inherent limitations: recommendations issued by the systems can only be as good as the guidelines they model, and as a result, flaws in the guidelines will unfortunately be systematically reproduced in the output of the system [6]. Clinical knowledge is always limited or partial and it is not unusual for certain guidelines to be revised or proved wrong. Ironically, in these situations, patients' health would actually benefit from errors of omissions which a CDSS would make less likely [7]. There is obviously little CDSS designers can do about this problem apart from updating the system whenever new guidelines are introduced or old ones revised.

Several systematic reviews [2,3,4,8,9,10] concluded that CDSSs had generally (although not always...) demonstrated some benefits on clinical behaviour, compliance and performance during clinical controlled trials, including drug dosing and prescribing systems, preventive care and other generic or disease-specific systems. Regarding patients outcomes, results are less clear cut, in part due to the sparsity of available studies in that respect. In contrast, these studies showed that diagnostic aids have generally demonstrated little benefit to clinical practice overall (see Taylor's insightful description of the issues surrounding health informatics for a potential explanation of the lack of successes of diagnostic aids in [11]).

## 2.2   Pre-operative Assessment

In the United Kingdom, a patient due to undergo surgery will typically be referred to a hospital by his family doctor and will then first get an appointment at an out-patient clinic at the hospital for initial screening. The patient will then undergo a pre-operative assessment consisting of: answering a clinical questionnaire, generally followed by a physical examination, certain laboratory tests and possibly referral to a specialist consultant. Patient screening can be performed in a variety of settings: face to face consultations, paper-based questionnaires, on the telephone, or through web-based forms on the internet.

García-Miguel et al. define pre-operative assessment as *"the clinical investigation that precedes anaesthesia for surgical or non-surgical procedures, and is the responsibility of the anaesthetist"* [12]. The primary goal of pre-operative assessment is to maximise a patient's fitness for a (i.e. surgical) procedure by:

- (i) ensuring that the patient is fully informed about the procedure and has provided informed consent.

- (ii) identifying early in the patient's health pathway potential risks of perioperative (*i.e. during*) and postoperative complications due to pre-existing conditions (e.g. cardio-vascular or respiratory conditions, chronic diseases, multiple co-morbidities, previous adverse events, etc.).

- (iii) requesting additional investigations (e.g. tests) or referral to a specialist.

- (iv) taking steps to improve patient fitness (e.g. referral to family doctor for smoking cessation, weight loss, chronic disease control and management, etc.)

- (v) allocating appropriate resources for the day of surgery (e.g. taking appropriate actions to deal with patient's allergies, booking specialist equipment or a bed in critical care unit, etc.).

- (vii) considering alternatives to surgery when the risks of surgery are considered too high for the patient's safety.

- (vi) reducing the overall risk of late surgery cancellation by ensuring that all feasible precautionary steps have been taken prior to surgery.

Due to the vast scope of pre-operative assessment, the clinical domain knowledge potentially relevant for assessment is virtually limitless. For this reason, a generic PA has traditionally focused on identifying common allergies, cardio-vascular and respiratory risks and pre-empting potential airway complications, such as difficult intubation during anaesthesia. Complex surgical procedures may require additional precautions or even have separate specific pre-operative protocols.

# 3   A Knowledge-Base Pre-operative Assessment System

## 3.1   System Design Considerations

In addition to general considerations relevant to all CDSSs reviewed in the previous section, this research project had several important specific requirements. During the design phase of the project, requirements identified included that the system:

- (i) be capable of capturing highly specific patient medical information in a structured and coherent, yet flexible (i.e. *adaptive*) manner.

- (ii) have the ability to use and combine heterogeneous sources of clinical information.

- (iii) be capable of making useful inferences based on available evidence-based pre-operative assessment medical knowledge.

- (iv) provide context specific explanations for these medical inferences, targeted at a variety of health professionals (i.e. nurse, doctor, anaesthetist).

- (v) provide some level of transparency regarding the mechanisms for reaching previous medical inferences.

- (vi) provide some mechanisms to conveniently update and maintain the system in the face of new requirements and advances in the availability of evidence-based medical knowledge.

- (vii) being compatible with earlier versions of the system, including handling data from legacy patient databases while providing the same level of decision support.

## 3.2  System Overview

Figure 1 gives an overview of the general principles behind the architecture of our knowledge-based pre-operative decision support system. The system is composed of five main elements:

- 1. A patient pre-operative medical history information collection module. This component can be designed to be adaptive to the medical context of the information collected for all new patients entered in the system (case 1.a.). The adaptive behaviour of the system is obtained by modelling medical relationships and dependencies in a questionnaire ontology [13,14]. For patients whose medical history is already stored in some legacy clinical databases (case 1.b.), the automatic generation of a medical history is obtained through a reverse-engineering mapping from the legacy database information model to the questionnaire ontology [15].

- 2. The previous steps results in the generation of a patient pre-operative medical history representation in OWL [16]. There is an important distinction to be

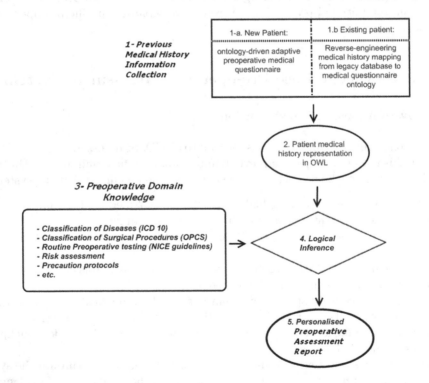

**Fig. 1.** Overview of Knowledge-based Pre-operative Decision Support System

made regarding this patient history: some of the information will have be obtained from clinical sources (e.g. examination by nurse or doctor, pre-operative tests, etc.) However, the information directly collected from the patient himself through the patient questionnaire is likely to be "coarse-grained", even if responding to the questionnaire is supervised by a pre-operative nurse, as is usually the case. Consider the following examples to illustrate this last point: a patient may know that he has *"diabetes"* but may be unable to qualify his condition any further than this. He may know that he is taking medication for a *"heart condition"* but may not be able to recall the exact name or the type of medication. While this unfortunately places some limitations on the accuracy and reliability of the patient's medical history, this is a consequence of the pre-operative process itself rather then a design flaw of the decision support system. What the system can do however is highlight which information was obtained through a reliable clinical source or was obtained from the patient himself, so a health professional can decide to make further investigation on a specific piece of information if necessary or relevant. This additional demand on the workload of the health professional could be somehow alleviated by obtaining the relevant information directly from the medical record at the patient's family doctor if available[1].

- 3. Because of the nature of pre-operative assessment in attempting to identify relevant risks of complications, the domain knowledge is potentially limitless. However, for practical reasons, pre-operative assessment has generally focused on generic risks (e.g. cardio-vascular, respiratory) unless the patient or surgical procedure require specific attention. We focused on designing a generic pre-operative assessment decision support ontology, including information on classification of morbidities using the ICD-10 International Classification of Diseases, classification of surgical procedures based on OPCS (Office of Population Censuses and Surveys) [2] and other relevant evidence-based pre-operative assessment medical knowledge, such as the NHS National Institute for Clinical Excellence (NICE) routine pre-operative tests guideline[3]. The pre-operative assessment ontology was developed in OWL along the principles of modularity developed by Rector et al. [17,18] for coherent and efficient knowledge update and management [19].

- 4. The personalised information representation obtained at step 2 is combined with the general domain knowledge of step 3 in order to make relevant logical inferences on this specific patient [20].

- 5. A personalised patient pre-operative report is compiled including (i) suggested pre-operative tests, (ii) risk assessments and (iii) suggested precaution protocols if relevant to the patient specific medical context.

---

[1] This is the subject of a research fellowship funded by the the Scottish Health Executive, Chief Scientist Office.

[2] http://www.connectingforhealth.nhs.uk/systemsandservices/data/ clinicalcoding

[3] National Health Service, National Institute for Clinical Excellence http://www.nice.org.uk/Guidance/CG3

# 4    System Implementation

## 4.1    System Development Tools

Prior to introducing semantic technology within the system, the pre-operative software was only composed of the following elements: user input, clinical data storage and a rule engine. The pre-operative risk assessment was then almost entirely based on the calculation of numeric scores. The introduction of semantic based technology in the system enabled adaptive information collection, high level semantic patient modelling and decision support based on patient classification rather than numeric rules only. This provides for a significant enhancement to the functionalities and capabilities of the system. Protégé-OWL was used as the main ontology development tool [21] and we used the web ontology language OWL [22], the java OWL application programming interface [23] and the Pellet reasoner [24].

## 4.2    Adaptive Medical Questionnaire

Context-sensitive adaptation is used to iteratively capture finer-grained information with each successive step, should this information be relevant according to a questionnaire ontology. The proposed method intends to replicate the investigating behaviour exhibited by clinicians when presented with items of information which may be cause for concern or require further attention. While the system has the potential to reduce the number of questions and thus save time and costs for healthy patients, the emphasis is rather on collecting *more* information *whenever* relevant so a proper informed patient risk assessment can be performed. The method is robust, scalable and highly configurable [13,14,16].

## 4.3    Medical Domain Knowledge of Pre-operative Assessment

As previously suggested, due to the nature of pre-operative assessment, the clinical domain knowledge relevant for assessment is potentially limitless. Some of the important knowledge resources introduced in the system included:

- Access to a knowledge base of approximately 1700 OPCS classification of surgical procedures. OPCS is the official classification of surgical procedures used by the NHS and is among other things used by hospital trusts to get reimbursed by the NHS for the procedures they carry out in the hospitals throughout the year (i.e. through the Payment by Results, PbR scheme by which funds are allocated according to levels of activity within the trusts). This feature is therefore of critical importance for integration of the application within hospital administrative information management systems. In addition, the OPCS knowledge base include unique OPCS code identifiers, detailed English clinical descriptions of surgical procedures, classification according to 16 major anatomical categories of procedures (e.g. vascular, thorax, abdomen, brain, etc.) and approximately 150 subcategories, allowing for fine grain classification of procedures. The repository also provides an overall surgery risk from grade 1 (minor) to grade 4 (major+) for each procedures.

- Access to the International Classification of Diseases, ICD-10 codes. This is a major feature in the application ability to classify elements of patient medical history into defined categories of morbidities (e.g. cardio-vascular, respiratory, renal diseases, unusual symptoms, etc.)

- Both of these previous features are critical to the integration of national and international pre-operative guidelines. We integrated the NICE pre-operative guideline, which is used to determine appropriate screening test investigations for a given patient, based on his individual medical context (medical history and planned surgical procedure). The guidelines are complex and are often not used in practice because healthcare professionals have neither the time nor the knowledge to apply them (see section 5).

### 4.4  Mechanism of Decision Support

In the system, decision support is usually provided in a 2 step process. The first step typically calculates risk scores using numerical formulas such as the Goldman and Detsky cardiac risk index [25]. or derives risk grades (e.g. ASA physical status classification grades[4]). The system does not use the decision support ontology at this stage but merely computes numeric values using an open source Java-based rule engine (JBoss Rules[5]). Once the risk grades and categories have been derived from the first risk calculation step, the system then performs decision support using the open-source java-based Pellet reasoner to reason on the decision support ontology given a patient OWL medical history profile. Examples of reasoning with the decision support ontology can be found in the next section as well as in [19,20].

## 5   Pre-operative Tests Recommendations and Reasoning

### 5.1  NICE Pre-operative Tests Guidelines

In the U.K., the effectiveness of pre-operative screening has been identified as a research priority by the National Institute for Health Research (NIHR) Health Technology Assessment (HTA) programme[6]. A systematic review on pre-operative screening tests was commissioned by HTA and published by Munro et al. in 1997 [26]. Following up on the previous work, the National Institute for Clinical Excellence (NICE) published in 2003 guideline recommendations on the use of routine pre-operative screening tests. These studies had important implications as they highlighted that: (i) the available evidence on the clinical benefits of routine pre-operative testing was mixed and the quality of studies was

---

[4] (American Society of Anaesthesiologists) ranging from ASA I (healthy patient) to ASA V (moribund).
[5] http://www.jboss.com/products/rules
[6] NIHR HTA, http://www.hta.ac.uk/

## ①②-GRADE IV SURGERY

| ③④ →  COMORBIDITY | ASA = 2 ③④ + Respiratory Comorbidity | | | | ASA = 3 ③④ + cardio-vascular Comorbidity | | | | ASA = 3 ③④ + Renal Comorbidity | | | |
|---|---|---|---|---|---|---|---|---|---|---|---|---|
| ↓ Preoperative  tests | AGE in years ⑤ | | | | AGE in years ⑤ | | | | AGE in years ⑤ | | | |
| | 16-40 | 40-60 | 60-80 | 80+ | 16-40 | 40-60 | 60-80 | 80+ | 16-40 | 40-60 | 60-80 | 80+ |
| Chest X-ray | | | | | | | YES | YES | | | | |
| ECG | | | YES | YES | YES | YES | YES | YES | | YES | YES | YES |
| Full Blood Count | YES | YES | YES | YES | YES | YES | YES | YES | YES | YES | YES | YES |
| Haemostasis | | | | | | | | | | | | |
| Renal function | YES | YES | YES | YES | YES | YES | YES | YES | YES | YES | YES | YES |
| Random glucose | NO | NO | NO | NO | NO | NO | NO | NO | | | | |
| Urine analysis | | | | | | | | | | | | |
| Blood gases | | | | | | | | | | | | |
| Lung function | | | | | NO | NO | NO | NO | NO | NO | NO | NO |

**Fig. 2.** NICE pre-operative guidelines. Investigations are recommended based on: (1) type and (2) risk grades of surgical procedure, (3) severity and (4) nature of co-morbidities and (5) patient age. There are 3 types of result for each test: "recommended", "not recommended" and "consider" (in amber).

weak overall and that (ii) there was no evidence to support the use of *systematic* routine pre-operative screening tests. In order to address the latest point, NICE set up a Guideline Development Group (GDG) to clarify the medical context in which the use of routine pre-operative tests was appropriate and justified by the available clinical evidence. After a thorough 13-steps guideline development process, including 2 parallel expert panels and a wide consultation exercise, NICE issued explicit guidelines for routine pre-operative testing. The guidelines made 3 types of recommendations: (i) test is "recommended", (ii) test is "not recommended" and (iii) "consider" the test based on 5 distinct input: (1) type of surgery (e.g. cardiac surgery, neuro-surgery or non-cardiac surgery) and (2) risk grades of surgical procedures, (3) severity (ASA grade) and (4) nature of co-morbidities (cardio-vascular, respiratory or renal) and (5) patient age. A sample of the guidelines is illustrated in Figure 2. The sample illustrates recommendations for various morbidities and severity levels. The guidelines were themselves summarised into various categories of surgery, surgery severity, morbidity categories and morbidity severity[7].

- *Type of investigations:* the guidelines include 9 potential investigations: Chest X-Ray, ECG (Electrocardiogram), Full Blood Count, Haemostasis, Renal Function, Random glucose, Urine analysis, Blood gases and Lung Function.

- *Type of recommendations:* there are currently 3 types of recommendations for each test: "test recommended", "test not recommended" and "consider test".

- *Factors Influencing recommendations:* There are 5 factors taken into consideration in order to find the relevant recommendations: the (i) age of the patient, (ii) his ASA grade, (iii) the type of co-morbidities the patient has (e.g. respiratory,

---

[7] http://www.nice.org.uk/nicemedia/live/10920/29090/29090.pdf

cardio-vascular, renal) (iv) the type of surgery (e.g. cardio-vascular surgery, neu-
rosurgery, etc.) (v) the risk grade of the surgery (from 1 to 4).

- *Number of cases in the guidelines:* the guidelines are summarised for pre-
operative health assessors into 36 tables such as the one illustrated in Figure 2.
There are different tables for different combinations of the 5 factors previously
described, including different tables for children under 16 years old and adults
over 16 years old. In total, there are at least 1242 different possible cases.

As the example in Figure 2 suggests, the guidelines recommend a chest X-Ray
test for patients over 60 years old *if* : they are scheduled for a surgical procedure
of severity risk grade IV and if they have a cardio-vascular co-morbidity of
severity level ASA=3. An ECG would be recommended for someone aged less
than 40 years old if they have a cardio-vascular co-morbidity of level ASA=3 but
not necessarily if they have a renal co-morbidity of level ASA=3 or a respiratory
co-morbidity of level ASA=2. If the patient is between 40 and 60 years old, an
ECG would be recommended if the patient has either a cardio-vascular or a renal
co-morbidity of level ASA=3 but again not necessarily if he has a respiratory
co-morbidity of level ASA=2, etc.

Perhaps not surprisingly, we found that in practice, pre-operative health assessors faced considerable difficulties in using the guidelines. The important number
of factors to take into consideration in order to find the correct table and then
the specific case within this table, combined with the significant number of tables meant that too much time was being spent by pre-operative health assessors
trying to refer to the correct case. In addition, the pre-operative health assessors would need to be able to categorise (i) the type of co-morbidities (ii) their
severity (e.g. for determining the patient's ASA grade) (iii) the type of surgical
procedures and (iv) their surgical risk grades: all of these steps being necessary
before being able to refer to the correct table. All of these tasks are obviously
highly knowledge intensive as well as being intellectually demanding. In addition, pre-operative health assessors typically see dozens of different patients a
day, each with a wide variety of health conditions and scheduled for various
types of surgical procedures. In practice, the consequences are that, if in doubt,
pre-operative investigations would probably be requested regardless of the guidelines (i.e. *better safe than sorry*) , thus defeating the purpose of the guidelines
in efficiently managing the allocation of pre-operative investigations within care
delivery.

## 5.2   Pre-operative Tests Recommendations

We combined the use of an ontology and reasoner in the pre-operative decision support system in order to automatically make recommendations regarding
the suitability of tests based on the NICE guidelines. The first step consisted
into transforming the NICE tables into rules. This enabled to considerably reduce overlap and redundant information in the current format of the guidelines. The 1242 different possible cases currently covered by the NICE guidelines
were reduced to around a hundred rules [20] (see Figure 3). Effectively we modelled instances of the NICE test recommendations as *super-test entities* in the

| Pre-operative TEST | RECOMMENDED TEST ENTITY | CONSIDER TEST ENTITY | NOT RECOMMENDED TEST ENTITY | TOTAL TEST ENTITIES |
|---|---|---|---|---|
| Blood gases | ✗ | ✓ 3 | ✓ 1 | 4 |
| Chest X-ray | ✓ 2 | ✓ 10 | ✓ 7 | 19 |
| ECG | ✓ 9 | ✓ 9 | ✓ 4 | 22 |
| Full blood count | ✓ 5 | ✓ 5 | ✓ 2 | 12 |
| Haemostasis | ✗ | ✓ 6 | ✓ 3 | 9 |
| Lung function | ✗ | ✓ 3 | ✓ 3 | 6 |
| Random glucose | ✗ | ✓ 6 | ✓ 3 | 9 |
| Renal function | ✓ 9 | ✓ 6 | ✓ 2 | 17 |
| Urine analysis | ✗ | ✓ 3 | ✓ 1 | 4 |
| TOTAL RECOMMENDATIONS | 25 | 51 | 26 | 102 |

**Fig. 3.** Categories of OWL Pre-operative test entities according to: (i) pre-operative tests and (ii) type of recommendations

ontology. Definition of these test entities are based: on the 3 dimensions of recommendations (recommended, not recommended, consider) and the 9 categories of pre-operative tests. Patient classification entities are then categorised as subclass entities of pre-operative test entities based on individual medical histories, a process which is described and illustrated in more details in the next section using a practical example. In Figure 3, certain tests have no *recommendation test entities* associated with them as the guidelines simply do not make explicit recommendations for these tests (but suggest instead that it may appropriate to consider the tests in a specific medical context, as described in Figure 2). Certain tests, such as chest X-ray, ECG, and renal function, are modelled using a larger number of test entities because their medical allocations is dependant on various specific combinations of the 5 input features described previously in 5.1. Others such as blood gases and urine analysis are more generic (e.g. the latter being potentially considered for *all* adults, regardless of surgery or morbidity). An important modelling consideration was that the test entities were not modelled in order to minimise the number of entities but rather to *facilitate the interpretation of the rules by health professionals*, as is illustrated in the next section.

## 5.3   Example of Pre-operative Tests Reasoning

Figure 5 provides an example of pre-operative test recommendation based on reasoning on the decision support ontology. The patient: (i) is between 40 and 60 years old, (ii) has unspecified angina pectoris of severity level (iii) ASA = 3 and is scheduled for a repair of mitral valve procedure (OPCS code K2580). The procedure K2580 is classified in the ontology as a cardio-vascular surgery of risk grade IV. Reasoning on the decision support ontology issues 16 test recommendations (some of which relating to duplicate tests) in total: 7 *recommended* tests

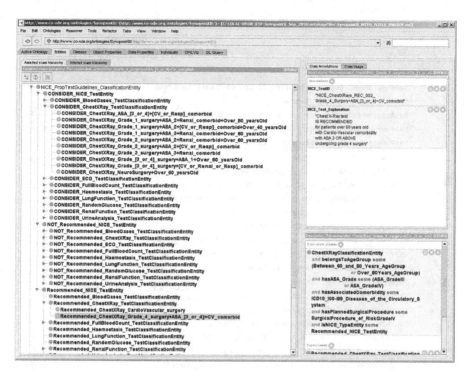

**Fig. 4.** The NICE guidelines as OWL entity rules as viewed through the Protégé-OWL User Interface

and 9 *consider* tests. The recommendations are made by the system based on the following reasons:

**Chest X-Ray, test recommended:** 1 recommended (patient is scheduled for *cardio-vascular* surgery) + 2 consider tests ({*grade IV* surgery and *cardio-vascular co-morbidity*} and {*ASA=3* and *cardio-vascular* co-morbidity}).

**ECG, test recommended:** 2 recommended ({*cardio-vascular* surgery} and {*cardio-vascular* co-morbidity})

**Full Blood Count, test recommended:** 1 recommended (*cardio-vascular* surgery).

**Renal function, test recommended:** 4 recommended ({ *cardio-vascular* surgery} and {*ASA=3* and *cardio-vascular* co-morbidity} and {*grade IV* surgery} and last 2 combined )

**Haemostasis, consider test** 2 consider ({*grade IV* surgery} and { *cardio-vascular* surgery})

**Blood Gases, consider test** 2 consider ({ *ASA=3* } and {*grade IV* surgery})

**Urine Analysis, consider test** 1 consider ( all *adults* over 16 years old)

**Random Glucose, consider test** 2 consider ({*grade IV* surgery} and {*cardio-vascular* surgery})

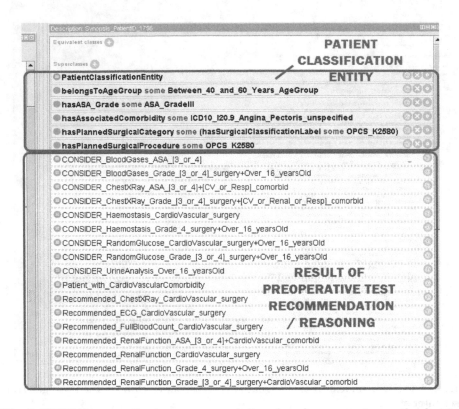

**Fig. 5.** NICE Pre-operative test recommendations based on reasoning on the decision support ontology

## 5.4   Pre-operative Tests and Multiple Morbidities

A significant development in current world health trends is the substantial increase in the prevalence of chronic diseases and multiple morbidities within the general population, both in developed and developing countries [27]. The NICE guidelines do not explicitly deal with the issue of *multiple co-morbidities* and this is an other aspect in which the system can provide additional decision support functionalities. In the case of a patient with multiple co-morbidities, a test may be recommended for multiple reasons, as illustrated in the previous example. In this case, the system can issue a *strong* recommendation alongside relevant explanations. Again, the recommendations in the guideline are not mutually exclusive, particularly not in the case of multiple co-morbidities. Thus, we implemented the system so one instance of a "recommended test" within a batch of test results would lead to a positive test recommendation regardless of all the other test recommendations. According to the same principle, if the system returns "consider test" instances along instances of "test not recommended", then the system issues a final "consider test" recommendation. Finally, the system issues a "test not recommended" advice only if all instances retrieved are negative for the specific test. We are hoping to introduce more sophisticated support for dealing with multiple conditions as clinical models of complex morbidities are being developed and guidelines become routinely available within the health services [28,29,30].

## 6    Conclusion and Future Work

We have presented our work to date on the design and implementation of a knowledge-based pre-operative assessment support system. We have discussed how pre-operative assessment is a generic clinical screening process, which purpose is to identify potential risks of complications prior to surgery. By its very nature, the clinical knowledge relevant to pre-operative assessment is potentially limitless. We have proposed some solutions to efficiently harness and manage pre-operative assessment clinical knowledge. The system was developed using semantic web technology including modular ontologies developed in OWL, the OWL API and an automated logic reasoner. This design has provided substantial improvements on earlier versions of the systems, including the ability to tailor patient information collection according to individual medical context, the ability to efficiently manage a vast repository of pre-operative assessment domain knowledge, including classification of surgical procedures and morbidities, and guidelines for routine pre-operative tests. An important modelling consideration was that test entities modelled in the system were not modelled in order to minimise the number of entities but rather to *facilitate and optimise the interpretation of rules by health professionals*. Future work will involve evaluating the use of the system in clinical settings, in particular with regards standardisation of pre-operative processes and compliances with the guideline recommendations suggested by the system.

**Acknowledgement.** Dr. Bouamrane is a Scotland Chief Scientist Office (CSO) funded postdoctoral fellow in health informatics and health services research working on the development of information systems in order to facilitate the integration of pre-operative assessment across primary and secondary care in Scotland.

## References

1. Reason, J.: Human error: models and management. British Medical Journal, BMJ 320, 768–770 (2000)
2. Johnston, M.E., Langton, K.B., Haynes, R.B., Mathieu, A.: Effects of Computer-based Clinical Decision Support Systems on Clinician Performance and Patient Outcome: a Critical Appraisal of Research. Annals of Internal Medicine 120(2), 135–142 (1994)
3. Hunt, D.L., Haynes, R.B., Hanna, S.E., Smith, K.: Effects of Computer-Based Clinical Decision Support Systems on Physician Performance and Patient Outcomes - A Systematic Review. Journal of American Medical Association, JAMA 280(15), 1339–1346 (1998)

4. Kaplan, B.: Evaluating informatics applications - clinical decision support systems literature review. International Journal of Medical Informatics 64(1), 15–37 (2001)
5. Bates, D.W., Cohen, M., Leape, L.L., Overhage, J.M., Shabot, M.M., Sheridan, T.: Reducing the Frequency of Errors in Medicine Using Information Technology. Journal of American Medical Informatics Association, JAMIA 8, 299–308 (2001)
6. Sim, I., Gorman, P., Greenes, R.A., Haynes, R.B., Kaplan, B., Lehmann, H., Tang, P.C.: Clinical Decision Support Systems for the Practice of Evidence-based Medicine. Journal of American Medical Informatics Association, JAMIA 8(6), 527–534 (2001)
7. Morris, A.H.: Decision support and safety of clinical environments. Quality and Safety in Health Care 11, 69–75 (2002)
8. Garg, A.X., Adhikari, N.K.J., McDonald, H., Rosas-Arellano, M.P., Devereaux, P.J., Beyene, J., Sam, J., Haynes, R.B.: Effects of Computerized Clinical Decision Support Systems on Practitioner Performance and Patient Outcomes. Journal of American Medical Association, JAMA 293(10), 1223–1238 (2005)
9. Kawamoto, K., Houlihan, C.A., Balas, E.A., Lobach, D.F.: Improving clinical practice using clinical decision support systems: a systematic review of trials to identify features critical to success. British Medical Journal, BMJ 330, 765(8p) (2005)
10. Randell, R., Mitchell, N., Dowding, D., Cullum, N., Thompson, C.: Effects of computerized decision support systems on nursing performance and patient outcomes: a systematic review. Journal of Health Services Research & Policy 12(14), 242–249 (2007)
11. Taylor, P.: From Patient Data to Medical Knowledge: The Principles and Practice of Health Informatics. BMJ, Blackwell Publishing (2006)
12. García-Miguel, F., Serrano-Aguilar, P., López-Bastida, J.: Pre-operative assessment. The Lancet 362(9397), 1749–1757 (2003)
13. Bouamrane, M.M., Rector, A., Hurrell, M.: Gathering Precise Patient Medical History with an Ontology-driven Adaptive Questionnaire. In: Proceedings of Computer-Based Medical Systems, CBMS 2008, Jyväskylä, Finland, pp. 539–541. IEEE Computer Society Press, Los Alamitos (2008)
14. Bouamrane, M.M., Rector, A., Hurrell, M.: Ontology-Driven Adaptive Medical Information Collection System. In: An, A., Matwin, S., Raś, Z.W., Ślezak, D. (eds.) ISMIS 2008. LNCS (LNAI), vol. 4994, pp. 574–584. Springer, Heidelberg (2008)
15. Bouamrane, M.M., Rector, A., Hurrell, M.: Semi-Automatic Generation of a Patient Pre-operative Knowledge-Base from a Legacy Clinical Database. In: Meersman, R., Dillon, T., Herrero, P. (eds.) OTM 2009. LNCS, vol. 5871, pp. 1224–1237. Springer, Heidelberg (2009)
16. Bouamrane, M.M., Rector, A., Hurrell, M.: Using Ontologies for an Intelligent Patient Modelling, Adaptation and Management System. In: Meersman, R., Tari, Z. (eds.) OTM 2008, Part II. LNCS, vol. 5332, pp. 1458–1470. Springer, Heidelberg (2008)
17. Rector, A.: Modularisation of domain ontologies implemented in description logics and related formalisms including OWL. In: Proceedings of the 2nd International Conference on Knowledge Capture, K-CAP 2003, Sanibel Island, FL, USA, October 23-25, pp. 121–128. ACM, New York (2003)
18. Rector, A., Horridge, M., Iannone, L., Drummond, N.: Use Cases for Building OWL Ontologies as Modules: Localizing, Ontology and Programming Interfaces & Extensions. In: Proceedings of 4th Int Workshop on Semantic Web Enabled Software Engineering, SWESE 2008, with ISWC 2008, Karlsruhe, Germany (October 2008)

19. Bouamrane, M.M., Rector, A., Hurrell, M.: Development of an Ontology of Pre-operative Risk Assessment for a Clinical Decision Support System. In: Proceedings of the 22nd IEEE International Symposium on Computer-Based Medical Systems, CBMS 2009, 4th Special Track on Ontologies for Biomedical Systems, Albuquerque, USA. IEEE Computer Society Press, Los Alamitos (August 2009)

20. Bouamrane, M.M., Rector, A., Hurrell, M.: A Hybrid Architecture for a Pre-operative Decision Support System using a Rule Engine and a Reasoner on a Clinical Ontology. In: Polleres, A. (ed.) RR 2009. LNCS, vol. 5837, pp. 242–253. Springer, Heidelberg (2009)

21. Knublauch, H., Fergerson, R.W., Noy, N.F., Musen, M.A.: The Protégé OWL Plugin: an open development environment for Semantic Web applications. In: McIlraith, S.A., Plexousakis, D., van Harmelen, F. (eds.) ISWC 2004. LNCS, vol. 3298, pp. 229–243. Springer, Heidelberg (2004)

22. OWL 2.: Web Ontology Language document overview (2009), http://www.w3.org/tr/owl2-overview/

23. Horridge, M., Bechhofer, S., Noppens, O.: Igniting the OWL 1.1 touch paper: the OWL API. In: Proceedings of the Third International Workshop of OWL Experiences and Directions, OWLED 2007, Innsbruck, Austria (2007)

24. Sirin, E., Parsia, B., Grau, B.C., Kalyanpur, A., Katz, Y.: Pellet: A practical OWL-DL reasoner. Journal of Web Semantic 5(2), 51–53 (2007)

25. Palda, V.A., Detsky, A.S.: Perioperative assessment and management of risk from coronary artery disease. Annals of internal medicine 127(4), 313–328 (1997)

26. Munro, J., Booth, A., Nicholl, J.: Routine pre-operative testing: a systematic review of the evidence. Health Technology Assessment 1(12), 76 (1997)

27. W.H.O: Preventing chronic diseases: a vital investment. World Health Organization, Geneva. Technical report (2005)

28. Fortin, M., Dubois, M.F., Hudon, C., Soubhi, H., Almirall, J.: Multimorbidity and quality of life a closer look. Health and Quality of Life Outcomes 5(52), 8 (2007)

29. Mercer, S.W., Smith, S.M., Wyke, S., O'Dowd, T., Watt, G.C.: Multimorbidity in primary care: developing the research agenda. Journal of Family Practice 26(2), 79–80 (2009)

30. Valderas, J.M., Starfield, B., Sibbald, B., Salisbury, C., Roland, M.: Defining co-morbidity: Implications for understanding health and health services. Annals of Family Medicine 7(4), 357–363 (2009)

# Enterprise Data Classification Using Semantic Web Technologies

David Ben-David[1], Tamar Domany[2], and Abigail Tarem[2]

[1] Technion – Israel Institute of Technology, Haifa 32000, Israel
davidbd@cs.technion.ac.il
[2] IBM Research – Haifa, University Campus, Haifa 31905, Israel
{tamar,abigailt}@il.ibm.com

**Abstract.** Organizations today collect and store large amounts of data in various formats and locations. However they are sometimes required to locate all instances of a certain type of data. Good data classification allows marking enterprise data in a way that enables quick and efficient retrieval of information when needed. We introduce a generic, automatic classification method that exploits Semantic Web technologies to assist in several phases in the classification process; defining the classification requirements, performing the classification and representing the results. Using Semantic Web technologies enables flexible and extensible configuration, centralized management and uniform results. This approach creates general and maintainable classifications, and enables applying semantic queries, rule languages and inference on the results.

**Keywords:** Semantic Techniques, RDF, Classification, modeling.

## 1 Introduction

Organizations today collect and store large amounts of data in various formats and locations. The data is then consumed in many places, sometimes copied or cached several times, causing valuable and sensitive business information to be scattered across many enterprise data stores. When an organization is required to meet certain legal or regulatory requirements, for instance to comply with regulations or perform discovery during civil litigation, it becomes necessary to find all the places where the required data is located. For example, if in order to comply with a privacy regulation an organization is required to mask all Social Security Numbers (SSN) when delivering personal information to unauthorized entities, all the occurrences of SSN must be found. That is when classification becomes essential.

Data discovery and classification is about finding and marking enterprise data in a way that enables quick and efficient retrieval of the relevant information when needed. Since classifying and tagging everything in the organization's data sources is time-consuming, inefficient and rarely done well, usually organizations need to choose what to classify, determining the relevant and important pieces of information that need tracking. The classical approach today is to start by examining the

P.F. Patel-Schneider et al.(Eds.): ISWC 2010, Part II, LNCS 6497, pp. 66–81, 2010.

purpose of the classification and how the classified data is going to be used. The process consists in manually examining the relevant laws or regulations, identifying which types of information should be considered sensitive and what are the different sensitivity levels, and then building the classes and classification policy accordingly. The main drawback of such an approach is that each time the requirements change (e.g., due to changes in existing laws or internal policies or the addition of new laws) it requires re-classification of the data.

Moreover, enterprise data can be stored in many varied formats: unstructured, semi-structured and structured data; and is typically stored in various data stores such as relational databases, file systems, mail servers, etc., sometimes even at different physical locations (sites) in the organization. Most existing solutions for automatic data classification apply to a single data type, mainly unstructured data, and can generally only identify predefined sets of known fields, depending on the domain. As a result, organizations wishing to classify their data typically need to deploy and configure several different products and store and manage their results separately. A major drawback of these traditional solutions is the lack of common interfaces.

The need to build a generic classification system with manageable results introduces several challenges. First, a common 'language' that can be used as input to all classifiers must be devised. This 'language' must enable an accurate classification process but also provide flexibility and extensibility to new types of classifications, classes and data types. In addition, a uniform format for the classification results must be created so as to maximize their usability and homogeneity.

In this paper we introduce a generic, automatic classification method that enables classifying data elements from all types, formats and sources based on a common classification scheme, and making it possible to use the results in a uniform manner. We suggest a method based on Semantic Web technologies for classifying any type of data from any domain, by modeling the information in question as an ontology. According to our approach, a model (or ontology) provides the means for describing the entities to discover, the classes to map them to, and information on how to discover them. Such an ontology can serve as input to several different classifiers, each for a different source of data. The ontology is then used again as a schema for representing the discovery results, thus unifying the results across the different sources.

The ontology used as input for the classification process can describe any type of content. It can be generic, containing information that may apply to many different organizations and/or requirements, such as knowledge on what identifies a person (PII); it can be specific to a certain domain, such as medications and treatments for the Healthcare domain; and it can be specific to a certain organization or even a particular application.

We use the Resource Description Framework (RDF) and its extensions (RDFS and OWL) to represent the classification models and results. By using RDF for these purposes we maximize modularity and extensibility and facilitate easy navigation between the results, the data models and the actual data in the data sources, as will be described in the following sections.

Using models, in combination with RDF and additional Semantic Web tools gives us several advantages. First, it enables decoupling the classification process from the intended usage, allowing more general and maintainable classifications with less dependency on changes. Using the model as a common input to all classifiers enables centralized management, auditing and change control. A uniform format for all classification results enables fast and easy location of the data when needed, thus increasing its usability. It is then possible to apply semantic queries on the results, feed them to inference engines to deduce additional relations, and even apply rule languages such as AIR [13] and TAMI [27] to enforce and verify compliance with existing policies.

Using an ontological model to describe the input and output of the different classification algorithms, each working on a different type of data source, enables decoupling the data descriptions from the actual implemented algorithms. This design pattern can prove beneficial for many different enterprise data management tasks, data classification being just one example.

The paper is organized as follows: Section 2 describes related work, both in the area of classification in general, and in the use of RDF in information management. Section 3 describes the general idea of classification models, as well as the PII model as a specific example. In Section 4 we discuss general requirements from classifiers and describe in detail a relational data classifier. In Section 5 we give a short description of our implementation and some results, and we conclude in Section 6.

## 2  Related Work

### 2.1  Classification

There are many existing algorithms for automatic classification of unstructured documents, using various techniqes such as pattern maching, keywords, document similarity and different machine learning techniques (Bayesian, Decision Trees, Resource Limited Immune Classifier System, k-Nearest Neighbor and more), as well as technologies such as Autonomy's Meaning-Based Computing Technology [2]. Any of the above can be used as part of the classification process.

In many existing works, the classes and the entities to discover are predefined, deriving from the specific use case and domain of the solution. There are several works that suggest a way to find and describe the classes. Miler [18] suggests a method for automatically enriching the classes based on the discovery. Ben-Dor et al. [3] suggest a way to find the meaningful classes from the content in the computational biology domain. Taghva at al. [24] suggest to use ontologies to describe the classes when classifying emails.

Our approach is to enable modelling any type of data to be classified, from any domain, and to use the same model to describe the classification schema for all input types, thus simplifying the creation and management of classification policies across the enterprise.

Most work related to classifying relational data deals with classifying or clustering records based on common patterns or associations. Much less has been done in the area of classifying based on metadata (tables, columns, types), according to specific categories. The same applies to semi-structured data such as XML files. One scenario in which we found some similar work is electronic discovery (eDiscovery)[1]. Butler et al. [7] developed a top level ontology to represent enterprise data for eDiscovery, and created a semantic mapping that manually maps the data models of heterogeneous sources into the ontology.

In addition, many existing classification techniques classify only whole documents or assets, and not single data elements. We suggest a method to automaticaly classify data from different sources according to a model, enabling accurate classification at the level of a single data element.

## 2.2 Use of Ontologies and RDF in Information Management

The use of Semantic Web methodologies in the world of information management, has lately gained much momentum.

The use of ontologies in information and knowledge management has long been recognized as a natural connection. According to Staab at al. [23], who present an approach for ontology-based Knowledge Management, ontologies open the way to move from a document-oriented view of Knowledge Management to a content-oriented view, where knowledge items are interlinked, combined, and used. Maedche et al. [17] also present an integrated enterprise-knowledge management architecture, supporting multiple ontologies and ontology evolution. Ontology-driven information and knowledge management in specific domains has also been investigated [9].

RDF specifically can also be used to unify data access across multiple sources. Langegger et al. in their work [16] presented a system that provides access to distributed data sources using Semantic Web technology designed for data sharing and scientific collaboration. Warren and Davies [26] studied a semantic approach to information management for integrating heterogeneous information as well as coping with unstructured information.

RDF has recently begun to be used in the context of data classification. Jenkins et al. [11], use RDF to express the results of classifying HTML documents using the Dewey Decimal Classification. In addition, Xiaoyue and Rujiang [28] showed how using RDF ontologies to perform concept-based classification of text documents significantly improved text classification performance.

We use RDF to express both the input and the output of the classification process, thus enabling high connectivity and easy navigation between the models, the results and the original data sources and providing a unified representation of all classified data sources.

## 3  Classification Models

The main goal of our solution is to enable organizations to annotate their knowledge bases with semantic meaning. Our approach offers to automate the classification process using classification models, which are ontologies that describe

---

[1] Discovery in civil litigation that deals with information in electronic format.

entities in the organization's knowledge base and their relationships. These models can be generic, domain-specific or even organization-specific.

The classification process is performed at the single data element level and results in bi-directional references between the individual data parcels and terms in the model. This method enables the "classify once, enforce many" approach for future application of policies and other semantic applications.

The Data Discovery and Classification process involves four steps:

1. Defining and identifying the data to discover (the valuable pieces of information to locate).
2. Building a classification scheme. One example of a common classification scheme is top secret, secret, confidential, etc.
3. Discovering where the valuable data is located.
4. Documenting the findings in a useful manner.

We offer to use the same model for several purposes throughout the discovery and classification process: the description of what entities to search for (step 1), the classification scheme to which the findings are associated (step 2), directives on how to search for each entity (input to step 3) and the schema for representing the discovery results (step 4).

Selecting RDF as the language in which to represent the classification model enables us to exploit existing and evolving tools for annotating, reasoning, querying, etc. the classified data. We can take advantage of common vocabularies (RDFS, OWL) to express stronger relations and properties, best suited for ontologies, and benefit from the powerful features of RDF that makes it easy to expand, merge and combine existing classification models, and generate new models for different purposes.

### 3.1  Models as Input

The classification models are used as input to an automatic classification tool (henceforth: *Classifier*), which is the software component that performs the actual mapping. Our goal is to have a general-purpose classifier, capable of handling any domain-specific knowledge, and so the input for the classifier must be designed to answer not only 'what to search for', but also 'how to search'. This means that in addition to functioning as an ontology, the classification model must also contain additional characteristics of the entities such as type, format and possible synonyms. This allows full automation of the classification process by use of natural language processing and information retrieval techniques.

To implement such a general-purpose classifier, a need emerged for a meta-model to describe the expected structure and contents of a valid classification model. An input model complying with this meta-model is regarded as an instance of the meta-model. The same meta-model serves for describing models for all data types; the specifics of searching in each data format are encapsulated in the classifier's implementation.

## 3.2   The Meta-model

We designed a preliminary meta-model for classification models used as input to our system implementation, which is depicted in Figure 1. The top-most element in the classification meta-model is the subject element that is the root node of any domain-specific ontology. A subject can possibly contain several categories, and each category can hold several fields, which are the basic terms linked by the classifier to data units in the organization's data sources. Each field can be associated with a data type and format (e.g., denoted by a regular expression), and one or more synonym elements, which represent keywords or phrases that can be used to describe the field. Each keyword (or token) in a synonym can be stated as optional, required or exact. These synonyms are processed by the classifier to create different search strings used to search for potential matching elements in the data stores. We elaborate on this matter in section 4.2.

The meta-model also supports mapping between a field in the classification model and one or more actual data elements. This mapping, obtained as the result of the classification process, is also represented in the model.

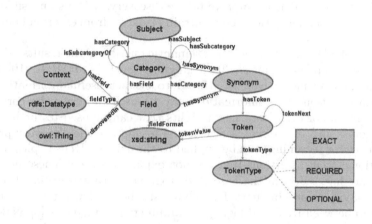

**Fig. 1.** The classification meta-model

## 3.3   Models as Output

The classification results, which are basically a mapping between fields in the model and specific data elements, are also saved in RDF format, as an extension of the models used to perform the search (i.e. using the same vocabulary defined in the meta-model). They are composed of triples, each triple representing a link between a classification field and a data element. The mapping uses the data element's URI, which differs according to the type of data and the manner in which it is represented in RDF. For example, in relational databases the mappings are to table columns, and the URIs used are those of the database server, schema, table and column. In XML files, the mappings are to specific XPaths, and the URIs are those representing the XML file location (e.g. in a file system or Content Management system) and the specific element's XPath. In unstructured data, the mapping is to a word or group of words in a document, and the URIs represent the location of the document and the location of the word/s in the document (e.g. chapter/page, paragraph, sentence, etc.).

The advantages in using RDF to represent the classification results are many:

1. Different results from different runs can be linked together to form one unified classification source containing all discovered entities (a concept demonstrated by Langegger at al. in [16]). If a model has been extended or a new model introduced, a partial search (only on the new parts) can be peerformed and the results combined with the existing ones. It is also possible to express information both on the location of the classified data (such as a database column or XPath), as well as the data itself (an actual value).

2. Thanks to the use of URIs to describe the different resources, it is very easy to navigate between the discovery results, the classification models and the RDF representations of the searched data sources. In this way, while accessing the results, additional information about the model or data source can be accessed to verify that the results are correct, add new links or add new entities to the models to improve future discovery. It is also possible to obtain additional information about certain resources from external knowledge-bases.

3. The use of RDF allows performing semantic queries on the results. Many existing query languages for the RDF language (such as SPARQL [19], RDQL [20] and many more) can be utilized to extract useful information from the classification. This information may be used to verify the correctness of the results or even learn new insights that can be used to further enrich the models and/or the search algorithms to improve the discovery process.

4. Inference (or reasoning) can be applied to the existing RDF triples to derive additional RDF assertions that are not explicitly stated. These new assertions are inferred from the base RDF together with any optional ontology information, axioms or rules. There have also been a few attempts to bridge the gap between RDF and logic programming and languages [6]. N3logic [4] provides a logical framework to express logic in RDF, enabling the use of rules to make inferences, choose courses of action, and answer questions.

5. Several policy and rule expression languages for RDF (such as AIR [13], KAoS [25], and Rei policy language [12], etc.) as well as tools for checking compliance with policies (such as TAMI [27]) exist, enabling the enforcement of privacy rules or any other business policies on the discovered data.

### 3.4  Example: PII Model

As a sample use case, we may consider the field of information security. In modern times, organizations retain an explosively growing volume of sensitive data to which access is available to an expanding user base. The growing public concern over the security and privacy of personal data has placed these organizations in the spotlight, and countries around the world are developing laws and regulations designed to support confidentiality.

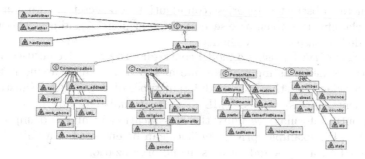

**Fig. 2.** Partial PII RDF Model

Personally Identifiable Information (PII) refers to information that can be used – whether alone or in combination with other personal or identifying information – to uniquely identify a single individual. For example, national identification numbers, vehicle registration plates and fingerprints are often used to distinguish individual identities and are clearly PII. Even less identifying characteristics such as city of residence, gender and age are considered PII for the potential that when joined together they might be linked with other public available information to identify an individual.

By successfully identifying and classifying PII in an organization's data, it is possible to apply virtually any privacy policy. Figure 2 displays a partial view of the PII model we have developed. The root element is Person, and it contains categories such as PersonName and Address. Each category contains fields such as firstName, middleName and lastName.

### 3.5   Advantages of Classification Using Models

There are several advantages for using ontologies in general, and RDF in particular, in the discovery and classification process:

- **Extensibility** and **Flexibility:** The model can be extended both by adding more entities to search for and by adding additional tips and hints to aid the discovery of existing entities. Since the information on how to search for the entities is included in the model that is used as input to the discovery algorithm, the model can be extended, edited or replaced without the need to change the discovery algorithm itself.
- **Linking to external knowledge bases:** The models used to perform the classification can themselves be derived from existing business models and/or domain- or industry-specific models and ontologies. Furthermore, additional knowledge from external ontologies or taxonomies can be linked into the model, enabling automatic extension of the model, addition of new synonyms, related terms, etc.
- **Resistance to policy changes:** By classifying data sources according to specific classes of information (first name, phone number...) as opposed

to general, abstract categories (confidential, top-secret...) we can use the same classification to enforce several different policies and no reclassification is needed in case of policy changes.

- **Homogeneity:** By using a model containing the sensitive entities and classification schema as input, the same model can be used as input to many different classifiers that classify different types of data sources (unstructured documents, forms, emails, XML files, relational data, etc.). By using the same output format, we unify the classification results and improve their usability. For example, privacy policies can be enforced in a unified manner across all data types and formats in the organization.
- **Centralization:** Everything is linked through the use of unique IDs. This allows the model to serve as a centralized point to manage all valuable information in the organization and enables easy location of all related pieces of data in one click. On the other hand, given the unique ID of data location, we can easily find its classification.

## 4  Classifier

Once the models describing the data the organization wishes to classify are designed, these models are used as input to a software component that performs the classification of the actual data stores. This component is referred to as the *Classifier*. In this section we will describe the approach in general and a relational data classifier in more detail.

### 4.1  Classifier Inputs

The classifier receives as input RDF files of two types: the first type is the RDF files representing the models described in the previous section; the second type is RDF files representing the data to be searched, for example, the schema of a database. The classifier performs a search for relevant items in the data schemas, refines these results according to several filters, and creates a new RDF file containing the search results.

**Using one or more models to perform the search.** New models can be created as needed for domain or organization-specific classifications and existing models can be extended. Any combination of models can be used as input when performing the search on a data source, and searches may be performed on different combinations of models to yield results for different purposes (such as privacy enforcement, retrieving specific information within a set timeframe in order to meet regulations, de-duplication and increasing data quality, etc.). Each model used as input to the classifier must be an instance of the classification meta-model described in Section 3.2.

**RDF representation of data source schemas.** Organizations today have many different data sources, which may include structured data (e.g., relational databases), semi-structured data (e.g., XML files), and unstructured data (e.g., e-mail archives and text documents). RDF provides a means to represent different types of schemas in a unified manner in order to perform semantic queries on them. This is what allows us to perform the classification in a uniform manner on all data sources, without the need for any prior knowledge about the specific underlying schemas.

Much work has been done in the domain of representing relational database metadata in RDF format. RelationalOWL [15] is an OWL-based representation format for relational data and schema components. The D2RQ API [5] provides access to relational database content from within the Jena and Sesame RDF frameworks.

Some work has been done in the direction of transforming XML schemas and instances to RDF format, however a completely generic technique for fully representing any XML file in RDF has still not been found at the time of writing this article. Existing work has been focused on either transforming "RDF-enabled" XML files to RDF or representating any XML document in RDF based on a simplified RDF syntax. These techniques, described in [22] and demonstrated in the XML2OWL[2], XSD2OWL and XML2RDF[3] tools, use XSLT to perform the transformations. Many additional implementations also exist. We believe that using a combination of such existing tools, possibly also extending them if needed, can be sufficient to provide adequate representations of almost all XMLs in an organization's data stores and enable meaningful classification of these.

Regarding unstructured data, the search techniques must of course differ, since there is no underlying data schema. A great deal of work has been done on classifying unstructured data: several tools, such as SystemT [14], InfoSphere Classification Module (ICM)[4], RSA Data Loss Prevention Suite[5] and Symantec Data Loss Prevention product[6] analyze and extract information from unstructured text such as documents and e-mails. Many new methods for efficiently classifying unstructured text have been examined [21], and some domain-specific techniques have been developed.

We believe that classifying unstructured data based on classification models can greatly enhance the existing techniques for the classification of this type of data, since it enables a clear separation between the data classification phase and the policy enforcement phase. Moreover, representing the classification results in RDF format enables high connectivity between the search results on the different data types, and increases the variety of possible uses for these results, a concept that has also been used in [1].

## 4.2 Performing the Search

The model used as input to the discovery process describes the entities for which to search and how to search for them. For each such entity there is a list of

[2] http://www.avt.rwth-aachen.de/AVT/index.php?id=524
[3] http://rhizomik.net/html/redefer/
[4] http://www-01.ibm.com/software/data/content-management/classification
[5] http://www.rsa.com/node.aspx?id=3426
[6] http://www.symantec.com/business/data-loss-prevention

synonyms that describe the concept, and for each synonym we can indicate which words (or tokens) in the phrase are required, which are optional and which need to appear in the same exact form. These synonyms are used to create several search strings to compare to the data store's metadata. In the example of relational databases, they are compared to the database table and column names. For example, for the term "family name", the additional synonyms "second name" and "surname" can be stated, and for the synonym "second name" we can indicate that the word "second" is required but "name" is optional. In addition an entity can have one or more types associated with it and additional information on the expected format of the data. These are used at runtime to verify that a column suspected of containing a certain data element from the model indeed matches the expected type and format for that element.

During the discovery process, for each data element in the input models, a list of search strings is created from the synonyms defined for that element. The aim is to cover all common abbreviations and hyphenations that may be used to describe that data element in the data sources. The algorithm uses syllabifying techniques to divide each token (which is not required to appear exactly) into syllables, creates different possible substrings representing each syllable according to specific rules, and then uses all possible combinations of the substrings and tokens to search the metadata representation for matching data. In this phase, in addition to syllabification, additional linguistic techniques can be employed, such as stemming and affix stripping.

## 5   Implementation

We developed a reference implementation for model based discovery and classification in relational databases. We used the PII model described above as our test case, but any model that complies with the meta-model can be used just as easily. We chose the RelationalOWL ontology (with some slight changes) and implementation to transform the database metadata to RDF format. The current implementation uses the syllabifying techniques described in Section 4.2, as well as type checking. Future extensions are planned to include the use of additional linguistic tools (such as stemming) and the use of sample content to verify the data's format.

### 5.1   NeOn-Based Implementation

As a basis for this implementation we chose to use the NeOn toolkit [10], which is an open-source ontology engineering environment based on the Eclipse platform. The toolkit provides an extensive set of plug-ins covering different aspects of ontology engineering. In addition we used the Eclipse Data Tools Platform (DTP)[7] for defining connections with local and remote databases, RelationalOWL [15] for creating an RDF representation of the database metadata, and the Jena

---

[7] http://www.eclipse.org/datatools

framework [8] for accessing and querying the different RDF representations. We implemented several Eclipse plug-ins that perform the actual discovery, integrate between the different components in the system and contribute some elements to the NeOn user interface.

In the classification tool, a user can create projects, create and edit models, import existing models (in both cases, the models are validated against the meta-model) and import or create database metadata RDF representations. The discovery process can be performed on any combination of models and databases, and the results are also displayed in the project. The results (as well as the models and database metadata) can be viewed in both a hierarchical view and a graph view, as depicted in Figure 3.

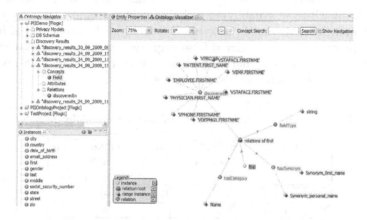

**Fig. 3.** A partial view of the discovery results in the NeOn-based tool

Thanks to the use of RDF to represent the discovery results, it is possible to navigate from any node in the result back to both the classification field in the model, and to the data field (column) in the database RDF representation. This easy navigation allows verifying the classification results, refining them (adding or removing triples), and enriching the model to be more accurate in subsequent runs. As all of the information created and used by the system (models, metadata RDF representations, results) is stored in an RDF store, it can also be exposed to additional existing and evolving Semantic Web tools that operate on RDF stores.

## 5.2   Some Results

We ran our discovery algorithm on two externally available databases: one representing employee records in an organization (taken from the sample database created by the DB2® software installation) and the other representing medical records of patients (taken from "Avitek Medical Records Development Tutorials" by BEA Systems, Inc.[8]). Both were run using our own PII model. The algorithm

---

[8] http://download.oracle.com/docs/cd/E13222_01/wls/docs100/
medrec_tutorials/index.html

was able to discover all columns that contained data from the classes in the PII model. For example, a person's first name was discovered in both databases under numerous forms such as FIRSTNAME, FIRST_NAME and FIRSTNME. Date of birth was found both as DOB and as BIRTHDATE. In the first round of the implementation, a few false positives were found. For example, a column called REGION matched with the class Religion. However after correcting the division into syllables for the word religion, this phenomenon was eliminated. This correction was done in an external resource file and not in the code itself.

We compared the results of running the discovery algorithm with different components of the search algorithm disabled. First we ran the algorithm without the use of additional synonyms for each field name. Then we ran it with synonyms but without the syllabifying techniques described in Section 4.2. Finally we ran it with the syllabifying techniques but without verifying the type of the column. Columns that fit a certain category in the input model but that were not discovered were considered false negatives, and columns that were discovered but did not actually match the category were considered false positives. A summary of these results is shown in Table 1.

**Table 1.** The results of running the classification using different algorithms

| Algorithm | Total no. of hits | No. of false negatives | No. of false positives |
|---|---|---|---|
| Original alg. | 37 | 0 | 0 |
| No synonyms | 35 | 2 | 0 |
| No syllabifying | 29 | 8 | 0 |
| No type matching | 41 | 0 | 14 |

The conclusion from our work is that the main advantages of our approach are the combination of different matching techniques and the extraction of most of the search logic to external files that can be easily changed, as opposed to changing the actual searching algorithms. In addition we suggest an application model in which there is an initial iterative "learning phase" on some sample databases in which the models are manually tweaked to optimize the discovery results. Using our NeOn-based tool, an administrator can easily navigate between the discovery results, the original models and the database representations, thus validating the results and refining the models. We believe that this manual effort is relatively minimal, and the advantages of our approach over any currently existing solution greatly outweigh the effort needed in order to optimize it.

## 5.3    Summary

We presented a model-based approach to data classification and demonstrated its advantages using a reference implementation for relational databases. We showed that this is a flexible solution that can be used to automatically classify information from different data types and sources, in a fine-grained manner, with no dependency on the existing or future policies that will be applied to the discovered data.

The model-based approach provides extensibility and flexibility in that the ontologies can be easily extended, edited or replaced without the need to change the discovery algorithm itself. The same ontology can be used as input to different classifiers that classify different types of data sources. In addition, all classification results are saved in the same format, providing a unified representation of all classified data sources and enabling uniform policy expression and enforcement across the enterprise.

Using RDF to represent both the ontologies and the results maximizes modularity and extensibility and facilitates easy navigation between the results, the models and the data sources. The ontology can serve as a centralized point to manage all valuable information in the organization and enables easy location of all related pieces of data in one click. The RDF representation of the classification results enables high connectivity between the search results on the different data sources, and increases the variety of possible uses for these results, also enabling the application of semantic queries.

We see this general meta-data representation approach as beneficial not only as a specific solution for enterprise data classification, but as a generic "blue print" that can be used to solve many problems in different data management scenarios. Such examples include developing Enterprise Data Catalogs for viewing data from back-end server applications (such as SAP and Siebel), data integration and quality assurance, creating different views of the data according to user roles, etc.

## 6 Conclusions and Future Work

There are many axes in which this work can be extended. One such axis is the variety of data types that can be classified. In the implementation presented above we classified relational (structured) data; doing so on unstructured or semi-structured data is not trivial because we would like to be able to link a certain class to a specific data element in the document and not to the entire document.

An additional axis is the development and combination of the ontologies themselves. We developed a PII model as the core knowledge for privacy purposes. However this is just an example. For different use cases in the various industries, additional models will be required, each with its specific domain knowledge and expertise.

Another direction is taking advantage of the RDF representation of the results to apply reasoning and additional Semantic Web technologies for policy enforcement, semantic data queries and more. There is a very rich set of existing and developing tools that can be applied to gain new insight into the organization's data and enable getting one step closer to the vision of the Semantic Enterprise.

One more aspect that can be investigated is improving the discovery algorithm itself, using additional available information sources, or even extending the method to learning based approaches, i.e. using supervised machine learning or other such techniques.

# References

1. Beliefnetworks - Semantically Secure Unstructured Data Cassification. White paper (2008), http://www.beliefnetworks.net/docs/classunstructdata.pdf
2. Autonomy - Meaning-Based Computing Technology (2009) (press release), http://www.autonomy.com/content/News/Releases/2009/0817.en.html
3. Ben-Dor, A., Friedman, N., Yakhini, Z.: Class discovery in gene expression data. In: Annual Conference on Research in Computational Molecular Biology, pp. 31–38 (2001)
4. Berners-Lee, T., Connolly, D., Kagal, L., Scharf, Y., Hendler, J.A.: N3Logic: A logical framework for the World Wide Web. TPLP 8(3), 249–269 (2008)
5. Bizer, C., Cyganiak, R.: D2RQ V0.2 - Treating Non-RDF Relational Databases as Virtual RDF Graphs. Tech. rep., School of Business & Economics at the Freie Universität Berlin (2004)
6. Boley, H., Forschungszentrum, D., Gmbh, K.I.: Relationships between logic programming and RDF. In: 14th Workshop Logische Programmierung (2000)
7. Butler, M., Reynolds, D., Dickinson, I., McBride, B., Grosvenor, D., Seaborne, A.: Semantic Middleware for E-Discovery. In: IEEE International Conference on Semantic Computing (2009)
8. Carroll, J.J., Dickinson, I., Dollin, C., Seaborne, D.R.A., Wilkinson, K.: Jena: Implementing the Semantic Web Recommendations. Tech. Rep., HP Laboratories (2003)
9. Castells, P., Foncillas, B., Lara, R., Rico, M., Alonso, J.L.: Semantic Web Technologies for Economic and Financial Information Management. In: Bussler, C.J., Davies, J., Fensel, D., Studer, R. (eds.) ESWS 2004. LNCS, vol. 3053, pp. 473–487. Springer, Heidelberg (2004)
10. Holger, P.H., Studer, L.R., Tran, T.: The NeOn Ontology Engineering Toolkit. In: ISWC (2009)
11. Jenkins, C., Jackson, M., Burden, P., Wallis, J.: Automatic RDF Metadata Generation for Resource Discovery. The International Journal of Computer and Telecommunications Networking 32, 1305–1320 (1999)
12. Kagal, L.: Rei: A Policy Language for the Me-Centric Project. Tech. rep., HP Labs (2002)
13. Kagal, L., Hanson, C., Weitzner, D.J.: Using Dependency Tracking to Provide Explanations for Policy Management. In: 2008 IEEE Workshop on Policies for Distributed Systems and Networks, pp. 54–61 (2008)
14. Krishnamurthy, R., Li, Y., Raghavan, S., Reiss, F., Vaithyanathan, S., Zhu, H.: SystemT: a system for declarative information extraction. ACM SIGMOD Record 37, 7–13 (2008)
15. de Laborda, C.P., Conrad, S.: RelationalOWL: a data and schema representation format based on OWL. In: Conferences in Research and Practice in Information Technology, pp. 89–96 (2005)
16. Langegger, A., Wöß, W., Blöchl, M.: A Semantic Web Middleware for Virtual Data Integration on the Web. In: Bechhofer, S., Hauswirth, M., Hoffmann, J., Koubarakis, M. (eds.) ESWC 2008. LNCS, vol. 5021, pp. 493–507. Springer, Heidelberg (2008)

17. Maedche, A., Motik, B., Stojanovic, L., Studer, R., Volz, R.: Ontologies for enterprise knowledge management. IEEE Intelligent Systems 18, 26–33 (2003)
18. Miller, D.J.: A Mixture Model and EM-Based Algorithm for Class Discovery, Robust Classification, and Outlier Rejection in Mixed Labeled/Unlabeled Data Sets. IEEE Transactions on Pattern Analysis and Machine Intelligence (2003)
19. Prud'hommeaux, E., Seaborne, A.: SPARQL query language for RDF. W3C recommendation, W3C (2008)
20. Seaborne, A.: RDQL query language for RDF. W3C member submission, W3C (2004)
21. Song, Y., Zhou, D., Huang, J., Zha, I.G.C.Z., Giles, C.L.: Boosting the feature space: Text classification for unstructured data on the web. In: ICDM. pp. 1064–1069 (2006)
22. Sperberg-McQueen, C.M., Miller, E.: On mapping from colloquial XML to RDF using XSLT. In: Extreme Markup Languages (2004)
23. Staab, S., Studer, R., Schnurr, H.P., Sure, Y.: Knowledge Processes and Ontologies. IEEE Intelligent Systems 16, 26–34 (2001)
24. Taghva, K., Borsack, J., Coombs, J., Condit, A., Lumos, S., Nartker, T.: Ontology-based Classification of Email. In: International Conference on Information Technology: Computers and Communications, p. 194 (2003)
25. Uszok, A., Bradshaw, J.M., Johnson, M., Jeffers, R., Tate, A., Dalton, J., Aitken, S.: KAoS Policy Management for Semantic Web Services. IEEE Intelligent Systems 19, 32–41 (2004)
26. Warren, P.W., Davies, N.J.: Managing the risks from information - through semantic information management. BT Technology Journal 25, 178–191 (2007)
27. Weitzner, D.J., Abelson, H., Berners-lee, T., Hanson, C., Hendler, J., Kagal, L., Mcguinness, D.L., Sussman, G.J., Waterman, K.K.: Transparent accountable data mining: New strategies for privacy protection. Tech. Rep., MIT-CSAIL (2006)
28. Xiaoyue, W., Rujiang, B.: Applying RDF Ontologies to Improve Text Classification. In: International Conference on Computational Intelligence and Natural Computing, vol. 2, pp. 118–121 (2009)

# Semantic Techniques for Enabling
# Knowledge Reuse in Conceptual Modelling

Jorge Gracia[1], Jochem Liem[2], Esther Lozano[1], Oscar Corcho[1], Michal Trna[1],
Asunción Gómez-Pérez[1], and Bert Bredeweg[2]

[1] Ontology Engineering Group, Universidad Politécnica de Madrid, Spain
{jgracia,elozano,ocorcho,mtrna,asun}@fi.upm.es
[2] Informatics Institute, University of Amsterdam, The Netherlands
{j.liem,b.bredeweg}@uva.nl

**Abstract.** Conceptual modelling tools allow users to construct formal representations of their conceptualisations. These models are typically developed in isolation, unrelated to other user models, thus losing the opportunity of incorporating knowledge from other existing models or ontologies that might enrich the modelling process. We propose to apply Semantic Web techniques to the context of conceptual modelling (more particularly to the domain of qualitative reasoning), to smoothly interconnect conceptual models created by different users, thus facilitating the global sharing of scientific data contained in such models and creating new learning opportunities for people who start modelling. This paper describes how semantic grounding techniques can be used during the creation of qualitative reasoning models, to bridge the gap between the imprecise user terminology and a well defined external common vocabulary. We also explore the application of ontology matching techniques between models, which can provide valuable feedback during the model construction process.

**Keywords:** Qualitative Reasoning, Semantic Grounding, Ontology Matching.

## 1 Introduction

The Qualitative Reasoning (QR) area of Artificial Intelligence (AI) researches conceptual representation of systems, and the prediction of their behaviour through reasoning. QR has been successfully applied in a variety of domains, e.g., environmental science [6,27], autonomous spacecraft support, failure analysis and on-board diagnosis of vehicle systems, automated generation of control software for photocopiers [7], etc.

Of particular relevance to this paper is the use of QR in science and education. QR models can be used as a means for learners to formally express and test their conceptual knowledge about systems in an educational context [3]. A desirable feature would be the possibility of uploading expert and learner models to a shared learning environment, and receiving feedback from the common knowledge contained in such a resource. This paper addresses the issue of how this environment can be created and used effectively.

In the current state of the art in qualitative modelling and simulation tools [5,4,28,17], modellers are free to choose their own domain vocabulary. However, this results in different modellers using different terms to denote the same concept (e.g. *death rate* and

P.F. Patel-Schneider et al.(Eds.): ISWC 2010, Part II, LNCS 6497, pp. 82–97, 2010.

*mortality*). Different languages and spelling variations further exacerbate the issue. This makes generating feedback based on a large set of models difficult, since the consensus and disagreement between models cannot be easily determined. We hypothesise that the application of Semantic Web techniques to describe and interlink QR models will be beneficial.

We call *grounding* the process of linking terms in models to concepts in a common vocabulary. Grounding transforms the set of models into a semantically enabled networked resource of scientific data that can be exploited both in the scientific and educational contexts. By allowing comparison between models, algorithms can be written to make modelling suggestions based on other models. Furthermore, when reusing model parts, knowledge can be more gracefully integrated, as equivalent knowledge already existing in a model can be reused. For finding these pieces of common information, *ontology matching* techniques can be applied to explore the similarities among models, with the purpose of getting valuable feedback during the model construction process.

The approach presented in this paper consists of the following steps (see Figure 1):

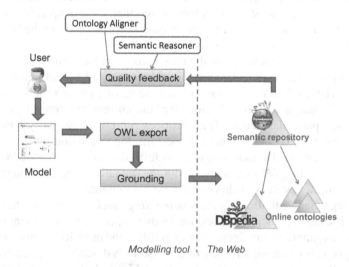

**Fig. 1.** Overview of the approach

1. *OWL Export*. The conceptual knowledge contained in QR models is extracted and expressed in OWL [20], to facilitate their ontology-based description.
2. *Grounding*. The terms from the QR models are linked into external vocabularies (e.g., DBpedia [2]). These grounded models are stored in a semantic repository.
3. *Quality feedback*. Alignment and reasoning techniques are applied to discover similarities and dissimilarities among models, and based on that to enrich the modelling process with adequate feedback and suggestions for knowledge reuse.

The output of this process is a networked pull of online aligned conceptual models (expressed as ontologies) anchored to common vocabularies, and representing specific scientific domains. This has the potential of being a valuable Web resource for scientific progress in general and for semantic guided learning in particular.

The rest of the paper is organised as follows. Section 2 introduces the topic of qualitative reasoning. In Section 3 a method for expressing QR models in OWL is presented. Section 4 describes the semantic grounding process. Quality feedback from stored models is described in Section 5. Section 6 describes our experimental results. In Section 7 some related work is presented. Conclusions and future work are discussed in Section 8.

## 2    Qualitative Modelling and Simulation

The functionality presented in this paper is implemented in the DynaLearn[1] Interactive Learning Environment (ILE) [4] (an evolution of Garp3[2] [5]), which implements a diagrammatic approach to modelling and simulating qualitative models.

DynaLearn allows modellers to capture their knowledge about the structure and the important processes governing their system of interest. Generic knowledge about processes, such as how a process causally affects quantities and when it is active, are represented in *Model Fragments* (MFs). MFs incorporate other model ingredients as either conditions or consequences, and thus form a rule that, for example, indicates that if a population has a biomass above zero, the production will increase the biomass, while the mortality will decrease the biomass (Figure 2(b)).

QR models can be simulated based on a *scenario*, which represents an initial situation of the system (i.e. a particular variant of the system and a set of initial values for its quantities). The result of the simulation is a state graph in which each state represents a qualitatively unique state of behaviour (i.e. the current structure of the system and quantities with particular values). The transitions represent how the system can change from one state of behaviour to others. To perform the simulation, MFs are sought that match the scenario (i.e. the model ingredients fulfil the conditions of the MF). The consequences of matching MFs are merged with the scenario to create an augmented state from which the next states of behaviour can be determined.

Model ingredient definitions, or *domain building blocks*, are instantiated in MFs and scenarios, and are of particular importance for this paper. These definitions include entities, agents, assumptions, configurations, quantities, and quantity spaces. *Entities* define the concepts with which the structure of the system is described, e.g. environment and population. Entities are organized in a taxonomy. Figure 2(a) shows an entity hierarchy. *Agents and assumptions* are also defined in taxonomies. Agents represent influences from outside the system (when a modeller decides these are not part of the system). Assumptions represent simplifying or operating assumptions about the system, such as the assumption that resources for primary producers is considered constant. *Configurations* define relationships with which the structural relations between entities are described. They are defined by their name (e.g. part of, contains, lives in). *Quantities* represent the features of entities and agents that may change during simulation, and are defined by their name and a set of possible quantity spaces. *Quantity spaces* represent the possible values a magnitude (or derivative) of a quantity can have, and are defined by their name and an ordered set of possible values.

---

[1] http://www.dynalearn.eu
[2] http://www.garp3.org

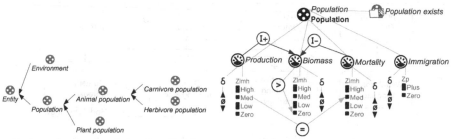

(a) The entity hierarchy of the plant growth resource model.

(b) The Population growth model fragment (from the plant growth model) incorporates the Population Exists model fragment (indicated by the folder with content icon) describing the population, its four quantities, and the inequalities. The model fragment introduces the production (I+) and mortality (I-) influences.

**Fig. 2.** The entity hierarchy and a model fragment of the *model of plant growth based on exploitation of resources* [27]

Next to the model ingredients defined by the modeller, there is also a set of predefined model ingredients called *generic building blocks*. These include causal relationships, correspondences, the operator relations plus and minus, value assignments, and inequalities.

## 3  Export of QR Models into OWL

To ease the ontology-based definition of QR models and its later semantic grounding, they are exported [24] to the Web Ontology Language (OWL) [20]. To determine how the QR models can be formalised as ontologies, an ontological perspective on QR is taken. Previous research distinguishes different types of ontologies based on the type of ontological commitments they make [29]. For example, the ontological commitments of a knowledge representation language consist of the domain independent concepts. However, a domain model created by a knowledge engineer using such a language defines new concepts based on the concepts in the knowledge representation language. We frame the QR knowledge representation on these different types of ontologies (Figure 3).

OWL provides the *representational ontology* we use to define the general model ingredients that can be used in a QR model (i.e. the QR vocabulary). We call the formalisation of the QR vocabulary in OWL the *DynaLearn QR ontology*[3]. This ontology defines the *generic building blocks* (e.g. the concepts entity, configuration and different kinds of causal relations and inequalities) that can be used in a QR model. The DynaLearn QR Ontology functions as our *generic ontology* that extends the ontological commitments made by OWL.

---

[3] http://staff.science.uva.nl/~jliem/ontologies/QRvocabulary.owl

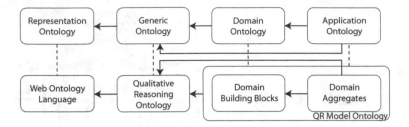

**Fig. 3.** Correspondences between the QR ontologies and ontology types based on the type of ontological commitments made

When modellers create QR models, they extend the QR vocabulary by defining domain specific model ingredients, called *domain building blocks*, such as entities, configurations, and quantities. Creating such a domain specific vocabulary can be seen as refining some of the generic building blocks in the generic ontology to define a *domain ontology*. Note that the domain building blocks correspond to the *model ingredient definitions* (Section 2). The generic building blocks in the DynaLearn QR ontology, and the domain building blocks in the QR model ontology (which are all represented as classes) are instantiated in model fragments and scenarios to represent specific situations and processes. We refer to these ontologies as *QR model ontologies*. The QR model ontologies refer to concepts in the DynaLearn QR ontology. In the rest of this paper, when we use the word *QR model* to refer to QR model ontologies.

## 4    Semantic Grounding

The text above details how QR models can be represented in terms of an ontological language. The next step is to link the unrestricted terminology utilized by users in the QR models into well defined external vocabularies. We refer to this process as *grounding*. Technically speaking, this is performed by an *anchoring* [1] process which connects model concepts to one or more equivalent concepts in a background knowledge ontology (or network of ontologies).

### 4.1    Grounding Process

From a user perspective, the grounding process follows a *semiautomatic* approach: for a given model term, a list of candidate ontology terms (representing the possible meanings of the model term) is automatically proposed to the user. Such a list is ranked according to the probability of being the right meaning. Then, the user can accept the first proposed ontology term (the most probable one) or may choose another one in the list, and move to the next model term to ground it.

In order to save time and effort, a more automatic way of operating is allowed, called *whole model grounding* (see Figure 4). This way, the whole model is grounded at once, and only the most probable grounding of each term is shown to the user (separated by types of model ingredients: entities, quantities, etc.). If the user is not satisfied with

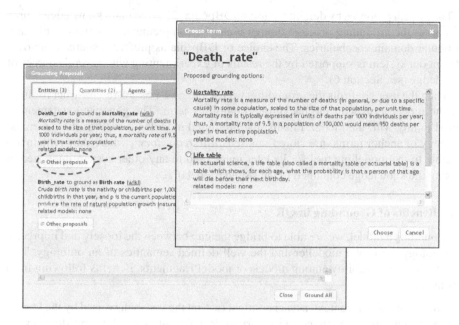

**Fig. 4.** Example of model grounding (left). When the user asks for alternative groundings for the term "death rate", the window on the right appears.

some default grounding, he/she can ask for other proposals and the whole list of candidate senses is shown.

In case that the term to be grounded is not well covered by the proposed groundings (the user is not satisfied, or no sense was found), two actions are possible:

1. We obtain from WordNet [25] syntactic variations of the initial word, as well as approximate forms coming from Yahoo Spelling Suggestion Service[4]. These alternative terms are offered to the user for grounding, thus increasing the range of possibilities.
2. The user can insert the "ungroundable" term anyway, hence generating a new ontology term that is added into an ontology of *anchor terms*. This way, the information is not lost and can be proposed for future groundings jointly with the other background ontology terms. The anchor terms may be related afterwards to terms in other ontologies (by other domain experts).

Different algorithms can be applied for ranking the list of candidate senses, taking into account the context where the model term appears (surrounding terms in the model) to determine the probability of being the right sense [19]. In our approach, the system promotes the reuse of already utilized groundings, which are shown first. A list of synonyms is maintained in the system (fed by the information accessible in the background ontologies, e.g., `rdfs:label`). This is used for expanding the list of candidate senses (when searching for a term we can also search for their synonyms).

---

[4] http://developer.yahoo.com/search/web/V1/spellingSuggestion.html

The system proposes by default the use of DBpedia [2] as the main knowledge source to support the grounding process, though it can be complemented by the use of other particular domain vocabularies. The choice of DBpedia as preferred source of knowledge in our system is supported by the results of experimenting with several sources of knowledge (see Section 6).

When the user confirms the grounding, we use the `owl:sameAs` construct for linking the model term with the background ontology term. The generated statement is stored jointly with the model. Finally, the grounded models (as well as the generated ontology of anchor terms) are stored in a *semantic repository*[5], where they remain accessible to the modelling tool for its later reuse (and to any other system interested in reusing the knowledge contained in the stored models)[6].

### 4.2  Benefits of Grounding in QR

By grounding a model, we are able to bridge the gap between the loosely and imprecise terminology used by a modeller and the well-defined semantics of an ontology. This facilitates interoperability among models or model fragments. Benefits following from this include:

1. In an educational context, a teacher might restrict the vocabulary used by the learner to the knowledge contained in a certain domain ontology, thus speeding up the training period required to learn that vocabulary.
2. New knowledge can be inferred using standard semantic reasoning techniques. For example, let us suppose that entities *whale* and *mammal* in a QR model are grounded to equivalent terms of the same background ontology. If this ontology asserts that *whale* is a subclass of *mammal*, then the same relationship can be inferred for the entities in the model. Other relations not explicitly declared in the model can be also inferred (such as *whale* is an *animal*).
3. Inconsistencies and contradictions between models can be more easily detected. Besides semantic inconsistencies (which can be discovered by applying a reasoner), other modelling issues can be detected. For example, suppose that a model asserts that the increasing *size* of a *population* increases the demand of *natural resources* of such a population, while another model establishes the opposite effect, that is, a growing *size* would decrease the demand of *natural resources*. If we are able to establish that both models are referring to the same concepts (*size, population, natural resources, etc.*), the contradiction between the shared concepts can be discovered and pointed out.
4. Additional knowledge and resources can be incorporated to the system. For example, DBpedia contains rich multilingual textual descriptions, links to pictures and web pages, etc. as part of a term description. This information can be imported if the term is grounded on that knowledge source, and shown to the user in the modelling tool.

Most of the previous features are exploited in our system for enabling knowledge-based feedback, as we will see in the following section.

---

[5] Based on Jena semantic framework (http://jena.sourceforge.net/) in our current prototype.

[6] A set of web services has been developed to support the communication between repository and modelling tool.

## 5   Ontology Based Feedback

As aforementioned, the repository of semantically grounded models created in our system is intended to support feedback during the model creation process. For such a purpose, we devise the use of ontology matching techniques [16], semantic reasoning, and QR specific comparisons between models. Depending on the particular technique, our system provides different types of feedback (see Figure 5). Notice, however, that these types of feedback are not mutually exclusive and can be combined.

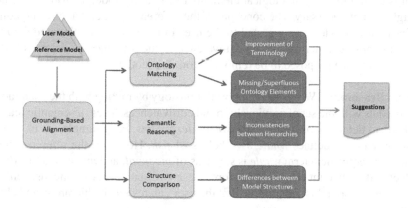

**Fig. 5.** Data flow diagram of the ontology-based feedback techniques

The input to the feedback process is a pair of QR models: one corresponding to the user model (under construction) and other corresponding to a reference model made by an expert and already stored in the repository (we do not enter here in the particular technique utilized for selecting the reference model, i.e., if manually or if based on a dynamic evaluation of relevance).

The first step in the process is to derive mappings from the shared groundings. Since the concepts of both models are grounded to a common vocabulary, we can use these relations to infer a preliminary set of mappings. For example, let us suppose that the user model has a concept labelled *Death* that is grounded to the DBpedia term *Mortality rate*[7], and the reference model has a concept labeled *Mortality* that is also grounded to the same DBpedia resource *Mortality rate*. In order to allow later inference, we determine that *Death* and *Mortality* are equivalent terms (expressed using owl:EquivalentClass). The next steps in the process depend on the particular technique:

**Ontology Matching.** The set of mappings inferred from the shared groundings are utilized, jointly with the user and reference models, as input of an ontology matching tool [18], to generate more pairs of equivalent terms. This enriched set of mapped terms are used to give two types of feedback:

---

[7] http://dbpedia.org/resource/Mortality_rate

*Improvements of terminology.* Two terms that have been deemed equivalent in the ontology matching process should share the same label and grounding. By comparing the user terms with their equivalent reference terms we are able to detect these differences and suggest a better option to the user. As an example of this, if a user has an entity labelled *Sustainable biomass* but the equivalent term in the reference model has the label *Carrying capacity*, the system suggests to the user the replacement of the current label by the one used in the reference model.

*Missing or superfluous ontological elements.* In this type of feedback, we use the set of mappings to find missing ontological elements in the user model, as well as elements that might be not necessary. The concepts of the reference model that have no equivalence in the user model are suggested to the user in order to enrich the model. On the other hand, those user terms with no equivalence in the reference model might be superfluous and hence proposed to be removed from the model.

**Semantic Reasoning.** We create a temporary ontology by mixing both the user and reference models with the set of previously found equivalences. Then, a semantic reasoner is applied[8] to detect *inconsistencies between hierarchies*. For example, let us suppose that the user model defines *whale* as subclass of *fish*. However, the reference model states that the equivalent term *whale* is subclass of *mammal*, and *mammal* and *fish* have been declared as disjoint classes. Therefore, these two statements are inconsistent. The system informs about this situation, so that the user can review the hierarchy and change it accordingly.

**Structure Comparison.** Besides ontology-based comparisons, we also exploit the particular semantics of the QR vocabulary to perform more QR-specific comparisons between the models. In fact, we can identify common model structures that are present in

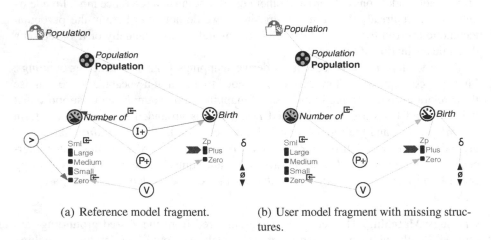

(a) Reference model fragment.          (b) User model fragment with missing structures.

**Fig. 6.** Example of feedback on missing model structures

---

[8] We use Jena built-in reasoner in our current prototype (http://jena.sourceforge.net/inference/), though any other reasoner can be used.

the reference model but not in the user model, thus revealing the *differences between model structures*. These missing structures can modify the final behaviour of the model. To detect them, we make a structural comparison between the models. First, patterns in the reference model are searched; then, by means of the set of mappings, we look for the same patterns in the user model. Once the mappings are established between the elements, the structure comparison process detects that some model structures are missing in the user model. Figure 6 exemplifies this. In the example, an inequality property in *Number of* quantity and the positive influence between the quantities *Birth* and *Number of* are pointed out to allow the user to make the corresponding changes.

## 6   Experimental Evaluation

To adequately ground QR model terms in an external vocabulary and be able to explore similarities between models for quality feedback purposes, specific concerns need to be addressed:

- Q1: Are Semantic Web resources suitable for grounding the specific domain vocabularies that QR models typically contain?
- Q2: Are the state of the art ontology matching techniques suitable for mapping QR models?

In this section, we present the description of the experiments carried out to answer our motivating research (and use) questions.

### 6.1   Grounding Experiments

In order to answer our first question Q1, we performed an experiment to study the *coverage* of different ontologies and semantic resources in specific domains. We measure the coverage as the amount of terms that a resource is able to describe semantically, divided by the total examined terms.

In a realistic usage scenario, the QR models are constructed on the basis of specific domain vocabularies. Therefore, we have focused in our experiment on a set of domain glossaries in environmental science developed by several universities[9]. These vocabularies have been specifically created to be used in the context of QR modelling, so that they constitute very valuable material for our purposes. Each glossary consists of a set of English words which covers seven topics: *Earth systems and resources, the living world, human population, land and water use, energy resources and consumption, pollution*, and *global changes*. We merged these glossaries and removed duplicated terms, obtaining a dataset of 1686 different words.

This unified dataset was used to explore the coverage of knowledge sources of different type: lexical resources such as WordNet [25], common knowledge ontologies such as DBpedia [2] and OpenCyc[10], and the large amount of online ontologies accessible in Watson [13].

---

[9] University of Brasília (Brazil), Tel Aviv University (Israel), University of Hull (United Kingdom), Bulgarian Academy of Sciences (Bulgaria), and University of Natural Resources and Applied Life Sciences (Austria).

[10] http://www.opencyc.org

**Table 1.** Coverage of knowledge sources

| Knowledge Source | Coverage |
|------------------|----------|
| DBpedia | 72% |
| OpenCyc | 69% |
| WordNet | 45% |
| Watson | 47% |

The first step of the experiment consisted in searching each word of the input dataset on each of the above external resources[11], obtaining (if available) a set of ontology terms from each resource that semantically describe the input word. Table 1 shows the different coverage degree obtained on each resource[12].

The immediate conclusion from the given results is that DBpedia has a better coverage than the other resources, closely followed by OpenCyc, for the utilized domain specific vocabularies.

We have analysed the uncovered cases in the experiment, noticing that most of them corresponded to complex multiword terms (e.g., "cultural habit", "distributed water governance"). There was also a reduced amount of misspelling errors in the glossaries (e.g., "fiter feeding" for "filter feeding") and some terms that do exist in the resource however in other syntactic variation (e.g., "meandering" for "meander"). In such cases (misspelling errors and variations), the grounding could be assisted by services like Yahoo! Spelling Suggestion. In order to measure that effect, we repeated the experiment with DBpedia but searching for alternative suggested forms when the term was not found in its initial form. The coverage of using DBpedia + Yahoo Spelling Suggestion service raised to a **78%**.

Though the reached 78% coverage indicates that DBpedia covers well the studied domain specific terminologies, other question arises, which is whether these proposed DBpedia groundings are acceptable according to human opinion or not. In order to answer that, we randomly selected 909 terms covered by DBpedia from the same glossaries used in the previous experiments. We asked 8 evaluators (experts in different fields of environmental science) to assess the correctness of the possible meanings given by DBpedia. Each evaluator assessed between 200 and 300 terms. The grounding of each term was double-evaluated.

We counted as positive groundings those terms for which there were at least one suitable meaning among the list of DBpedia results for such a term. We define *accuracy* as the amount of positive groundings divided by the number of evaluated groundings. The obtained average accuracy in the experiment was **83%**. The observed inter-evaluator agreement was 85%, and Cohen's kappa of inter-evaluator agreement [10] was 0.47, which can be considered as "moderate" and gives us an idea of the difficulty of the task (if another expert does not fully agree with me, why a computer should?).

Finally, although DBpedia exhibits a good coverage, we wonder whether it can be further improved with the addition of other resources. Table 2 shows the result of

---

[11] The search mode was "exact match".
[12] The experimental data can be found in
http://delicias.dia.fi.upm.es/~jgracia/experiments/dynalearn/groundingcoverage.html

**Table 2.** Combined coverage of knowledge sources

| Knowledge Source | Coverage |
|---|---|
| DBpedia and OpenCyc | 87% |
| DBpedia and Watson | 73% |
| DBpedia and WordNet | 72% |
| DBpedia, OpenCyc, Watson, and WordNet | 88% |

combining DBpedia with the other resources[13]. From the data we conclude that the combined use of OpenCyc and DBpedia increases the coverage significantly (while the further addition of other sources has only a minor effect).

**Discussion.** The results of the grounding experiment show a high coverage degree of DBpedia (78%) when used to ground domain specific terminologies, as well as a high accuracy (85%) of the covered terms according to human opinion. These results support the use of DBpedia as preferred source of knowledge for the grounding of the vocabulary involved in QR modelling. Notice that although the coverage of OpenCyc and DBpedia are comparable (see Table 1), there are other good reasons for choosing DBpedia, such as its multilingual capabilities: DBpedia contains data in up to 92 languages while OpenCyc is monolingual in English, thus reducing its potential usage in a multilingual modelling environment. Nevertheless, the combined use of both ontologies further improves the coverage of English terms up to a 87%, as we have found out empirically.

### 6.2 Ontology Matching Experiment

In our approach, we use ontology matching techniques for reconciliating models coming from different authors but modelling similar domains. Thus, the question that we posed above (Q2) emerges naturally. Our target is to study the applicability of already existent alignment approaches in the context of QR modelling.

We have tested the use of two state of the art ontology matching systems: Falcon-AO [21], and CIDER [18] (a complete evaluation of the large amount of alignment systems currently available is out of the scope of this paper). We chose these two systems owing to their good behaviour in previous Ontology Alignment Evaluation Initiative (OAEI) competitions[14] and to their complementary nature: Falcon is more focused on structure-based matching and incorporates efficient partition-based mechanisms, while CIDER relies on a context-based similarity measure and applies lightweight semantic reasoning. The evaluation was conducted as follows:

1. First, a golden standard[15] was defined by human experts[16]. They created specific QR models for this experiment and identified semantic equivalences between them.

---

[13] Notice that this time the experiment was run without applying Yahoo Spelling Suggestion service.

[14] http://oaei.ontologymatching.org/

[15] It can be found at
   http://delicias.dia.fi.upm.es/~jgracia/experiments/dynalearn/omtechniques.html

[16] Three researchers in biological science from University of Brasília (Brazil).

As result, eight QR models grouped in pairs were created, and a reference alignment file was produced for each pair, containing a total of 85 equivalences between the model terms[17].

2. Then, each ontology aligner system was run separately. Each one received the two QR models as input, and produced an alignment file as result, containing the found correspondences between the two models. Note that the models were not previously grounded, to allow a raw comparison.

3. Finally, each produced alignment was compared to the golden standard, and *precision* and *recall* computed.

In this study we focused on comparing certain types of QR model ingredients: entities, quantities, and configurations (those more closely identifiable as ontology elements). Table 3 shows the experimental result.

**Table 3.** Averaged results for the QR Model matching experiment

|       | Precision | Recall |
|-------|-----------|--------|
| CIDER | 92%       | 95%    |
| Falcon| 67%       | 95%    |

Notice that the lower precision given by Falcon is in part owing to the fact that Falcon also aligns the imported ontologies (thus, it aligns the QR vocabulary imported in the models). If the alignment is post-processed and these unnecessary alignments removed, precision also reaches 92%.

**Discussion.** Although these results are not conclusive, they are indicative that the use of traditional ontology matching techniques perform well for giving the similarity between QR models. Only minor adaptation were required to reuse such techniques for the purposes described in this paper. Notice that the experiment is not intended to study the differences between the compared matching tools, but to determine whether these systems are suitable for the proposed task or not (a question that we have answered positively).

# 7   Related Work

To our best knowledge, the approach described in this paper has no counterpart in the field of qualitative modelling [5,17,28,14]. Other modelling and simulation tools, such as Betty's brain [23] or Stella [11] neither ground terms to a common vocabulary, nor get quality feedback from other models.

With regard to other conceptual modelling techniques beyond QR, CmapTools [9] is a software for representing knowledge using such as *concept maps* [26]. Although CmapTools is also intended for collaborative use, it neither relies on Semantic Web standards to maximize its interoperability, nor uses common shared vocabularies to

---

[17] We expressed the alignments in the Alignment Format [15], to facilitate their later processing.

minimize the semantic gap between models. In [8] a method for suggesting concepts during concept map modelling based on Web mining techniques is proposed, though an effective grounding with external concepts is not performed.

Regarding the use of semantic techniques to enhance collaborative learning (one of the goals of DynaLearn), there have been some specific efforts, such as the work described in [22], where DEPTHS (Design Patterns Teaching Help System) is introduced. DEPTHS system establishes a common ontological foundation for the integration of different existing learning tools and systems in a common learning environment. Though our techniques differ, the motivation is along the same lines as our work. Nevertheless, DEPTHS focuses on the particular scenario of software engineering education, and supports recommendation more than quality feedback.

## 8 Conclusions and Future Work

In this paper we propose a method for the semantic enhancement of the modelling process in the field of Qualitative Reasoning. Our goal is to support the creation of semantically networked models as a means to share and reuse conceptual knowledge. In our approach, QR models are first exported into an ontological language and grounded to an external common vocabulary. Then, ontology matching techniques are used for getting quality feedback, by identifying pieces of common knowledge across models.

Our experiments show that the coverage of DBpedia, when used for grounding specific domain terminologies, is above other studied resources. We have also shown that 85% of the covered terms in DBpedia contain suitable meanings according to human opinion. Finally, our experiments indicated a good behaviour of the state of the art ontology matching systems when applied for the alignment of QR models.

As future work, we will especially focus on running usability studies on our ontology-based feedback functionalities. We will also enrich our ontology matching-based techniques with other advanced metrics that measure the agreement and disagreement degree between ontologies [12]. We devise also the application of collaborative filtering techniques for model recommendation based on the community of users of the system. We also plan to specifically use our system in the academic domain to support semantic-guided learning. In this real usage context, more human-based evaluations will be possible for the improvement of our approach.

Finally, although the semantic data that we generate is accessible on the Web by different means, a natural improvement of our system is to "open" this semantic data to the web of Linked Data, by adhering to the Linked Data principles. This will be favoured also by the preferred use of DBpedia (one of the most important nodes in the cloud of Linked Data [2]) in our system.

**Acknowledgments.** We thank Thanh Tu Nguyen for her valuable assistance in the implementation of the experiments. We thank Andreas, Michaela, Petya, Richard, Ian, Dror, Paulo, Isabella, and Gustavo for their kind help with the grounding evaluation. The work presented in this paper is co-funded by the EC within the FP7 project DynaLearn (no. 231526, http://www.DynaLearn.eu) and by a PhD research grant of Universidad Politécnica de Madrid

# References

1. Aleksovski, Z., Klein, M.C.A., ten Kate, W., van Harmelen, F.: Matching unstructured vocabularies using a background ontology. In: Staab, S., Svátek, V. (eds.) EKAW 2006. LNCS (LNAI), vol. 4248, pp. 182–197. Springer, Heidelberg (2006)
2. Bizer, C., Lehmann, J., Kobilarov, G., Auer, S., Becker, C., Cyganiak, R., Hellmann, S.: Dbpedia - a crystallization point for the web of data. Web Semantics: Science, Services and Agents on the World Wide Web 7(3), 154–165 (2009)
3. Bredeweg, B., Forbus, K.: Qualitative modeling in education. AI Magazine 24(4), 35–46 (2004)
4. Bredeweg, B., Liem, J., Linnebank, F., Bühling, R., Wißner, M., Gracia, J., Salles, P., Beek, W., Gómez-Pérez, A.: Dynalearn: Architecture and approach for investigating conceptual system knowledge acquisition. In: Aleven, V., Kay, J., Mostow, J. (eds.) Intelligent Tutoring Systems. LNCS, vol. 6095, ch. 42, pp. 272–274. Springer, Heidelberg (2010)
5. Bredeweg, B., Linnebank, F., Bouwer, A., Liem, J.: Garp3 - workbench for qualitative modelling and simulation. Ecological Informatics 4(5-6), 263–281 (2009)
6. Bredeweg, B., Salles, P.: Qualitative models of ecological systems – editorial introduction. Ecological Informatics 4(5-6), 261–262 (2009); Special Issue: Qualitative models of ecological systems
7. Bredeweg, B., Struss, P.: Current topics in qualitative reasoning (editorial introduction). AI Magazine 24(4), 13–16 (2003)
8. Cañas, A., Carvalho, M., Arguedas, M., Leake, D.B., Maguitman, A., Reichherzer, T.: Mining the web to suggest concepts during concept mapping: Preliminary results. In: Proc. of the First International Conference on Concept Mapping, Pamplona, Spain (2004)
9. Cañas, A.J., Hill, G., Carff, R., Suri, N., Lott, J., Gómez, G., Eskridge, T.C., Arroyo, M., Carvajal, R.: Cmaptools: A knowledge modeling and sharing environment. In: Proc. of the First International Conference on Concept Mapping, Pamplona, Spain, pp. 125–133 (2004)
10. Cohen, J.: A coefficient of agreement for nominal scales. Educational and Psychological Measurement 20(1), 37–46 (1960)
11. Costanza, R., Voinov, A.: Modeling ecological and economic systems with stella: Part iii. Ecological Modelling 143(1-2), 1–7 (2001)
12. d'Aquin, M.: Formally measuring agreement and disagreement in ontologies. In: Proc. of the 5th International Conference on Knowledge Capture (K-CAP 2009), pp. 145–152 (September 2009)
13. d'Aquin, M., Baldassarre, C., Gridinoc, L., Angeletou, S., Sabou, M., Motta, E.: Characterizing knowledge on the semantic web with Watson. In: 5th International EON Workshop, at ISWC 2007, Busan, Korea (November 2007)
14. Dehghani, M., Forbus, K.: QCM: A QP-based concept map system. In: Žabkar, J., Bratko, I. (eds.) The 23nd International Workshop on Qualitative Reasoning (QR 2009), Ljubljana, Slovenia, pp. 16–21 (June 2009)
15. Euzenat, J.: An API for ontology alignment. In: McIlraith, S.A., Plexousakis, D., van Harmelen, F. (eds.) ISWC 2004. LNCS, vol. 3298, pp. 698–712. Springer, Heidelberg (2004)
16. Euzenat, J., Shvaiko, P.: Ontology matching. Springer, Heidelberg (2007)
17. Forbus, K.D., Carney, K., Sherin, B.L., UreelII, L.C.: Vmodel: A visual qualitative modeling environment for middle-school students. In: Proc. of the 16th Conference on Innovative Applications of Artificial Intelligence, San Jose, California, USA, pp. 820–827 (July 2004)
18. Gracia, J., Mena, E.: Ontology matching with CIDER: Evaluation report for the OAEI 2008. In: Proc. of 3rd Ontology Matching Workshop (OM 2008), at ISWC 2008, Karlsruhe, Germany, vol. 431, pp. 140–146. CEUR-WS (October 2008)

19. Gracia, J., Trillo, R., Espinoza, M., Mena, E.: Querying the web: A multiontology disambiguation method. In: Sixth International Conference on Web Engineering (ICWE 2006), Palo Alto, California, USA, pp. 241–248. ACM, New York (July 2006)

20. Grau, B., Horrocks, I., Motik, B., Parsia, B., Patelschneider, P., Sattler, U.: Owl 2: The next step for owl. Web Semantics: Science, Services and Agents on the World Wide Web 6(4), 309–322 (2008)

21. Hu, W., Qu, Y.: Falcon-ao: A practical ontology matching system. Journal of Web Semantics 6(3), 237–239 (2008)

22. Jeremic, Z., Jovanovic, J., Gasevic, D.: Semantic web technologies for the integration of learning tools and context-aware educational services. In: Bernstein, A., Karger, D.R., Heath, T., Feigenbaum, L., Maynard, D., Motta, E., Thirunarayan, K. (eds.) ISWC 2009. LNCS, vol. 5823, pp. 860–875. Springer, Heidelberg (2009)

23. Leelawong, K., Biswas, G.: Designing learning by teaching agents: The betty's brain system. Int. J. Artif. Intell. Ed. 18(3), 181–208 (2008)

24. Liem, J., Bredeweg, B.: OWL and qualitative reasoning models. In: Freksa, C., Kohlhase, M., Schill, K. (eds.) KI 2006. LNCS (LNAI), vol. 4314, pp. 33–48. Springer, Heidelberg (2007)

25. Miller, G.A.: Wordnet: A lexical database for english. Communications of the ACM 38(11), 39–41 (1995)

26. Novak, J., Gowin, D.: Learning How to Learn. Cambridge University Press, Cambridge (1984)

27. Nuttle, T., Salles, P., Bredeweg, B., Neumann, M.: Representing and managing uncertainty in qualitative ecological models. Ecological informatics 4(5-6), 358–366 (2009)

28. Soloway, E., Pryor, A.Z., Krajcik, J.S., Jackson, S., Stratford, S.J., Wisnudel, M., Klein, J.T.: Scienceware's model-it: Technology to support authentic science inquiry. Technological Horizons on Education 25(3), 54–56 (1997)

29. van Heijst, G., Falasconi, S., Abu-Hanna, A., Schreiber, G., Stefanelli, M.: A case study in ontology library contruction. Artificial Intelligence in Medicine 7(3), 227–255 (1995)

# Semantic Technologies for Enterprise Cloud Management

Peter Haase, Tobias Mathäß, Michael Schmidt,
Andreas Eberhart, and Ulrich Walther

fluid Operations, D-69190 Walldorf, Germany
firstname.lastname@fluidops.com

**Abstract.** Enterprise clouds apply the paradigm of cloud computing to enterprise IT infrastructures, with the goal of providing easy, flexible, and scalable access to both computing resources and IT services. Realizing the vision of the fully automated enterprise cloud involves addressing a range of technological challenges. In this paper, we focus on the challenges related to intelligent information management in enterprise clouds and discuss how semantic technologies can help to fulfill them. In particular, we address the topics of *data integration, collaborative documentation and annotation* and *intelligent information access and analytics* and present solutions that are implemented in the newest addition to our eCloudManager product suite: The Intelligence Edition.

## 1 Introduction

Cloud computing has emerged as a model in support of "everything-as-a-service" (XaaS) [9]. Cloud services have three distinct characteristics that differentiate them from traditional hosting. First, cloud services are sold on demand, typically by the minute or the hour; second, they are elastic – users can have as much or as little of a service as they want at any given time; and third, cloud services are fully managed by the provider (while the consumer needs nothing but a personal computer and Internet access) [13]. Significant innovations in virtualization and distributed computing, as well as improved access to high-speed Internet and a weak economy, have accelerated interest in cloud computing.

While the paradigm of cloud computing is best known from so called public clouds, its promises have also caused significant interest in the context of running enterprise IT infrastructures as private clouds [11]. A private cloud is a network or a data center that supplies hosted services to a limited number of people, e.g. as an enterprise cloud. As with public clouds, the goal of enterprise clouds is to provide easy, scalable access to computing resources and IT services [14].

The emergence of cloud offerings such as Amazon AWS or Salesforce.com demonstrates that the vision of a fully automated data center is feasible. Recent advances in the area of virtualization make it possible to deploy servers, activate network links, and allocate disk space virtually via an API rather than having to employ administrators who physically carry out these jobs. Note that

P.F. Patel-Schneider et al.(Eds.): ISWC 2010, Part II, LNCS 6497, pp. 98–113, 2010.

virtualization is not limited to CPU virtualization – virtualization can be defined as an abstraction layer between a consumer and a resource that allows the resource to be used in a more flexible way. Examples can be drawn from the entire IT stack. Storage Area Networks (SANs) virtualize mass storage resources, VLAN technology allows using a single physical cable for multiple logical networks, hypervisors can run virtual machines by presenting a virtual hardware interface to the guest operating system, and remote desktop software such as VNC virtualizes the screen display by redrawing it on a remote display.

Realizing the vision of the fully automated data center – the enterprise cloud – involves addressing a range of technological challenges, touching the areas of infrastructure management, virtualization technologies, but also distributed and service-oriented computing. In this paper, we focus on the challenges related to intelligent information management in enterprise clouds and discuss how semantic technologies can help to fulfill them. In particular, we address the topics of *data integration, documentation and annotation,* and *intelligent information access and analytics.* We present solutions that we have implemented in the newest addition to our eCloudManager product suite: The Intelligence Edition. In the following, we give a brief overview of the contributions in each of the dimensions.

**Data Integration.** Clearly, being able to automate data center operations via low level APIs is the prerequisite for achieving the requirements listed above. The challenge lies in the proper integration of data received from infrastructure components and the orchestration of subsequent actions as a response to events such as user requests or alarms. Many layers play a role in this picture and one is faced with a large set of provider APIs ranging from storage to application levels. The situation grows even more complex when products and solutions from different vendors coexist in the data center. In fact, CIOs tend to mix hardware from different vendors to avoid vendor locks, in order to benefit from price competitions among the individual vendors. Hence, in the end they often face a mix of technologies acquired over several years, where products from different vendors and sometimes even different product versions differ vastly in syntax and semantics of the data supplied and functionality offered via APIs.

Semantic technologies have been designed for these real-world situations. In our solution, we use RDF as a data model for integrating semantically heterogeneous information sources in order to get a complete picture across the entire data center, both horizontally – across different product versions and vendors – and vertically – across storage, compute units, network, operating systems, and applications. The RDF-based integration offers the flexibility needed to integrate new sources in the presence of heterogeneity and dynamics in data centers.

**Collaborative Documentation and Annotation.** Data integration is a key aspect of running data centers and clouds in an efficient and cost effective way. For this purpose, cloud management software is fed with data from provider APIs. This data contains technical information about the infrastructure and the software running on it. In order to have a complete picture available, organizational and business aspects need to be added to the technical data. Consider the following examples: The decision whether to place a workload on a redundant cluster with highly available storage is strongly affected by the service level the system needs to meet, data center planning tools must take expiring warranties of components into account, and having a relatively mild punishment for SLA violations may lead a cloud operator to take a chance and place workloads on less reliable infrastructure.

In order to collaborate efficiently, data center operators need to document procedures and log activities. Proper knowledge management is essential in order to avoid a problem having to be resolved repeatedly by different staff members. Activities are usually managed via a ticketing system, where infrastructure alerts and customer complaints are distributed and resolved by operators.

The examples above show that business and organizational information must be addressed in a unified way. When information about systems or customers is stored or documentation about a certain hardware type is written, it should be possible to cross reference information collected from infrastructure providers. In our solution we apply Semantic Wiki technology to satisfy these requirements. Operating on an RDF base that is fed by infrastructure providers, operators can extend this data by documenting and annotating the respective items.

**Intelligent Information Access and Analytics.** Efficient management of a data center requires providing data center managers with the information they need to make intelligent, timely and precise decisions. The range of specific information needs that should be supported is very diverse, including the generation of reports about status and utilization of data center resources over time, the visualization of key performance metrics in dashboards, the search for specific resources etc. Many of these information needs require multi-dimensional queries that span across both technical, IT-related aspects and business aspects, and therefore cannot be answered by a single data source alone.

Enabled by an integrated view on the data, we support queries that overcome the borders of data sources. Apart from predefined queries that drive reports and dashboards, a clear benefit is also the ability to perform expressive *ad hoc* queries. As it is desirable to hide the complexity of the underlying data model and query languages from the end user, we use approaches of schema-agnostic semantic search that combine the expressiveness of structured queries with the ease of use of keyword driven interfaces. Novel approaches to the visualization of structured data as well as visual exploration of resources enable new forms of interaction with the resources and provide insights into previously hidden relationships.

**Structure of the Work.** We start with a brief overview of the overall solution in Section 2. Next, Section 3 introduces our ontology for the domain of enterprise cloud management, which serves as the core of the data integration and management tasks within the eCloudManager. We then discuss the specific uses of semantic technologies: semantics-based integration in Section 4, collaboration support in Section 5 and intelligent information access in Section 6. Subsequently, we report on practical experiences in Section 7 and, after a discussion of related work in Section 8, conclude with an outlook to future work in Section 9.

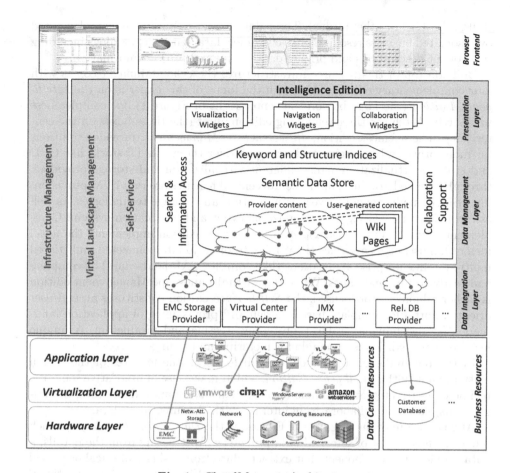

**Fig. 1.** eCloudManager Architecture

## 2   Solution Overview

The eCloudManager Product Suite is a Java-based software solution that is targeted at the management of enterprise cloud environments. Starting with the big picture, we first discuss the eCloudManager's overall architecture depicted in Figure 1. The bottom of the figure shows the two dimensions of information relevant to the eCloudManager, namely *Data Center Resources* and *Business Resources*. The data center resources are divided along the IT stack into (i) a *Hardware Layer* that consists of physical storage, network and compute infrastructure, (ii) a *Virtualization Layer* built on top of the hardware layer that is made up of hypervisors with appropriate management capabilities (enabling virtual clusters, live migration etc.), and finally (iii) the *Application Layer* built on top of the virtualization layer, comprising applications and landscapes on top of the virtualized resources (where the term *landscape* refers to a set of enterprise systems plus optional auxiliary systems that enable network access using VPN,

RDP, SAP GUI or other means). These data center resources are complemented by associated business resources, like customer data, hardware catalogs including component descriptions and pricing data, or related project information.

Built on top of this resource infrastructure, the eCloudManager comes with four complementary editions. Although the focus of this paper is on the *Intelligence Edition* (the large box on the center right of Figure 1), for completeness we shortly sketch the three remaining editions first (the three boxes on the center left). The leftmost edition, *Infrastructure Management*, provides solutions for monitoring and managing tasks ranging over the whole IT stack, like CPU and storage virtualization across different virtualization and storage providers, through a unified graphical interface. Its central features are rapid VM provisioning utilizing multi-vendor server virtualization and storage technologies, as well as error detection through a customizable event and notification system. Complementarily, the *Virtual Landscape Management* edition implements the novel idea of Landscape-as-a-service (LaaS), i.e. to offer up interconnected multitiered multi-system enterprise applications (like complex SAP landscapes) as complete and ready-to-run landscapes. While the Virtual Landscape Management edition is rather administrator-focused, the *Self-Service* edition constitutes an end-user oriented portal for template-based on-demand provisioning of application landscapes for development, value prototyping, testing and production. A unique feature of this edition is a module for metering and billing.

In the remainder of this paper, we will focus on the features of the fourth edition, namely the *Intelligence Edition*, which makes use of innovative semantic technologies to integrate available resources into a semantic data store, investigate this data, and collaboratively interact with the integrated data.

At the bottom of the Intelligence Edition is the *Data Integration Layer*, which relies on the concept of so-called *data providers*. Abstracting from the details, a data provider is a component that extracts data from a single physical or logical resource (e.g. an EMC storage device, a VMware Virtual Center, or a relational database), converts it into RDF and integrates the resulting RDF data into the central repository. It is crucial that several providers can – and typically do – coexist and that data from different providers is automatically interlinked by use of common URIs, ultimately providing a unified and integrated view on all data center and business resources (for more details see Section 4).

The central repository where the provider data is written to is settled in the *Data Management Layer*. Technically, it is realized as a Sesame [6] triple store that adheres to a predefined (yet extendable) OWL ontology (cf. Section 3). In addition to the repository, the layer provides components for search and intelligent, semantics-based information access (cf. Section 6), whose efficiency is made possible by keyword and structure indices over the data. A central component in this layer are also semantic wiki pages that are associated with the resources contained in the repository; they offer an entry point to the eCloudManager users, allowing to add new and complement existing information (cf. Section 5).

The uppermost layer in the Intelligence Edition is the *Presentation Layer*. Located on top of the Data Management Layer, it comes with a predefined set of widgets with varying functional focus, e.g. offering support to display wiki pages, visualize the underlying data using charts and diagrams, navigate through the underlying RDF graph, and collaboratively annotate resources in the database using both semantic annotations as well as free-text documentation. Ultimately, the combination of these widgets results in a customizable user-interface – realized as an Ajax-based browser frontend – which is flexibly adjustable to fit the needs of different user types (like data center administrators or CIOs).

# 3 The eCloudManager Ontology

Having presented the overall architecture, we now take a closer look at eCloud-Manager data model and introduce a conceptual model for the domain of enterprise cloud management in form of an ontology that abstracts from vendor-specific representations, data sources and management APIs. The ontology has been modeled as an OWL 2 ontology, consisting primarily of a class hierarchy and property definitions with domain, range and cardinality restrictions. While in the current application many of the expressive features of OWL 2 are not yet required, we opted for OWL 2 in order for the ontology to serve as a reference ontology for the data center domain and to perspectively enable reasoning-driven tasks in the application (cf. the discussion of future work in Section 9).

Figure 2 surveys the main concepts and relationships of the eCloudManager ontology (using the UML profile for modeling OWL ontologies [5]). Each concept carries a number of data properties (attributes), which capture information about properties and status of the respective resource; for space limitations, we included only the most important subsets of attributes. We next describe the four major subareas of the figure in more detail, namely *Storage Infrastructure*, *Compute Infrastructure*, *Application-level* and *Business-level* resources.

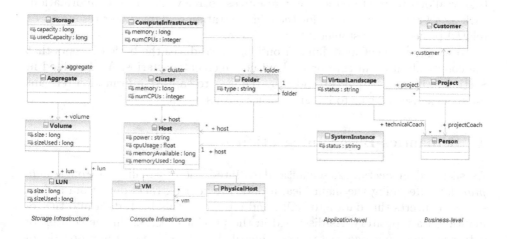

**Fig. 2.** Overview of the eCloudManager Ontology

*Infrastructure Resources.* Infrastructure resources include storage and compute resources. In our enterprise cloud solutions, storage virtualization is realized through Network Attached Storage (NAS), in which virtual disks are managed as so-called LUNs. This allows to maintain complete systems in a central place (the Storage filer) and dynamically relocate between hosts. Within a Storage filer, LUNs are typically grouped and managed in Volumes as logical container, which can further be grouped in Aggregates.

Specific solutions for compute virtualization environments are modeled as ComputeInfrastructure, representing a certain technology such as FlexFrame, AWS, XEN or VMware. Within an ComputeInfrastructure, a Cluster represents a collection of compute resources, specifically a group of tightly coupled computers that work together closely and in many respects can be viewed as though they are a single resource. A Folder is a logical set of resources that is grouped together, assigned resources, managed etc. Examples are VMware resource pools. A Host represents an actual – physical or virtual – compute system (PhysicalHost or VM), where one PhysicalHost can host multiple VMs. As shown in the figure, the primary connection between compute and storage resources is given by the lun relationship between Hosts and the logical storage units (LUNs) they are associated with.

*Application-level Resources.* In a virtual environment, complex application landscapes can be managed as a VirtualLandscape, which is assigned a Pool of virtualized infrastructure resources. A virtual landscape typically comprises a set of applications that run as SystemInstances on dedicated hosts. Examples of a virtual landscape would be a landscape of SAP system instances as applications, e.g. consisting of a CRM, ERP and enterprise portal.

*Business-level Resources.* In addition to the technical aspects of the data center (which is typically populated automatically from management APIs), the ontology allows to relate data center resources to relevant business information, such as a Person responsible for the administration of data center resources, or related project and customer information etc.

To wrap up, the eCloudManager ontology introduces the basic concepts that are commonly used in enterprise cloud management scenarios. As described in Section 5, it can easily be extended by the user to capture information relevant for specific use cases or to integrate other data sources.

## 4   Semantics-Based Integration

As sketched in Section 2, semantic data integration is realized through *data providers*. Recalling the main idea, a data provider extracts data from a data source, converts this data into RDF, and integrates this RDF data into the semantic data repository. As discussed in the previous section, in the context of enterprise cloud management we are primarily interested in physical properties of hardware components like storage devices, available CPUs, physical hosts, as well as the properties and the current state of software components like hypervisors and virtual machines. Addressing these information needs, the eCloudManager comes with a broad set of predefined data providers for integrating data from hard- and software components typically encountered in the context of enterprise cloud management. To illustrate the concept by example, let us consider a small data center employing a single Xen hypervisor that runs a couple of virtual machines over a set of physical hosts and two storage systems, say one EMC Symmetrix and one NetApp system. In this scenario, data integration in the eCloudManager would be realized through the following providers.

- A Symmetrix provider extracts physical properties from the Symmetrix storage unit such as number and size of physical disks contained in the system.
- A NetApp provider analogously extracts the physical properties from the NetApp storage system.
- Complementarily, a Xen provider gathers the properties hidden behind the Xen hypervisor, such as the number of the virtual machines running on top, configuration details of the individual virtual machines, information about underlying physical hosts, disk occupation of the virtual disks, etc.

Technically, the data providers are predefined modules that can be instantiated for existing data sources and, given only meta information like their network address and login information as input, use available drivers or APIs to gather the relevant data. This data is then translated into RDF, thereby using the key attributes of resources to generate URIs that uniquely identify these resources. To give an example, the key for the Symmetrix storage device is its IP address, so its URI might look like `http://my.datacenter/Symmetrix/192.168.0.55`. The crucial thing here is that both the Symmetrix provider and the Xen provider will generate this URI when storing information related to the Symmetrix system. This way, the data that is generated from different providers is automatically interlinked when integrating it into the global repository (cf. Figure 1).

When instantiating the providers, one can also configure advanced properties like the interval for periodically refreshing data, or whether to store old provider snapshots for recovery purpose. Ultimately, the provider framework serves as an abstraction layer for data integration that allows to connect data sources in a plug & play fashion to the central semantic repository, while abstracting from the technical details and the APIs of the physical and logical components accessed by the provider. If the data center is extended by, say, another Symmetrix storage system, all that needs to be done is instantiating a fresh Symmetrix provider that collects its data (which can be done by an administrator using the eCloudManager Web frontend). A second benefit of the provider concept is the high degree of reuse, e.g. the NetApp provider can be employed in every data center using a NetApp storage system. Taken all these properties together, our data provider concept enables a fast and simple data integration process (cf. Section 7).

Apart from the integration of technical data directly related to the enterprise cloud data center, the provider concept also makes it possible to integrate data from other sources, such as additional documentation or customer information (e.g. in the form of spreadsheets or relational databases). To this end, the eCloudManager offers predefined, customizable providers, e.g. to extract RDF data from external sources, SOAP-based Web Services, SNMP- and SSH-connected devices, relational DBs or tabular CSV files. In addition, there are script-based providers that allow users to integrate data from virtually every accessible source. The underlying idea is to support data integration in a pay-as-you-go fashion: while data center information is typically directly aligned with the eCloudManager ontology presented in Section 3, for other data sources it is up to the user how much effort she wants to spend on data integration: In some cases, only little effort is needed and generic, predefined providers can be used to integrate data sources quickly, leaving it to the user to semantically annotate and interlink this data afterward (cf. Section 5); in other cases, the user may put more effort in data integration and write his own provider that aligns the integrated data directly with existing data and the eCloudManager ontology. In the end, this concept gives users a high degree of flexibility in data integration.

## 5   Collaborative Documentation and Annotation

As motivated in the Introduction, in order to collaborate efficiently data center administrators need to be able to document and share knowledge. Wikis have long been used as a tool for knowledge sharing in administration environments, yet a major obstacle of traditional wikis is that they again are silos, with largely unstructured information that is poorly connected with other data sources. As outlined in Section 2, in our solution we tightly integrate Semantic Wiki functionality, where the wiki pages are directly associated with the structured data objects. Our wiki implementation follows the ideas of the Semantic MediaWiki [8], providing features like semantic annotations or built-in support for visual display of information (e.g. as charts or bar diagrams). Semantic annotations are automatically extracted from the wiki pages, converted into RDF triples and persisted in the repository. In the user interface, the structured and unstructured information then is presented in a seamlessly integrated way. In the following paragraphs, we discuss uses cases around user-driven documentation and annotation, describing how they are supported in our solution.

*Documentation.* Technical documentation is a central use case in data center management. Examples include customized installation instructions or best practices for handling errors on a host system of the cloud that is known to frequently cause problems. Such documentation, typically provided in form of unstructured text (possibly with some internal and external links), can directly be entered in the wiki pages associated with the respective resource in the central repository.

*Interlinkage of Resources.* In addition to unstructured documentation, the user can use semantic annotations to establish new relationships between resources. For example, one may link responsible administrators directly to hardware devices, or associate customers with the concrete virtual landscapes they are using.

*Completion of Missing Information.* In some cases, data providers may not have access to all information relevant to the user, which ultimately leads to incomplete information in the central repository. Our approach here is to use the eCloudManager ontology to identify missing information: whenever displaying a resource of a given type, we compare its properties against the properties specified in the ontology for the respective type, to identify properties that are currently not populated. For instance, the ontology may define that the warranty status of storage resources should be provided, and this property may not have been specified for some storage system. Using structured tabular templates that are automatically generated from the ontology, the user can easily detect properties with missing values and, in response, directly fill in the missing data.

It is worth mentioning that the wiki is equipped with a revision management system to track the provenance of changes. For the structured data, the provenance is maintained down to the level of single triples, i.e. for every information item it is possible to see when and by whom (either a user or a data provider) it was changed. Via RSS feeds users can get notified in the case of changes, e.g. when the status of an object is updated, or a corresponding wiki page has been modified. With these functionalities, the system offers a wide range of possibilities to support collaborative working processes in a data center environment.

# 6    Semantics-Based Information Access

Having discussed the collaborative annotation and documentation features in the previous section, we now present our techniques and paradigms to semantics-based information access. We divide our discussion into three areas, namely *presentation of resources*, *search and querying* and *exploration and analytics*.

**Presentation of Resources in the UI.** The user interface of the Intelligence Edition follows a resource centric presentation scheme, i.e. every resource has exactly one page associated that aggregates resource-related data in a transparent way. The UI is composed from a set of widgets with different functional focus, like visualization, navigation or collaboration. Some of these widgets are generic and displayed for all resources, while others are specific for a fixed type of data (or, alternatively, data with certain properties). To give some concrete examples, the widgets include a browsable graph view, a tabular view to edit structured data, the semantic wiki to create documentation and semantic annotations, as well as widgets displaying charts or bar diagrams. Figure 3 shows an example of a page for a virtual landscape, including a tabular view, the data graph and a dashboard displaying statistics.

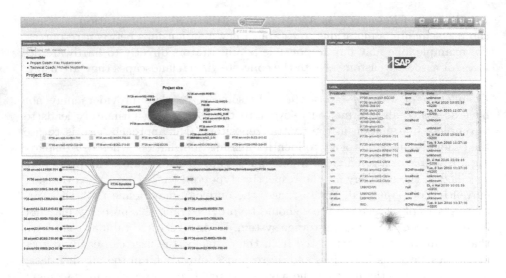

**Fig. 3.** Screenshot Displaying a Virtual Landscape Resource

**Search and Querying.** As argued in the introduction and solution overview, one particular strength of having data center and business resources integrated in a single repository is the ability to perform queries across the borders of data sources. Following a pay-as-you-go approach, we implement a variety of semantics-based paradigms to searching and querying the integrated data.

*Keyword Search.* The simplest search paradigm from a user's perspective is keyword search. In contrast to classical keyword search, in our scenario search is not document- but entity-centric: When answering a keyword query, we investigate both the data properties and the associated unstructured data (i.e., the wiki pages) for each entity and, if the keywords are matched, return the respective entity. Hence, keyword search is well-suited to quickly locate entities with known properties and/or annotations. Technically, our search engine uses the Lucene[1] library to maintain the keyword indices for RDF literals and wiki pages.

*Structured Queries using SPARQL.* Advanced users may also be interested in answering more complex queries, e.g. asking for all projects of a certain costumer together with billing contact information of this costumer. In such cases, the RDF query language SPARQL[2] can be used to encode requests directly.

While the SPARQL query language is very expressive (i.e., relationally complete [3]), formulating SPARQL queries requires knowledge about the query language and the schema of the underlying data. To equip users with the high precision of structured search without having detailed knowledge about the schema, we support a variety of means to perform ad hoc queries in a schema-agnostic way [12], as dicussed in the following.

---

[1] http://lucene.apache.org/
[2] http://www.w3.org/TR/rdf-sparql-query/

*Form Based Search.* We developed a form based search interface that supports the user in formulating SPARQL queries by presenting templates the user has to complete. This way the user can apply the expressiveness of a structured query language without directly being confronted with its concrete syntax. The templates are directly derived from the ontology. For instance, when searching for entities of type "Virtual Landscape", the dynamic form automatically searches for and presents properties having domain `VirtualLandscape`. The user can fill in values for the properties she is interested in, such as e.g. the virtual landscape's status or associated resource pools.

The form based approach is one way to bridge the gap between easy-to-use keyword search and precise structured search. It should be noted that, while greatly facilitating the access to precise structured search, form based search also restricts the expressiveness of SPARQL, since only a subset of the language is mapped to form structures. Therefore, another paradigm is supported:

*Query Translation.* An innovative approach to make structured search usable in practice is the automated translation of a keyword query into a set of structured candidate queries. Technically, this is realized by mapping every keyword to elements of the structured data and deriving possible structured queries that lead to "reasonable" results, letting the user select the query that comes closest to his information needs. Matching of keywords happens on both instance and schema level, which means that a query is suggested if all the keywords match on elements of a subgraph, no matter if these elements are RDF entities or links between these entities. We omit the technical details here, referring the interested reader to [12] for the theoretical background of this search approach.

As an example, in the enterprise cloud context the translation of the keywords "LUN size NetApp" could lead to a structured query looking for all LUNs on a NetApp filer, together with their respective size. While query translation resembles the form based search paradigm, it is much more flexible and may help to discover relations between data items the user did not even know existed.

**Exploration and Analytics.** We offer different ways to visualize and explore the results obtained from user-defined queries. Going beyond widgets that display the results in form of simple result tables, we also provide widgets for displaying the results in form of charts, diagrams or timeplots (e.g. for historic data). Queries and their visualizations can be stored and attached to resource pages, making it possible to create expressive dashboards and reports that visualize different dimensions of the underlying resources. It is even possible to create query-based page templates for individual resource types, to be instantiated dynamically when loading the page for an instance of the respective type.

*Visual Data Exploration.* The eCloudManager Intelligence Edition features Microsoft PivotViewer[3] as a widget for data exploration. This widget opens up a whole set of new data interaction paradigms, allowing the effective visual exploration of very large datasets by filtering the results with faceted search, drilling down a heterogeneous intermediate result set by any attribute and grouping a

---

[3] http://www.getpivot.com/

result set by different aspects. A data center provider may start with an overview of all the Virtual Machines in his data center, then decide only to display those VMs with the power state "on", apply another filter to only see those VMs having between two and five LUNs attached, and finally group these VMs by their state to see the set of VMs that at the moment are in an erroneous state. Such an approach could bring new correlations to light: The data center provider might notice that all the erroneous VMs have attached LUNs on the same physical host. This host should probably be checked, it could well be the source of the problems.

*Analysis of Historical Data.* One key feature of the Intelligence Edition is the support for historical data management, which is made possible by regularly storing snapshots of the system data in a queryable history database. With this feature at hand, admins and CIOs can track developments and changes in the data center over time. Given that the historical data is fully queryable, one could e.g. ask for hosts with constantly low CPU usage, and in response redistribute CPU resources in order to optimize the data center's resource utilization.

## 7    Experiences and Lessons Learned

The eCloudManager has been in use productively since 2008 in various enterprises. For instance, an early setup was done at a large software development organization of about 1500 developers to provision a sufficient number of test systems and landscapes for all required development configurations and scenarios. Previously this had been impossible in terms of an overwhelming number of configurations and changes to be administered, with systems regularly needing more than 4 weeks to set up, and "shot" systems often needing a week to be reverted to a functioning status. With the eCloudManager it was possible to use large-scale rapid provisioning of complete application landscapes in minutes.

In another deployment at SAP Value Prototyping, which maintains data centers in both Walldorf (Germany) and Cupertino (USA), in one data center alone it has become possible to support more than 875 virtual machines concurrently, delivering a total of 198 SAP Virtual Landscapes to customers.

Clearly, the most immediate and measurable benefits of using the eCloudManager platform result from the management, provisioning and automation features that allow cutting down on provisioning time and reducing the complexity of the support environment. As these features were not the focus of this paper, we refer to [1], where we have reported on experiences and results in these dimensions. Instead, we here want to discuss experiences related to the use of semantic technologies within the eCloudManager Intelligence Edition.

The first aspect we would like to discuss are the experiences with our data integration approach. The flexibility of the RDF-based integration proved to be a significant advantage in the heterogeneus environments of the enterprise data centers. Using the standard data providers of the eCloudManager, a typical setup including an integration of the relevant data center resources could be performed by the data center administrator with little manual effort in less than a day. In previous deployments, the initial integration of custom sources, which required the instantiation of custom data providers and extensions to the ontology, have been performed by the developers of the eCloudManager. While an integration based on a fixed schema and an a priori limited set of supported APIs would have required significant custom development, with our approach custom sources could be integrated within a matter of a few hours rather than days, primarily because additional data providers could be scripted without changes to the code base. Based on the experiences so far, we expect that in the future also data center administrators themselves will be able to integrate custom sources.

Concerning the interaction with the data center, the users reported several qualitative improvements. While previous management solutions let the resources in the data center appear as unconnected silos, for the first time users are now able to see and explore the data center as an integrated whole, manifested in a connected data graph that can be browsed and explored without boundaries. The ability to perform structured queries was perceived a huge efficiency gain, as previously creating integrated reports across management APIs and databases was considerable effort that involved performing queries against and compiling the results manually. The most enthusiastic reactions were received in response to the visual exploration with the Pivot interface, which provides an unprecedented user experience. With respect to user-generated content, we found that semantic wikis provide an adequate interaction paradigm to generate documentation augmented and linked with structured data. As administrators are used to maintain wikis with documentation anyway, the barrier to add structured annotations is low, while at the same time the benefits are immediately visible.

The final observation concerns scalability and performance of the system. In our current projects – which already include rather large enterprise cloud environments – we have to deal with up to 5000 data center objects per installation, where for each object the number of statements is in the range of tens. When also managing historic and aggregated data, the amount of data increases roughly by a factor of 10. This results in a total of $10^5$-$10^6$ statements – an amount of data that has not posed a serious challenge for the underlying data store nor any other component of the system. As a result, we were able to realize real time ad hoc queries and analytics with sub-second response times.

## 8  Related Work

In the domain of data center (and more recently enterprise cloud) management solutions, major vendors have commercial offerings, such as IBM Tivoli, HP OpenView or VMware vCenter, yet most commercial systems are far from having a truly integrated data source and appear more like a bundle of individual standalone products. To our knowledge, the fluidOps eCloudManager product suite is the first commercial offering based on semantic technologies to enable an open, vendor-independent integration of heterogeneous data center resources.

The idea of using wikis to improve collaboration and knowledge management in the enterprise is of course not new (see e.g. [2]). Recent works also explore the use of semantic wiki technology in the corporate context. For instance, [7] presents an approach to the collaborative management of systems monitoring information based on Semantic MediaWiki. From a more general, wiki-independent perspective, the benefits of semantic technologies for enterprise data management have been exemplarily discussed in the domain of customer data management in [10]. All these works are complementary to our work, which has its focus on the domain and challenges of enterprise cloud management.

The topic of intelligent analytics for data center operations – or *data center intelligence* – has recently attracted significant interest. As an example, CIRBA[4] offers analytics software that enables organizations to support intelligent planning and management of physical and virtual infrastructure. It helps answer the questions of where to place workloads and how to allocate and configure resources. While the task of data center planning goes beyond the currently available features (cf. future work), the eCloudManager provides integrated access to the relevant technical and business data, including the historical data.

# 9   Conclusions and Future Work

We have shown how semantic technologies are applied in addressing some of the key challenges in managing enterprise cloud environments. Summarizing the main results, the approach of RDF-based data integration allows us to deal with the highly heterogeneous and changing set of resources encountered in enterprise data centers. Semantic wikis provide an end-user oriented interface for creating structured and unstructured annotations, supporting the main use cases for documentation and knowlegde management, seamlessly integrating automatically obtained data with user-generated content. This data can be searched, explored and analyzed without system boundaries, supported by state-of-the-art techniques of semantics-based information access. Having summarized the results, we now conclude with a discussion of items that remain on the research agenda.

**Policies.** Cloud operators typically define business policies on a high level, which are subsequently monitored by software or taken into account when decisions are being made. In order to be successful, cloud providers need to be able to adapt policies upon changing market requirements or new competition coming into the equation. Changing policies quickly is only possible when the gap between high level business policies and low level implementation is not too big. Future work may consider how to map business policies into rule based systems.

**Reasoning.** Recalling the discussion in Section 3, we currently use the eCloud-Manager's OWL 2 ontology primarily to define domain, range and cardinality restrictions. As explained earlier, this schema information is then used at runtime, e.g. for guided completion of missing information (cf. Section 5) or schema-agnostic search (cf. Section 6). Going beyond the pure use of schema information, we are planning to investigate in how far the use of OWL (and possibly, advanced features) and OWL-based reasoning can help to monitor system health and detect misconfigurations in the data center.

---

[4] http://www.cirba.com/

**Complex Event Processing.** A closely related area where we expect interesting opportunities for future work is complex event processing. Using our provider concept, we periodically receive data from sources distributed across the whole enterprise, and it is a challenging task to define, identify and react to globally meaningful event patterns. The topic seems particularly interesting in the context of reasoning and data semantics, where logic-based approaches like in [4] may be a good starting point for future work.

# References

1. Whitepaper: Making the case for the private sap cloud. Technical report, fluid Operations (2010), http://www.fluidops.net/download/VLM-Whitepaper.pdf
2. Andersen, E.: Using wikis in a corporate context. Handbuch E-Learning (2005)
3. Angles, R., Gutierrez, C.: The expressive power of sparql. In: Sheth, A.P., Staab, S., Dean, M., Paolucci, M., Maynard, D., Finin, T., Thirunarayan, K. (eds.) ISWC 2008. LNCS, vol. 5318, pp. 114–129. Springer, Heidelberg (2008)
4. Anicic, D., Fodor, P., Stojanovic, N., Stühmer, R.: An approach for data-driven and logic-based complex event processing. In: DEBS (2009)
5. Brockmans, S., Haase, P., Hitzler, P., Studer, R.: A metamodel and uml profile for rule-extended owl dl ontologies. In: Sure, Y., Domingue, J. (eds.) ESWC 2006. LNCS, vol. 4011, pp. 303–316. Springer, Heidelberg (2006)
6. Broekstra, J., Kampman, A., van Harmelen, F.: Sesame: A generic architecture for storing and querying rdf and rdf schema. In: Horrocks, I., Hendler, J. (eds.) ISWC 2002. LNCS, vol. 2342, pp. 54–68. Springer, Heidelberg (2002)
7. Kleiner, F., Abecker, A., Brinkmann, S.F.: Wisymon: managing systems monitoring information in semantic wikis. In: Int. Sym. Wikis (2009)
8. Krötzsch, M., Vrandecic, D., Völkel, M.: Semantic mediawiki. In: Cruz, I., Decker, S., Allemang, D., Preist, C., Schwabe, D., Mika, P., Uschold, M., Aroyo, L.M. (eds.) ISWC 2006. LNCS, vol. 4273, pp. 935–942. Springer, Heidelberg (2006)
9. Lenk, A., Klems, M., Nimis, J., Tai, S., Sandholm, T.: What's inside the cloud? an architectural map of the cloud landscape. In: CLOUD 2009: Proceedings of the 2009 ICSE Workshop on Software Engineering Challenges of Cloud Computing, Washington, DC, USA, pp. 23–31 (2009)
10. Ma, L., Sun, X., Cao, F., Wang, C., Wang, X., Kanellos, N., Wolfson, D., Pan, Y.: Semantic enhancement for enterprise data management. In: Bernstein, A., Karger, D.R., Heath, T., Feigenbaum, L., Maynard, D., Motta, E., Thirunarayan, K. (eds.) ISWC 2009. LNCS, vol. 5823, pp. 876–892. Springer, Heidelberg (2009)
11. Sotomayor, B., Montero, R.S., Llorente, I.M., Foster, I.: Virtual infrastructure management in private and hybrid clouds. IEEE Internet Computing 13(5), 14–22 (2009)
12. Tran, T., Mathäß, T., Haase, P.: Usability of keyword-driven schema-agnostic search. In: Aroyo, L., Antoniou, G., Hyvönen, E., ten Teije, A., Stuckenschmidt, H., Cabral, L., Tudorache, T. (eds.) The Semantic Web: Research and Applications. LNCS, vol. 6089, pp. 349–364. Springer, Heidelberg (2010)
13. Vaquero, L.M., Rodero-Merino, L., Caceres, J., Lindner, M.: A break in the clouds: towards a cloud definition. SIGCOMM Comput. Commun. Rev. 39(1), 50–55 (2009)
14. Winter, M.: Data center consolidation: A step towards infrastructure clouds. In: Jaatun, M.G., Zhao, G., Rong, C. (eds.) CloudCom 2009. LNCS, vol. 5931, pp. 190–199. Springer, Heidelberg (2009)

# Semantic MediaWiki in Operation: Experiences with Building a Semantic Portal

Daniel M. Herzig and Basil Ell

Institute AIFB
Karlsruhe Institute of Technology
76128 Karlsruhe, Germany
{herzig,basil.ell}@kit.edu
http://www.aifb.kit.edu

**Abstract.** Wikis allow users to collaboratively create and maintain content. Semantic wikis, which provide the additional means to annotate the content semantically and thereby allow to structure it, experience an enormous increase in popularity, because structured data is more usable and thus more valuable than unstructured data. As an illustration of leveraging the advantages of semantic wikis for semantic portals, we report on the experience with building the AIFB portal based on Semantic MediaWiki. We discuss the design, in particular how free, wiki-style semantic annotations and guided input along a predefined schema can be combined to create a flexible, extensible, and structured knowledge representation. How this structured data evolved over time and its flexibility regarding changes are subsequently discussed and illustrated by statistics based on actual operational data of the portal. Further, the features exploiting the structured data and the benefits they provide are presented. Since all benefits have its costs, we conducted a performance study of the Semantic MediaWiki and compare it to MediaWiki, the non-semantic base platform. Finally we show how existing caching techniques can be applied to increase the performance.

## 1 Introduction

Web portals are entry points for information presentation and exchange over the Internet about a certain topic or organization, usually powered by a community. Leveraging semantic technologies for portals and exploiting semantic content has been proven useful in the past [1] and especially the aspect of providing semantic data got a lot of attention lately due to the Linked Open Data initiative. However, these former approaches of semantic portals put an emphasis on formal ontologies, which need to be build prior to the application by a knowledge engineer resulting in formal consistent and expressive background knowledge [1,2]. This rather laborious process yields further efforts when changes and adjustments are required. Beside this disadvantage, [3] points out that versioning of the structured knowledge is missing and the community features are essential,

P.F. Patel-Schneider et al.(Eds.): ISWC 2010, Part II, LNCS 6497, pp. 114–128, 2010.

but insufficient. Recently, [4] showed how the popular content management system *Drupal*, which will support semantic data from version 7, can be applied for building semantic applications. We pursue an alternative approach, leveraging communities for the creation and maintenance of data.

One of the most successful techniques to power communities of interest on the web are wikis. Wikis allow users to collaboratively create and maintain mainly textual, unstructured content. The main idea behind a wiki is to encourage people to contribute by making it as easy as possible to participate. The content is developed in a community-driven way. It is the community that controls content development and maintenance processes. Semantic wikis allow to annotate the content in order to add structure. This structure allows to regard the wiki as a semi-structured database and to query its structured content in order to exploit the wiki's data and to create various views on that data. Thus wikis become even more powerful content management systems. Moreover due to the semantic annotations, structured content becomes available for mashups with semantic content residing outside the wiki, for example as Linked Open Data.

In this paper we describe how we use the semantic wiki *Semantic MediaWiki*[1] [5,6] (SMW) for creating a portal for our institute, which can be accessed at `http://www.aifb.kit.edu`. The portal manages the web presence of the AIFB institute, an academic institution with about 150 members. The portal is a semantic web application with about 16.7k pages holding 105k semantic annotations. Table 1 gives an overview in numbers of the portal. While wikis provide free, wiki-style semantic annotations with the complete freedom regarding the adherence to any vocabulary, users can be guided to adhere to use certain vocabularies by providing form-based input. The importance of the right balance between unstructured content, which is better than no content, and structured content, which is more efficient to use, was studied already by [2]. However, this approach focussed on automatic crawling of structured data and did not regard the user as the primary provider.

This paper is structured as follows: In Section 2 we report on design and development decisions and in particular discuss the free, wiki-style editing versus guided user input. Further, we report on the development efforts and on the subsequent usage and maintenance. In Section 3 we show the advantages and features made possible by the semantics of the portal. And finally, we report in Section 4 on performance tests and compare Semantic MediaWiki to its non-semantic base platform MediaWiki, before we conclude in Section 5.

## 2   Designing and Developing the Portal

The most common and original application of Semantic MediaWiki and wiki systems in general is collaborative knowledge management, e.g. for communities such as `semanticweb.org`. In this section we present the portal we built using Semantic MediaWiki and in particular its features exploiting semantic technologies.

---

[1] `http://semantic-mediawiki.org/`

## 2.1    Free Annotations versus Guided Input

A wiki provides the users with the means of rather easy and unconstrained adding and changing of content, in the sense that they just need to know the simple wiki markup and have a web browser. Further, the users can publish content themselves without the assistance of a webmaster. One aim when designing the portal was to have low barriers for the institute's members to contribute, extend and maintain the content. Hence, we considered a wiki as an appropriate choice. In contrast to regular wiki systems, Semantic MediaWiki allows to semantically annotate the content. This free and independent annotation paradigm has the advantage of being flexible, and expandable. Moreover, it does not require the knowledge of a predefined schema. The underlying notion is that more annotations are in general better than less annotations even if they are not well organized and do not follow a predefined vocabulary or ontology. However, when using *inline queries*, see Section 3.1, one has to know the exact property names, since formal queries are strictly sensitive and minor derivations are not tolerated. The same is true for many applications building on top of the structured data. They are often build on a specific schema or vocabulary. Thus, one has to find the right balance between a predefined schema and keeping it flexible and expandable at the same time. For the case of Semantic MediaWiki, *templates* and *forms*[2] allow to restrict the user to a predefined set of annotations. A template defines the logic and the appearance of a part of a page. It keeps placeholder variables, which are filled by the instantiating page. Inserting annotations in the template entails the annotation of all pages using the template with the same annotations. Consequently, changing the annotation inside a template cascades this change to all pages and thereby allows a flexible modification of the structured data. Forms provide a graphical user interface for using templates correctly and do not even require the usage of wiki markup. Thereby the combination of forms and templates allows to have a set of predefined annotations.

For the portal, we created about 30 templates and corresponding forms for all major, reoccurring resources, like *people, lectures, publications*, and so on. Figure 1 shows an example form for editing a page about a project. Forms can consist of different types of fields, e.g. for text, dates, choices, etc. Behind each field is an annotation, i.e. a property. By entering a value in a field, the value is assigned to the corresponding property. It is good practice to import these properties from already existing vocabularies[3], e.g. FOAF[4] for persons, if applicable. In order to keep the possibility for free, unconstraint annotations, the forms can contain text areas, which can contain text with arbitrary annotations. Thereby, we tried to find a balance between guided input with predefined annotations and the possibility to have free annotations.

The advantage of this mixture of guided input and open annotations is that the structure of the data can evolve dynamically, which we report on in Section 2.3.

---

[2]  http://www.mediawiki.org/wiki/Extension:Semantic_Forms
[3]  http://semantic-mediawiki.org/wiki/Help:Import_vocabulary
[4]  http://xmlns.com/foaf/spec/

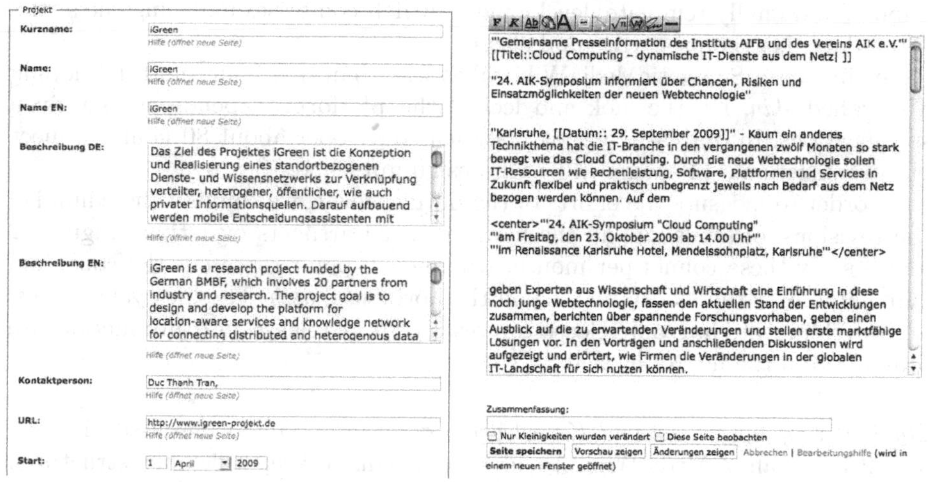

**Fig. 1.** The left side shows guided input via a form. The form consists of several different input fields. The content of each field is assigned to a predefined annotation. The form holds also free text areas, which can contain text with arbitrary annotations. The right hand side shows entirely free, wiki-style editing without any constraints regarding predefined annotations.

## 2.2   User Roles

When we proposed a wiki system for the portal, the first concern from colleagues was the fear that *everybody* can edit it, even *anonymously*. This is of course not the case. We used MediaWiki's internal user right management[5] to create four different groups: The anonymous web surfers can only read regular pages, i.e. those in the main namespace. The authenticated users may also read pages in other namespaces and in addition are allowed to edit pages, except for pages in the template and form namespace. The latter can only be manipulated by admins. The fourth group are bureaucrats, which have the same rights as admins, but in addition they can appoint and withdraw the admin right.

Since having an extra user account might impose a barrier for people to participate, we used the *Lightweight Directory Access Protocol* (LDAP) extension[6] for MediaWiki and an SSL encrypted connection between the portal and the LDAP server. This allows to use already existing user accounts for the authentication at the portal.

## 2.3   Development and Dynamics of the Structured Knowledge Representation

The development efforts of the portal can be broadly separated into four different areas: system setup, visual design and custom function development, data

---

[5] http://www.mediawiki.org/wiki/Manual:User_rights
[6] http://www.mediawiki.org/wiki/Extension:LDAP_Authentication

import, and finally template development, which comprises modeling the structured data, i.e. the properties and classes.

Setting up a SemanticMediaWiki takes less than one hour[7] and developing a so-called *skin*, i.e. the look and feel of the platform, depends on the given specifications. In our case, it took a student developer about 80 hours to meet our organization's 148 pages in length design guideline.

In order to measure the efforts of the development, we counted the edits, i.e. the revisions, of artifacts and the newly created artifacts over time. Figures 2 and 3 show these counts per month, where each point represent the count accumulated over the month marked on the horizontal axis. The life cycle phases, i.e. development, internal release for testing, and release into production are illustrated in the figures as well.

*Dynamics of the Structured Knowledge Representation.* As discussed in Section 2.1, Semantic MediaWiki provides the means to keep a flexible, structured data schema consisting of properties and classes. Figure 2 shows how these elements changed over time and how often new elements were added. Categories group pages and correspond to classes in the structured representation. Regarding the manipulation of classes and properties, one can see that most of the structured data layout was done at the very beginning of the project in April 2009. In particular, the classes involved were known right from the beginning and relatively few changes were needed during the subsequent phases. The same holds for the properties in an alleviated form. Still, one can see that a small, but steady number of properties and classes were added or changed over the course of the project with the exception of the peaks in March 2010. In this month, the annual institute report about events, publications, and people was prepared. The data for the report was exported from the portal. Since the editors requested changes and additions to the data, e.g. splitting names into first and last name or adding the location of publication to some publication types, we needed to change the structure in the portal accordingly. In particular, the class structure underwent refactoring, e.g. splitting the class *employee* into former and active members.

All these adjustments were done in an agile way driven only be requirements and demands. In particular, one has to keep in mind that all changes happen on the application level. Touching the underlying database was never necessary nor taking the system offline for modifications. Furthermore, the wiki provides a versioning system, which tracks all changes, also those of properties and classes, a crucial capability for semantic portals [3].

*Data Migration and Template Development.* In order to populate the platform, we used the Pywikipediabot framework[8], a Python application for manipulating wiki pages via scripts. The loading of the existing data into the platform explains the high peak on the left plot of Figure 3. Creating the templates was the most

---

[7] http://www.mediawiki.org/wiki/Manual:FAQ#How_do_I_install_MediaWiki.3F
[8] http://meta.wikimedia.org/wiki/Pywikipediabot

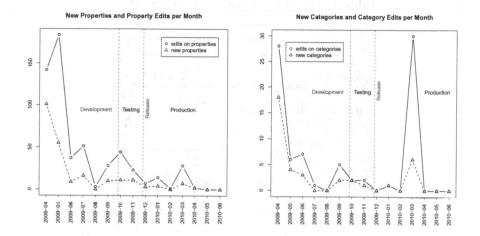

**Fig. 2.** The plot on the left side shows the number of new property types and edits on property types per month. The plot on the right side show the number of new categories and edits on categories. Categories correspond to classes of the structured data. Since properties and classes are the elements of the structured data, these plots show the evolution of the structured data over time.

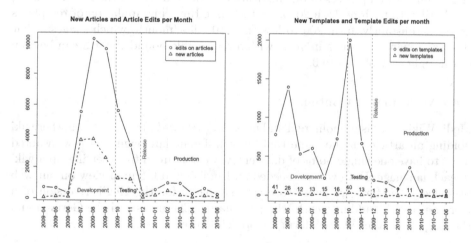

**Fig. 3.** The plot on the left side shows the number of new articles and edits on articles per month for the different periods from development to production. The high numbers during the development phase are due to automatic batch jobs populating the portal with content. The plot on the right side shows new templates and edits on templates. Since these can only be edited by admins, this plot allows to estimate the development effort as well as the maintenance effort after the release. The peak in March 2010 was the result of implementing the annual reporting, see Section 2.3.

**Table 1.** The portal in numbers as of June 2010

| | |
|---|---|
| pages | 16.716 |
| templates | 219 |
| forms | 30 |
| uploaded files | 1.773 (1.2 GB) |
| users (total) | 142 |
| active users (last 91 days) | 83 |
| annotations (property instances) | 104.182 |
| property types | 191 |
| categories (classes) | 40 |
| OWL/RDF | 238k triples |
| code base | 132 MB |
| database | 99.5 MB |

time consuming task. However, the peak at the start of the testing period is solely due to the tidy visual requirements.

*Usage and Maintenance.* Since the release, we observe a steady user participation with an average of 550 edits/month on articles and about 195 new pages/month, as shown in the production period in left plot of Figure 3. About 83 users or 66% of the full time employees of our institute contributed within the last 3 months, i.e. April 18th to June 18th 2010. At the same time, manipulation of templates declined constantly, from 200 to less than 10 edits/month, which suggests that the maintenance by the admins is within reasonable bounds, which can be seen in the right plot of Figure 3.

### 2.4   Multilingual Content

MediaWiki per se is monolingual and uses *interwiki-links* to point to another wiki holding an article on the same topic in a different language. Since we wanted users to have one single point of data entry, we abstained from setting up a wiki in each language. However, we needed an English and German view on our web presence. Therefore, we chose to create subpages for the English version of a German page by adding */en* to the page name. The users add the German and the English content via one form for predefined resources.

### 2.5   Challenges during Design and Development

The biggest challenge during the development was the creation of templates and trimming them to the strict design guidelines. Whereas it is acceptable for most wiki applications to have little blemish and accept the free and sometimes untidy appearance, an official web presence should avoid it, e.g. all empty variables in templates needed to be hidden. Moreover, due to the tidy appearance requirement, some annotations contain markup, e.g. for italic font style or font size, which is desirable in the structured data representation. Furthermore, the templates combine the description of the appearance of a page and the logic at the same time, which makes them complex and overcharged and require advanced knowledge of the wiki markup for further development and maintenance. Therefore, template manipulations are restricted to admins in our portal.

# 3 Where Semantics Help - Features of Semantic MediaWiki

In the previous section we reported on the development process and the dynamics of the structured data. In this section, we show the features taking advantage of the structured data.

## 3.1 Inline Queries

The biggest advantage of SMW, beside its flexible annotation paradigm, is the possibility to reuse data across the platform by querying it from other pages. These *inline queries* allow to request sets of data or just single property values and display them on a page in various result formats, such as tables, list, charts, maps, etc. This reuse of data avoids data redundancy, e.g. the information about a person, like name, email, or phone number, is entered once on the page about this person and then later this information is queried and displayed on pages about projects, publications, etc., where this person is involved in. If the data changes on the source page, the data on the requesting page changes accordingly when the inline query is executed again. Inline queries create dynamic pages. Figure 4 illustrates an example of an inline query and its results as it appears on the requesting page.

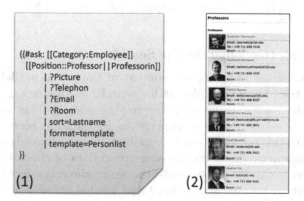

**Fig. 4.** An inline query requesting all employees, which are professors, and information about them (1) and its result representation (2)

## 3.2    Querying Linked Open Data Sources

We created an extension that allows querying external sources using the simple syntax of inline queries [7]. This mediation-based approach allows for either
displaying or importing externally retrieved data from the Linked Open Data
source Freebase, other SMWs, or from CSV files, in order to enrich the wiki's
content with external data. In the first two cases a mediator translates an inline
query into a query in the query language supported by the remote source, which
is MQL in the case of Freebase. Figure 5 illustrates an example. Translation is
not a task of solely syntactical transformation but also involves ontology mapping. The mappings are stored in the wiki as annotations. Thus they can be
contributed and maintained by users.

In our portal we query the SMW of semanticweb.org in order to retrieve
events, such as conferences or workshops and present them on a timeline in
order to offer visitors of our page an interactive conference radar with up-to-date
information. Moreover, we are using Freebase to retrieve location information
about the institute's industrial and academic partners, in order to be able to
sort them by region.

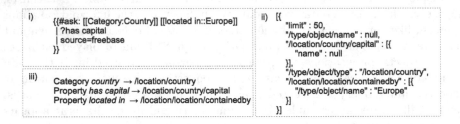

**Fig. 5.** Using the Freebase mediator an inline query such as in i) is translated into an
MQL query such as in ii) by using the mapping information such as in iii)

## 3.3    Exploiting the Semantics for Search

One certain advantage of having the content of the portal available in a structured form is the ability to exploit it for search. [8] presents an approach for
semantic search in wikis, which we apply for the portal[9]. This approach allows
to use keywords as the means to express an information need, because most
users are used to this common search paradigm. These keywords are then transformed into interpretations using the structured data of the wiki as the search
space. The interpretations are shown to the user, who can select the interpretation fitting best to his information need and further refine it in the next step.
Figure 6 shows an example search over the structured data for employees, their
email addresses, and office location. In contrast to the inline queries, which use a
simple, but formal query syntax and are therefore inadequate for ad-hoc search,
*Ask The Wiki* is suitable for end users and exploits the semantic annotations.

---
[9] http://www.aifb.kit.edu/web/Spezial:ATWSpecialSearch

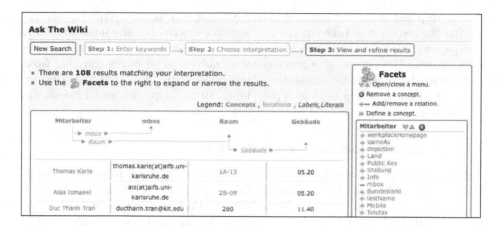

**Fig. 6.** This figures shows the result of a search for all employees, their emails, room numbers, and corresponding building numbers. The facets menu on the right hand side allows to refine the result based on the structured data.

## 4 Performance

MediaWiki, the platform powering Wikipedia, runs on many sites and is well known for being scalable and fast. Although the usefulness of the features provided by Semantic MediaWiki get the interest of many potential users, often skepticism about SMW's resource requirements, its stability and scalability are brought forward. In this section, we report on stress tests conducted on both Semantic MediaWiki and MediaWiki with the data from the portal in order to allow for their comparison.

*Test Environment.* We performed the tests with a common desktop computer, which has one CPU with 2GHz, 2 GB memory and runs on Debian 5.0[10]. The wiki runs on an Apache web server with PHP 5 and uses a MySQL database[11]. The load tests are conducted with Apache JMeter and the server was monitored by sysstat[12]. The client sending the requests was connected through the same 100 MBit backbone to the server.

This system configuration is a common single machine setting and by no means laid out for high-performance. However, it allows to compare the two systems, MediaWiki 1.15.3 and Semantic MediaWiki 1.5. The wiki holds the data of the AIFB portal, see Table 1 for an overview. The data contains annotations in the SMW syntax. If SMW is not enabled, MediaWiki interprets these statements just as text. SMW allows to restrict the usage of inline queries, e.g. the maximal number of conditions, query depth, and maximal retrieved results, in order keep reasonable bounds. However, these settings were all set to unlimited for the tests.

---

[10] AMD Athlon 64 3200+, 2.6.26-2-amd64 kernel.
[11] Apache 2.2.9, PHP 5.2.6-1+lenny3 with APC 3.0.19, MySQL 5.0.32.
[12] Apache JMeter 2.3.4, sysstat 7.0.

The response time at the client side, i.e. from sending the request until receiving the response, is taken as the performance metric. For the measurements, 310 pages ($\sim$ 2% of all pages) accessible from the main page within 2 clicks were chosen. The pages are a representative subset of all pages, ranging from pages with little semantic annotations and queries to pages that make heavy use of these features. On average a page holds 10 inline queries and 12 semantic annotations.

### 4.1 MediaWiki vs. Semantic MediaWiki

A test consisted of $N$ parallel users requesting the 310 pages in random order. Figure 7 shows a box-plot illustrating the response times in milliseconds for MediaWiki (MW) and Semantic MediaWiki (SMW), and Semantic MediaWiki with Caching (SMW+C) for $N = \{1, 10, 25, 50\}$ parallel users. It shows that the response times are linear with respect to the number of parallel users. This linear behavior becomes apparent in Table 2, which shows the throughput, i.e. the number of served requests per second. The throughput is constant at about 4.7 requests/sec for MW and 4.1 requests/sec for SMW, which means that using SMW costs about 13% in performance compared to MW during operation.

However, it is unexpected that the spread of the response times over the 310 pages is so little, as illustrated in the box-plot of Figure 7. Especially in the case

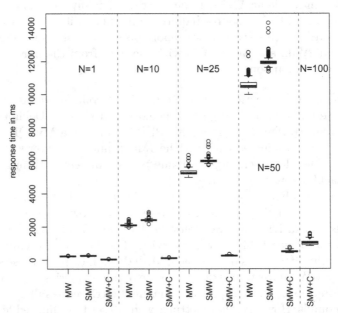

**Fig. 7.** Box-plot illustrating the response time per page for MW, SMW, SMW+Cache with N parallel users requesting 310 pages in random order.

**Table 2.** Throughput (requests/sec) for $N$ parallel users. The percentages are compared to the MediaWiki (MW) baseline. When applying the cache (SMW+C) the server's limits were not met.

| N | 1 | 10 | 25 | 50 | 100 |
|---|---|---|---|---|---|
| MW | 4.36 | 4.75 | 4.75 | 4.73 | n/a |
| SMW | 3.83 (-12 %) | 4.10 (-14%) | 4.13 (-13 %) | 4.13 (-13%) | n/a |
| SMW+C | > 25.68 (+489%) | > 90.80 (+1810%) | > 96.78 (+1930%) | > 96.31 (+1930%) | > 95.01 |

of SMW, the pages contain semantic annotations and in particular inline queries, which need to be parsed and processed. The response time should depend on the number and complexity of these queries. This low spread is due to the implicit caching. The web server has a build-in PHP code cache (APC), which MW and therefore also SMW exploits. Also, a page is not rendered for each request, but only when necessary. In addition, the database caches requests (InnoDB). All these build-in caches absorb most of the additional overhead by SMW during regular operation and make it possible to run SMW at a constant cost of about 13% in performance decrease compared to the non-semantic MediaWiki, see Table 2. The bottle neck resource during these tests was the CPU for MW and SMW for all runs with more than one user. The CPU was consumed for about 95% by the web server and 5% by the database.

In order to avoid the implicit caching behavior and to assess the actual resource requirements, we performed a *cold* test run, i.e. we restarted the machine after each page was requested once, and repeated this for 10 times. Figure 8 shows the average response time over the 10 runs for each page. The pages are sorted by number of inline queries in ascending order, which is displayed on the horizontal axis, and subsequently by the number of templates. It can be seen that

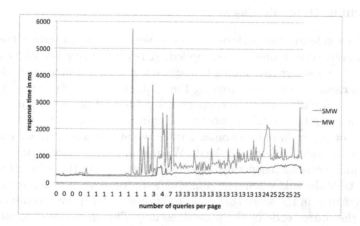

**Fig. 8.** Response times (cold) for 310 pages sorted by increasing number of inline queries, which is shown on the horizontal axis. The high peaks in the center of the plot are due to inline queries involving image operations.

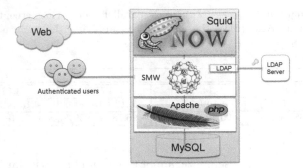

**Fig. 9.** The infrastructure stack of the portal. Anonymous readers get the content served from the Squid cache, if available. Authenticated users are directly connected to the web server.

the response time for MW increases slightly with the number of templates per page. In the case of SMW, one can say that more queries cause a higher response time in general. However, the response time depends mostly on the particular query, which can be seen on the high peaks. The highest peak is a page presenting a list of all people through an inline query, which retrieves an image for each person and creates a custom sized thumbnail for it. The same holds for the other high peaks. These pages all contain one query, which involves operations on images. Queries retrieving only textual information are far less expensive, e.g. pages containing 20 and more inline queries take all less than 2 seconds to serve, if no images are involved. Since images are static content, which can easily be cached, we applied a cache, which is discussed in the following section.

## 4.2   Caching Dynamic Pages

In order to accelerate the performance of a web site and to reduce the load of the web server, reverse proxies are applied. A reverse proxy is a cache installed in front of a web server responding to requests, if the requested content is available in the cache, or otherwise routing the request to the web server. A popular web cache is Squid[13], which is supported by MediaWiki, see Figure 9 for an overview of the setup. While it works well with static content, such as HTML documents or image files, it becomes harder when dynamic content comes into play. In the context of Semantic MediaWiki, dynamic content is foremost produced by inline queries requesting data from other sites of the wiki or from other sources outside the wiki when querying via mediator. The page holding the inline query is dynamic in the sense that its appearance and displayed content changes although the source code of the page does not. Therefore, we encountered the problem that the `Last-Modified` entry in the HTTP header remained the same, because the web server did not recognize a change. Since this entry is used by

---
[13] http://www.squid-cache.org

the cache to determine whether a page is still fresh, we needed to modify the caching mechanism. We chose an aggressive caching strategy by suppressing the `Last-Modified` entry and set a hard maximal expiration time of 3 hours for pages. Thereby, implicit changes of a dynamic page will be updated within this period at the latest. When a page is edited directly, it is immediately purged from the cache. Images and other static content is cached for longer periods. Applying the cache yields a huge performance increase to about 90 requests/sec as shown in Table 2 and Figure 7. However this is by far not the limit, since the CPU was used to about 30%, even for 100 parallel users. One needs to setup multiple physical clients sending requests to asses the actual limit when using a cache, which was beyond our scope.

## 4.3   Performance in Operation

The portal is online since more than six month now with only one interruption due to an DoS attack, which we addressed by restricting the number of connections to the web server. Therefore, we regard the solution as stable and quite robust. There are between 60k and 120k hits per day, which results in an average CPU usage of 6% and an average load of 0.1 on our production web server, which has 2 CPUs[14] and 2GB memory. The median response time, in this case the time between arrival of a request at the server and the sending of the response, is about 2 ms on average, where cache hits take slightly less than 2ms and cache misses take about 200ms to serve on average. The cache has a request hit ratio of about 80% and a size of about 550MB on disk.

## 5   Conclusion

In this paper we have shown how to apply the wiki paradigm of collaborative editing to a web portal using semantic technologies. We discussed how free, unconstrained annotations can be combined with predefined annotations in order to allow flexible and expendable structured data. Further, we reported on how the structured data made available by semantic annotations evolved over time and that it was possible to extend and change it during operation without touching the underlying database. How the semantic data is used and taken advantage of by Semantic MediaWiki's features is illustrated by several examples. Finally, we evaluated the performance and compared it to its non-semantic alternative and showed how caching can be applied to boost the performance. Taking everything in consideration, one can see how Semantic MediaWiki can be used as a successful portal platform providing the advantages of semantic technologies.

## Acknowledgments

Special thanks go foremost to Martin Zang, whose dedication to this project contributed a big part to its success. Also Nicole Arlt and Fabio Garzotto are

---

[14] Intel(R) Xeon(R) CPU E5450 @ 3.00GHz.

thanked for their commitment as well as Philipp Sorg and the IT team at AIFB for their valuable feedback and technical support. Work presented in this paper has been funded by the EU IST FP7 project ACTIVE under grant 215040.

# References

1. Maedche, A., Staab, S., Stojanovic, N., Studer, R., Sure, Y.: Semantic portal - the seal approach. In: Fensel, D., Hendler, J., Lieberman, H., Wahlster, W. (eds.) Spinning the Semantic Web, pp. 317–359. MIT Press, Cambridge (2003)
2. Hotho, A., Maedche, A., Staab, S., Studer, R.: Seal-II the soft spot between richly structured and unstructured knowledge. Journal of Universal Computer Science 7(7), 566–590 (2001)
3. Lara, R., Han, S.K., Lausen, H., Stollberg, M., Ding, Y., Fensel, D.: An evaluation of semantic web portals. In: IADIS Applied Computing International Conference, pp. 23–26 (2004)
4. Corlosquet, S., Delbru, R., Clark, T., Polleres, A., Decker, S.: Produce and consume linked data with drupal! In: Bernstein, A., Karger, D.R., Heath, T., Feigenbaum, L., Maynard, D., Motta, E., Thirunarayan, K. (eds.) ISWC 2009. LNCS, vol. 5823, pp. 763–778. Springer, Heidelberg (2009)
5. Krötzsch, M., Vrandecic, D., Völkel, M., Haller, H., Studer, R.: Semantic wikipedia. Journal of Web Semantics 5(4), 251–261 (2007)
6. Krötzsch, M., Vrandecic, D., Völkel, M.: Semantic mediawiki. In: Cruz, I., Decker, S., Allemang, D., Preist, C., Schwabe, D., Mika, P., Uschold, M., Aroyo, L.M. (eds.) ISWC 2006. LNCS, vol. 4273, pp. 935–942. Springer, Heidelberg (2006)
7. Ell, B.: Integration of external data in semantic wikis. Master's thesis, Hochschule Mannheim (2009)
8. Haase, P., Herzig, D.M., Musen, M., Tran, D.T.: Semantic wiki search. In: Aroyo, L., Traverso, P., Ciravegna, F., Cimiano, P., Heath, T., Hyvönen, E., Mizoguchi, R., Oren, E., Sabou, M., Simperl, E. (eds.) ESWC 2009. LNCS, vol. 5554, pp. 445–460. Springer, Heidelberg (2009)

# A Case Study of Linked Enterprise Data

Bo Hu[1] and Glenn Svensson[2]

[1] SAP Research
[2] BTS EMEA, SAP AG
{bo01.hu,glenn.svensson}@sap.com

**Abstract.** Even though its adoption in the enterprise environment lags behind the public domain, semantic (web) technologies, more recently the linked data initiative, started to penetrate into business domain with more and more people recognising the benefit of such technologies. An evident advantage of leveraging semantic technologies is the integration of distributed data sets that benefit companies with a great return of value. Enterprise data, however, present significantly different characteristics from public data on the Internet. These differences are evident in both technical and managerial perspectives. This paper reports a pilot study, carried out in an international organisation, aiming to provide a collaborative workspace for fast and low-overhead data sharing and integration. We believe that the design considerations, study outcomes, and learnt lessons can help making decisions of whether and how one should adopt semantic technologies in similar contexts.

## 1 Introduction

Thus far, the Linked Data (LD) initiative has demonstrated its value through a variety of projects aiming at improving data accessibility for primarily public and academic users [2]. The success stories certainly have not slipped the attention of large enterprises. Cautious attempts were made to experiment the LD principles and to evaluate the benefits, leading to the so-called "linked enterprise data" paradigm, the counterpart of LD in the business domain [14].

The motivation behind linking enterprise data is evident. Nowadays, with the deepening of globalisation, more and more non-mission-critical businesses are outsourced away from the home countries to for example design teams in Europe, manufacturers in China and service support in India. Fluctuation and risk in local markets, especially volatile ones, therefore becomes more manifested at the global level. This phenomenon has drawn more attention to business agility and continuity, a common ingredient of both being the easy access to data facilitating coordination and collaboration across different geographical locations. Businesses must be able to optimise their internal enterprise data landscape and explore such a landscape at the speed of thought so as to react to the rapidly changing market. Executives must be timely and comprehensively informed so that they can make decisions to counteract the threats to business revenue. More importantly, everyone needs to have ready and immediate access to information/data that enable her to carry out the allocated tasks.

P.F. Patel-Schneider et al.(Eds.): ISWC 2010, Part II, LNCS 6497, pp. 129–144, 2010.

Accessing data in an enterprise context, though not a new research area, is not a topic that we can comfortably mark as "solved" [10]. Enterprise data management has become a prevalent challenge with the rapidly plummeted storage and digitising cost resulting in an unprecedent amount of artefacts available in electronic form[1]. Linked Data initiative was proposed to deal with exactly this problem in the public domain, i.e. removing the barriers to data access and sharing. Intuitively, it seemed that we can just borrow the concepts having been so successfully implemented and recreate the stories in the enterprise environment. Our experience, however, prove otherwise. Indeed, enterprise data has many characteristics that resemble the data from public domains [6]. It, at the same time, presents unique requirements that put into test the principles and assumptions that are widely enjoyed when linking public data sets. The differences are demonstrated in the following aspects. Firstly, enterprise data is normally tied closely with the business processes. Peeling off the contexts wherein the use of such data takes place might render the data linking effort less fruitful. Secondly, it becomes increasingly important to link to data sets outside organisational boundaries. This is evident in use cases such as supply chain management and pre-sale where data from public domain significantly enrich internal data sets. We, therefore, see a mixture of public, partner, and proprietary data complicating data transparency and accessibility. To our best knowledge, none of the existing efforts have addressed the process driven requirement unique to enterprise data.

Inspired by the misalignment between the requirements of enterprise data and existing LD efforts, we carry out studies with real users to investigate how the linked data principles and concepts can assist customer account executives and team members when they need comprehensive and real-time access to internal and external data sets. We first elaborate on the differences (Section 2) between enterprise data and public data. Bearing these differences in mind, we discuss certain design considerations and the system architecture in the context of a customer information portal (CIP) project (Section 3). This is followed by three real-life use cases demonstrating the value of CIP (Section 4). We then discuss the lessons learnt (Section 5) and conclude the paper in Section 6.

## 2    Why Corporate Data Is Different?

As a collaborative and international effort, Linked Data has gained good publicity in the academic and to some extent the public sector communities [2]. With all the exciting success stories of massive development effort in linked data projects, we now face the question regarding the applicability of "linked data" principles in the corporate sub-domain.

It is evident that enterprise data lend themselves as both an opportunity and a challenge. On the one hand, enterprise data have well-defined boundaries with rigid protocols regulating the transition across the boundaries. They present less heterogeneity and diversity comparing with public data from the Internet.

---

[1] http://www.thegoldensource.com/component/attachments/download/36

Furthermore, even though divided into different departments focusing on different areas, modern enterprise normally reinforces a common corporate culture, which fosters a common, shared corporate "language", i.e. domain vocabulary. In many cases, this vocabulary may even impinge on communities beyond corporate boundaries. A good example is the jargons and acronyms used by the global SAP customer network. Finally, enterprise data are normally well documented and preserved either formally as white-papers, official publications, etc. or informally in e-mails, task log data, wiki pages, etc. Different from public data from the Internet, enterprise data are normally subject to internal review, for the purpose of auditing and quality control, or, at minimum, created with good intentions to fellow workers. We can therefore enjoy a much smaller amount of noise compared to general public data.

On the other hand, enterprise data still present significant research challenges. Simply connecting different islands together in an archipelagic data landscape will not be convincing enough. "Process-driven" is a unique feature that one has to bear in mind when migrating the LD concepts into the enterprise environment. Meanwhile, the relatively small size and homogeneous nature of enterprise data suggest that superficial connection of in-house data may not generate a good enough business value. In many cases, internal data alone is not sufficiently rich to satisfy diverse business requirements and thus incorporating external data sources is inevitable. How and what data should be exploited, however, can only stem from real-life scenarios. It is, therefore, salient to align with end users to understand and demonstrate the "return of value" of linking enterprise data. We will discuss these points further in this section.

## 2.1    Process-Driven

Currently, there are roughly two approaches to fulfill the LD vision, namely data-driven and community-driven. Data-driven starts with a set of core data and tries to establish connections with as many relevant data sets as possible to emerge patterns not possible to individual data sets alone, while community-driven tries to fulfill the data request of a community, e.g. movie fans, gene researchers, etc. Both approaches may find themselves struggling in the enterprise environments.

Data management in an enterprise environment always has one ultimate purpose: improving the efficiency of a company's core business. However, linking data together does not necessarily mean that the implications, with which data are generated and leveraged, automatically become explicit to those linked in. The business implications can only be understood when we situate data into their original business processes. Therefore, different from the dominant data-driven nature when linking data from the Internet, linking enterprise data demonstrates a strong process-driven characteristic. That is the connections among data can and should only be revealed within the context of business processes where such data are consumed. Similarly, links among data should not be arbitrarily created independent from business processes. Aligned with companies' mission-critical businesses, linking enterprise data from both inside and outside a company can be rightfully leveraged in decision making.

We would also argue that the successful community-driven approach (c.f. [16]) is not strictly applicable in the enterprise environments. Such communities are normally self-organised by common interests and loosely regulated, mainly self-disciplined. Misconduct and inappropriate behaviours do not result in the same consequence as in enterprise environments. Meanwhile, members of the community are organised in a rather flat structure with equal access to resources, which is a freedom that is not valid in companies. To the best interests of employees, taking a process-driven approach to data linking therefore can guarantee the alignment between personal interests and organisational policies.

## 2.2 External Data

At the beginning of the CIP project, our intuition was that in an international organisation, the internal data alone should present enough challenges and offer sufficient business value for the LD paradigm. This was proved partially wrong during the discussion with end users. Internal data, although distributed across a large geographical region, are well-regulated and to some extent aligned attribute to common corporate cultures and operational regulations. Making internal data compliant with LD principles is more an organisational and motivational effort than a technical challenge.

The real challenge comes from defining good scenarios that can meaningfully link data together to satisfy needs of everyday businesses. For such a purpose, internal data can only tell part of the story. Very frequently, employees refer to external data sets for essential information that is not available from within the corporate boundaries. For example, the latest volcanical ash disturbance resulted in changes of project execution, project management decisions, and customer relationship management; natural disasters (e.g. the earthquake in SiChuan Province, China) can lead to major changes in supply chain management. The importance of such external data will not be fully demonstrated if they are not combined with internal enterprise data and consumed in real-time business decision making. The linking of public, partner and proprietary data should conform to the following guidance. External data should not interfere with internal ones. Where conflict observed, organisational protocols should be consulted to resolve the inconsistences. Meanwhile, it should follow existing organisational policies: this again points back to the process-driven aspect.

## 2.3 Personal Space

The most controversial argument that we would like to put forward, which can be deemed against the total "openness" of the LD initiative, is that when linking enterprise data, the personal comfortable zone in data sharing should be respected. For organisations of different sizes, cultures, and structures, there is a long standing tendency of information *disintegration* attribute to a lack of trust in fellow workers, feeling of insecurity, and fear of disgrace [12]. We did not plan to deal with such motivational issues. Rather we acknowledge the existence of such barriers and try to accommodate user requirements that stem therefrom. Obsering such a requirement allows users to more comfortably position themselves in data sharing initiatives. This is, however, done at the price of sacrificing fundamental LD principles to a certain extent.

# 3    Customer Information Portal

The concept of linked enterprise data is materialised in a pilot study that is meant to facilitate data integration and data sharing in a geographically distributed international organisation. When a company operates in more than one locations, it is not surprised to find different regional representatives approaching the same customer with different stories. The representatives sometimes are caught totally unprepared with questions regarding latest business and technical development and, even worse, regarding technical proposals and sales offers made from other units or even within the same units. A simple and effective remedy to such a problem is to create a portal for all the data concerning a customer. It can serve as a briefing tool for any one working on a customer so as to avoid the aforementioned embarrassment. We take advantage of the CIP project as a platform for understanding benefits and constraints of applying LD concepts in the enterprise environment.

## 3.1    Design Decisions

During the definition of this pilot project, we try to address the unique characteristics of enterprise data (as discussed in previous sections).

Process-driven is given particular emphasis. Projected on design decisions, this implies the ability to answer "what data should be accessed by whom at what stage?". Based on business processes, one is prescribed to navigate the internal resources, employee profiles, and external data only specified in the business processes. Doing so will ensure that enterprise data are linked in line with organisational policies and strategies. Business processes can be standard ones or created for personal needs using predefined building blocks. We provide a list of exemplary business processes that are modelled and executed using in-house software (e.g. SAP Netweaver BPM) due to practical considerations. The in-house software is well understood by all the end users that reduces the learning curve. Meanwhile, in order to ensure a smooth integration with internal data sets, we try to avoid unnecessary disturbance to the platforms wherein such data sets are used. The in-house business process management system offers adapters compliant with J2EE Connector Architecture[2] and thus can seamlessly integrate with Java-based semantic systems.

The privacy concerns are addressed by maintaining a clear separation between data sets that are available to everyone and those that are only visible to the selected few. When creating an online article, a new business process, or uploading

---

[2] http://java.sun.com/j2ee/connector/

a document, people can opt-in to share or not share such resources. Effectively this is tantamount to linking private data space with the public one. Regardless of whether or not the resources are made public, an excerpt is produced to inform others of the contents.

We try to accommodate the general LD principles as followings. Using URI for resource identification can be easily satisfied—all internal resources (including documents and people) are uniquely identifiable through URIs. When this is not the case, we annotate data sets with uniquely identifiable labels based on RDF-coded ontologies. Links among internal resources are implemented as ontology properties among annotated resources. For internal data, syntactic and semantic mismatch does not present as a problem due to the existence of well-defined common vocabularies normally exercised by large organisations. Semantic interoperability becomes more of an issue when linking to external data sets. We adopt a simple but effective solution: embedding a Wikipedia link in concept definition. For instance, the "Course" Wikipedia article (URI) is introduced as a super-concept of concept Training_Course. The benefit is seen in two aspects: explanation and alignment. With links pointing to Wikipedia articles, we can easily extract the natural language based explanation of a concept. This is, in many cases, the first paragraph of the article. This explanation can be displayed to human readers for better understanding of the concept. Nearly all end users find this helpful. On the other hand, Wikipedia serves as a good reference point for aligning external resources with internal ones, for instance, through DBPedia. For those that are not currently covered by DBPedia, we leverage existing ontology mapping tools [7].

RDF representation is used exclusively in the background. We would argue that any efforts to make the underlying RDF representation transparent to the end users are likely to create more questions than answers in an enterprise environment. The following observations underpin our contentions. A majority of the corporate users are not semantic-web minded. More precisely, they do not care whether the data provision is facilitated by traditional technologies or semantic technologies, as long as data are provided in a timely and accurate manner. Such end users are for instance executives, sales and pre-sale personnel, service support and human resource. Understanding semantic technologies is certainly not a competence that they intent to develop. Ironically, the end users who will benefit from linked enterprise data is likely to enjoy such benefits only when the semantic technologies totally disappear from the user interface. A direct design consequence is that we had to improve user experience through good visualisation techniques (c.f. [4]) and RDF adaptors for conventional programming languages (c.f. [15]) for intuitive RDF data manipulation.

## 3.2   System Architecture

CIP is a multiple-layered data/information integration platform (see Figure 1). At the bottom, there is the Data Layer. We clearly distinguish data sources that are only available to internal users and those in the public domain due to

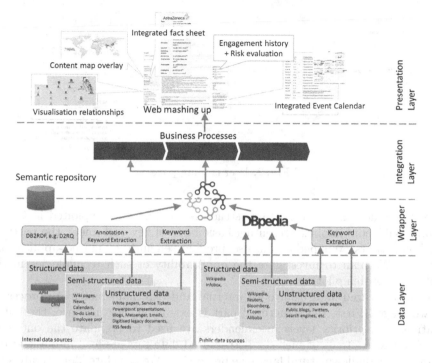

**Fig. 1.** Customer Information Portal Architecture

data safety and privacy concerns. We also differentiate data that are properly structured (e.g. databases), semi-structured (e.g. wiki pages, calendars, to-do lists, etc.), and un-structured (e.g. e-mails, blogs, and legacy documents).

Structured data from internal sources are mapped directly to the ontologies via for example manually/semi-automatically crafted D2RQ scripts[3]. Note that semi-automatically identifying correspondences between database schemata and ontologies is not a disadvantage. In our case, the internal databases are specialised for managing certain types of mission-critical data where consistence and stability is observed. We do not expect the schemata to be frequently updated/upgraded. Therefore, the DB2RDF mapping, once defined, has a knock-on effect on data migration. On the other hand, data stored in such databases capture critical information of the company's core business. In order to support sensible and accurate decision making, such data have to be faithfully presented. We evaluated several automatic database to ontology mapping toolkits and none of them produced satisfactory results. Human intervention and verification is inevitable and, we believe, is more cost-effective if introduced in the early stage of mapping. String similarity was leveraged to produce recommendations and based on our experience string similarity or a combination of its variants is by far the most effective method. We leverage DBpedia to align structured data

---

[3] http://www4.wiwiss.fu-berlin.de/bizer/D2RQ/spec/

**Fig. 2.** User landing page

from public domain. At this stage, structured public data exploited in CIP is mainly Wikipedia infobox presenting basic facts of key customers, the partners and competitors. Wikipedia can also provide semantic enhanced applications (*c.f.* [5]). We plan to investigate the applicability of such technologies in the next phase of this pilot.

Semi-structured data from both internal and external sources are processed in two stages. First, the structured part is extracted. For instance, the dates, locations and priority levels in Calenders are used to populate the ontology. The free-text contents of such semi-structured data sets are feed into a keyword extractor for shallow natural language processing. We use Gate [3] to create such extractors.

Processed data are stored in a semantic repository and are consumed by a business process management system residing in the integration layer. End users of the CIP do not assume equal privilege of internal as well as external data sets. What data sets should be linked is entirely decided by use cases and thus essentially driven by the business processes associated with the use cases. For instance, if the use case is to establish new sales opportunities, one needs to access potentially full customer engagement history in the Customer Relationship Management (CRM) data. On the other hand, if the use case is cost reduction, one focuses on product life-cycle management data, supply-chain management data, etc. Process driven is facilitated by providing predefined use cases at the personal landing page (Figure 2) of the CIP tuned against one's profile (role, area of working, professional responsibilities, etc.).

## 4    Use Cases

The value of linked enterprise data can only be fully appreciated if it supports the real needs from end users. In the context of the CIP project, we carried out workshops with different stake-holders to elicit their requests. Out of the discussion with end users, we identify a list of interesting web mashing up scenarios. In this section, we elaborate on three exemplary ones.

## 4.1 Meeting the Customers

Nearly all the modern sales 101 courses emphasise on "focusing on the prospect's point of view". Meeting with the prospects is always the best way to establish mutual trust and to understand their needs. The information portal facilitates this through linking external and internal data showing major events that the prospect is likely to participate and how events overlay with internal events (from e.g. internal event calendars).

Finding the prospect's events presents a technical challenge. We tackled this in the following steps. Firstly, we extract event information from the prospect's home page. Such pages can be easily found since almost all large enterprises maintain event calendars of various details. With little variants, entries in the event calendars are normally in the form of ⟨Date, Type, (Location), Description⟩ and can be easily processed with text analysis tools. The second data source is the recurrent past events identifiable in the internal customer engagement record. This shows where positive contacts were established before and are likely to happen in the future. Keywords from the past events (e.g. titles) are used to search and retrieve the date and location of the next event in the series from the Internet. We also identified several event portals as auxiliary data sources. Such portals are domain specific and can only be identified on a per customer basis. For pharmaceutical industries, exemplary web portals include pharmiweb.com, pharmaceutical-int.com, etc.

Data from the above three sources are used to create instances of the CIP domain ontology. We define seven different event types, namely conferences, exhibitions, trade fairs/expos, media/press events, training courses, unconferences, and the unspecified, while Unspecified is used to collect those of unknown or unconcerned types. Equation 1 is fragments of event type Training_Course: where $Course_{wpd}$ refers to the corresponding Wikipedia article via its URL. Denoted in Turtle notation[4], an event instance is as follows:

```
<http://www.***.com/EventsCalendar.mvc/EventDetail/32831>
      rdf:type #Training_Course ;
      rdfs:label "GCP"^^xsd:string ;
      #starts "07/06/2010"^^xsd:string ;
      #ends "07/09/2010"^^xsd:string ;
      #location "Costa Mesa"^^xsd:string ;
      #participants #NovoNordisk , ... , #Pfizer ;
      ...
```

$$Training\_Course \equiv Event \sqcap Course_{wpd} \sqcap =_1 starts.xs:date \sqcap =_1 ends.xs:date$$
$$\sqcap =_1 location.xs:string \tag{1}$$
$$\sqcap \forall participants.Organisation \sqcap \dots$$

We used simple domain heuristics to recognise types of events. In majority of the cases, types of events are either explicitly specified (e.g. in AstraZeneca event

---

[4] http://www.w3.org/TeamSubmission/turtle/

page), indicated in the titles (e.g. names of conferences), or given in event description. For instance, in the above example, we looked for keywords such as "course", "educational", "learning", etc. Such keywords are manually compiled and so far have produced good results: an F-measure value of 64.77% with respect to the six named event types. This value is obtained by comparing to the classification from human experts.

We use web crawlers to regularly harvest events from the Internet and populate the RDF repository accordingly. End users can then choose from several visualisation options: a list of next events, map overlay of event locations, and conventional calendars. A typical map overlay (implemented with GoogleMap) is illustrated in Figure 3, showing the location of events and one's current location (marked with "L" and retrieved from the employee's directory).

**Fig. 3.** Visualisation of events

### 4.2 What Has Happened to the Project?

Public news can lead to major decisions on the customer relations and thus impinges on account activities. In the CIP project, we compile multiple news sources and present to the end users in a coherent story.

We source news from mainly the following categories: i) internal news bulletin, ii) press releases from targeted customer, and iii) public news websites, e.g. FT.com and Bloomberg.com. Harvesting from the first two categories is straightforward and is constrained by the role of the requestor in the organisation. The third requires fine-tuning. News from public domain can be easily retrieved with the current capacity of general web search engines. We, however, would like to go one step beyond simply retrieving and presenting the news to the end users. We have done this by combining customer specific news together with other major events coinciding at the same location. In many cases, apparently irrelevant events happening in the same geographic area might significantly influence on the decision making regarding sales and long-term customer relations. Therefore, keeping end users up-to-date is crucial. This is done as follows:

1. identify customer's headquarters and important branch offices through internal customer profile,
2. use extracted locations to search in public data sets for major events (denoted as $\mathcal{E}$), e.g. festivals, natural disasters, urban uprising, etc.

3. use the geographic scales of $\mathcal{E}$ to analysis whether known partners or competitors with respect to the aforementioned customer are affected.
4. federalise $\mathcal{E}$ with news stories from internal and external sources.

Strategic locations of an organisation can be found from the organisation's homepage with shallow text analysis and simple domain heuristics. In some cases, this will require manual extraction for new customers. Deciding the scale of major events is simply done with shallow text analysis to extract location names. The connections between events and organisation locations is done via ontology properties. In CIP, news is introduced as a sub-concept of Event and is linked to organisations through location property.

Summarisation of collected new stories are presented to the end users ranked by significance. So far the best news summarisation technique is simply extracting the first paragraph of the news article based on the observation that the baseline algorithm, extracting the first $n$ sentences, has outperformed most of the "smarter" algorithms [11].

## 4.3   Where Are We with the Customer?

Customer accounts are in different stage of maturity. Moreover, one customer can be of different maturity with respect to different technical solutions. The content of the customer information pages should reflect such a diversity and put emphasis on different aspects accordingly by way of page layout, highlighting, etc. For instance, for a potential customer in pharmaceutical industries, the information page can focus on the solutions of competitors, key facts from similar customers, rules of engagement, etc. that will facilitate smooth initial contact of the account team. The emphasis will for example dynamically change to pre-sale, sales, and supports according to the status of the account. This is guided by high level business processes of general customer engagement.

Meanwhile, we support linking data based on more specific business processes for real-time decision making. For instance, a customer-facing project, $P$, may be divided into several tasks each having milestones and checkpoints. Team members working on $P$ use the dedicated customer page for keeping up with the progress of the project. It could be the case that the news of recent volcanic ash cloud raise concerns of potential disturbance to air-travel that can in turn impinge on project execution coinciding with the affected areas. Task managers can then use widgets on the CIP to adjust progress indicator (Figure 4(a)). The impact of such a change is two-fold: the disturbance can be propagated along task dependency links (through ontology properties) and cause the status of other tasks to change accordingly; management will be informed if the effects reach a certain level. Semantic technologies can facilitate such a scenario through modelling and reasoning of task dependency and the alignment between tasks and (news) events (as illustrated in Figure 4(b)). It is evident that the connections are established by extracting locations from external news stories, which are then mapped to the locations of customer organisations. At this moment, connecting news with projects cannot be fully automated. News that has potential to impinge projects are first crawled from selected news agencies (normally as regional headline stories). Harvested news stories are processed and presented as potential threats to tasks/projects that can be affected. Project managers or task owners will be summoned to confirm or reject such connections. If he/she opts for accepting, the page content is then updated accordingly.

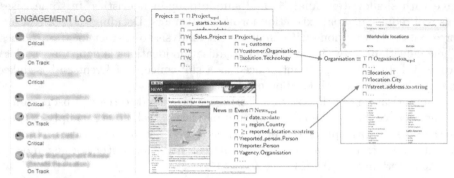

(a) Project monitoring                    (b) Linking project and news data

## 5    Discussions

Even though semantic technologies have been around for many years, the introduction of them into a well-established, high-tech organisation is not entirely hassle-free. In this section, we report some of the findings acquired when carrying out the pilot study. We believe our experience can be beneficial to those projects in similar settings.

### 5.1    Motivational Barriers

One of the major barriers to successfully exploiting the LD concepts in an enterprise environment is the lack of incentive. In the past, we witnessed the ups and downs of similar initiatives (e.g. Enterprise 2.0 [9]). The initial excitement slowly fades off when the attention from management has been deviated to other businesses and when "try-out" has become work routine. Unless such tools become an integral part of one's daily working environment, it is not likely to maintain the same level of enthusiasm in the long term. In order to convince the end users, we reckon it is important to demonstrate the benefit from two aspects: showing added value in the business context and showing improved work efficiency.

**Business context:** "Providing better access to data" has been a cliche when persuading end users with semantic technologies. In an enterprise environment, this will have to be made tangible in terms of business applications. Our experience indicates that the presentation is as important as, if not more important than, the underlying technologies. The merit of new technologies can only be delivered and well-accepted if they are presented in the end users' language. In our case, this is achieved by situating semantic technologies in the core business of an organisation. Linking enterprise data is then guided with the use cases derived from everyday work activities to avoid over exposure of data.

**Improved work efficiency:** We worked closely with the end users, customer-facing teams, to concretise the benefit of CIP in terms of saving on capital expenses and operational expenses. More specifically, we observe how employees work with the current technologies and how many short cuts they can enjoy with the help of semantic technologies. We estimated the time saved as per employee per customer with respect to mission-critical businesses and then summed up across the entire department. We also take a practical approach restricting "short cuts" to those that will cause minimum disturbance to employees' work routine and those that request only minimum investment in terms of labour and monetary resources. Able to demonstrate the improvement through financial gain increases the chance of obtaining management endorsement—this is a unique characteristic differentiate us from public social web-sites.

**Individual participation:** The barrier commonly seen in Enterprise 2.0 applications [9] were not observed in the CIP pilot. Collaborative and social network platforms have gained popularity in both public and corporate domains. The failure of certain initiatives does not deny their values but emphasises how the contents are organised, presented, and delivered. Again, we situate linked enterprise data in everyday work routine and bind it closely with a company's core businesses. We, therefore, experienced a very low level of reluctance from individuals and management.

## 5.2   Lightweight Ontologies and Incremental Approaches

Introducing ontologies was proved to be a more difficult task than we originally had expected, even though alternative names e.g. "vocabularies" and "taxonomies" were used. The hesitancy towards ontologies is seen from the availability and cost of domain expertise, the threshold of comprehending representation formalisms, and the misunderstanding of ontology commitment. In practice, we took an incremental and application-driven approach. Instead of constructing the ontology once for all, we started with a selected application (news integration for example) for a small subset of the end users. The ontology is made modular with only the most essential entities. For each entity, we did not make effort to cover every aspect that defines a concept for conceptual perfectness, but only those that are necessary to enable the application.

We adopted the "Scrum" agile development principle with end users' involvement throughout the project (with different intensities at different stages). Our experience is that through small and manageable projects, we can demonstrate the merit of semantic technologies and thus establish the initial trust among end users. The "teaser" applications can then be gradually extended with one concrete and tangible improvement at a time. By doing this user commitment and involvement are kept to minimum. In the CIP project, this approach was proved to be effective.

## 5.3   Minimum Disturbance

The importance of minimum disturbance to existing infrastructure was address already (c.f. [1]). Our experience underpins such a conclusion and further extend it with two other principles. Firstly, introducing semantic technologies should not manifest at the user interface level. Secondly, new technologies should not alter established protocols.

Semantic technologies worked better when they have totally disappeared from the user interface, blending into everyday work environments. The value of linking enterprise data is best shown in areas where timely delivery of data is deemed important. It is exactly such areas where concerns were raised regarding the potential risk of not meeting key performance indicators while staff are trying to gain proficiency of new technologies. We confined the semantic technologies to the background and worked closely with end users on the foreground (interface). Meanwhile, we based our development on platforms (e.g. confluence wiki[5] and Jive[6]) currently in use to ensure a smooth learning curve. We observe the integrity and access control of all legacy data. For instance, even though we maintain a link to existing CRM database, what are shown to the end users depends entirely on his/her access right granted based on business processes. Meanwhile, we did not migrate legacy data. Instead, semantic annotation and mapping are established on-the-fly with the help of tools such as CROSI [7,13]. In general, it is impractical (if not impossible) to remodel all the legacy data. It is equally difficult to abandon existing relational database (RDB) implementations to switch entirely to RDF repositories. In fact, there was a major discussion regarding the benefit and disadvantage of RDF. The end users' main concern appears not on the change of mindset, but at the programming cost and extra learning efforts.

Obeying the Linked Data principles, however, should not compromise the intact of existing data repositories and established work processes of a company. Customer information is confidential with multiple levels of clearance. As an international organisation, majority of the data is continuously accessed by different departments across the globe in different time zones. Applying semantic technology should not alter existing data models causing disruption to normal business. Linking different data sets should not break existing access restrictions. In order to maximise the value of linked enterprise data, data sets with restricted access are handled as follows. We introduced a transition layer on top of raw data. Based on users' access privilege, the transition layer either populates ontology with data from such databases or presents a demilitarised summary of what data could have been accessed. Maintain a transition layer on top of existing data seems redundant. The extra cost, however, is marginal comparing to interrupted businesses.

Considering individual motivation, the same minimum disturbance requirements exist. The true value of semantic technologies can only emerge when a large amount of data is ready to be consumed which, in turn, relies on the

---

[5] http://www.atlassian.com/software/confluence/
[6] http://www.jivesoftware.com/

willingness of involvement. Less disturbance to people's everyday work routine is likely to encourage their participation.

The requirement of minimum disruption also suggested that any development should be based on existing platform instead of introducing new ones. Fortunately, we were aiming at employee-only communities, for which modern organisations tend to impose less strict regulations, encourage employees to experiment the benefits of new technologies, and deploy at minimum some collaboration platforms [8]. Many of such platforms can be easily extended with web widgets encapsulating extended functionalities.

## 6   Conclusions

Linked Data has demonstrated noticeable value in the public domain [2]. Whether the same principles and the same outcomes are true in a "semi-closed" and proprietary domain has not been properly investigated. In this paper, we report on a pilot study carried out in an international high-tech company. Even though the project was not initiated as a proof-of-the-concept for semantic technologies, we discovered the real value of giving it a semantic touch. The advantages of doing this are not much different from the ones reported previously [1]. However, due to the characteristics of our domain, we made the following arguments that are unique to the studied domain. We would argue that linking enterprise data should be derived from processes faithfully reflecting the core business of a company. Deviation from this principle may lead to "yet another Enterprise 2.0" toy that does not bring real values to the company as well as its employees. The second principle emphasises on the interplay between external and internal data. Thus far, reaching a semantic consensus across different departments has not been a real challenge due to well-defined organisational boundaries and a common corporate culture, leading to well-understood common vocabulary. Counterintuitively, linking internal data sometimes can be made easier through references to external ones than solely based on internal links. Such a phenomenon might eventually encourage more companies to make their non-mission critical data sets available to the public to savor the payback. Finally, in many cases, linking enterprise data presents as less a technical challenge than a psychological one. How to motivate corporate employees to consume as well as actively contribute worth further investigation. We briefly discussed some of the motivational issues, for which we only gleaned the tip of the iceberg.

The crux of our further work is on performing larger scale evaluation with users invited from different departments and different geographic regions. It is also important to identify more application scenarios that can show values to a diversity of users including pre-sales, sales, education, technical support, etc.

## Acknowledgements

This work is partially supported by the European Union IST fund through the EU FP7 MATURE Integrating Project (Grant No. 216356).

# References

1. Alani, H., Dupplaw, D., Sheridan, J., O'Hara, K., Darlington, J., Shadbolt, N., Tullo, C.: Unlocking the potential of public sector information with semantic web technology. In: ASWC 2007 and ISWC 2007. LNCS, vol. 4825, pp. 701–714. Springer, Heidelberg (2007)
2. Bizer, C., Heath, T., Berners-Lee, T.: Linked data - the story so far. International Journal on Semantic Web and Information Systems 5(3), 1–22 (2009)
3. Cunningham, H., Maynard, D., Bontcheva, K., Tablan, V.: GATE: A framework and graphical development environment for robust NLP tools and applications. In: Proceedings of the 40th Anniversary Meeting of the ACL (2002)
4. Deligiannidis, L., Kochut, K.J., Sheth, A.P.: Rdf data exploration and visualization. In: Proceedings of the ACM First Workshop on CyberInfrastructure: Information Management in eScience, pp. 39–46. ACM, New York (2007)
5. Gabrilovich, E., Markovitch, S.: Computing Semantic Relatedness Using Wikipedia-based Explicit Semantic Analysis. In: Proceedings of the 20th IJCAI, pp. 1606–1611 (2007)
6. Ghani, R.: Research challenges in enterprise information retrieval (2008), http://videolectures.net/active09_ghani_rdekm/
7. Kalfoglou, Y., Hu, B., Reynolds, D., Shadbolt, N.: Semantic integration technologies. 6th month deliverable, University of Southampton and HP Labs (2005)
8. Matuszak, G.: Enterprise 2.0 - The Benefits and Challenges of Adoption (2007), http://www.kpmg.com/Global/en/IssuesAndInsights/ArticlesPublications/Enterprise-fadfuture/Documents/Enterprise-2.0-The-benefits-and-challenges-of-adoption.pdf
9. McAfee, A.P.: Enterprise 2.0: The dawn of emergent collaboration. MIT Sloan Management Review 47(3), 21–28 (2006)
10. Munkvold, B.E., Päivärinta, T., Hodne, A.K., Stangeland, E.: Contemporary issues of enterprise content management: the case of statoil. Scand. J. Inf. Syst. 18(2), 69–100 (2006)
11. Nenkova, A.: Automatic text summarization of newswire: lessons learned from the document understanding conference. In: AAAI 2005: Proceedings of the 20th National Conference on Artificial Intelligence, pp. 1436–1441. AAAI Press, Menlo Park (2005)
12. Riege, A.: Three-dozen knowledge-sharing barriers managers must consider. Journal of Knowledge Management (3), 18–35 (2005)
13. Roset, R., Lurgi, M., Croitoru, M., Hu, B., Lluch i Ariet, M., Lewis, P.: A visual mapping tool for database interoperability: the healthagents case. In: Proceeding of the 3rd CS-TIW Workshop (2008)
14. Servant, F.-P.: Linking enterprise data. In: Linked Data on the Web Workshop at the 17th International World Wide Web Conference (2008)
15. Völkel, M., Sure, Y.: Rdfreactor - from ontologies to programmatic data access. In: Poster session at the International Semantic Web Conference (2005)
16. Zhao, J., Miles, A., Klyne, G., Shotton, D.: Linked data and provenance in biological data webs. Briefings in Bioinformatics (2), 139–152 (2009)

# Linkage of Heterogeneous Knowledge Resources within In-Store Dialogue Interaction

Sabine Janzen[1], Tobias Kowatsch[2], Wolfgang Maass[1,2], and Andreas Filler[1]

[1] Furtwangen University, Robert-Gerwig-Platz 1, 78120 Furtwangen, Germany
`{sabine.janzen,wolfgang.maass,andreas.filler}@hs-furtwangen.de`
[2] University of St. Gallen, Dufourstrasse 40a, 9000 St. Gallen, Switzerland
`{tobias.kowatsch,wolfgang.maass}@unisg.ch`

**Abstract.** Dialogue interaction between customers and products improves presentation of relevant product information in in-store shopping situations. Thus, information needs of customers can be addressed more intuitive. In this article, we describe how access to product information can be improved based on dynamic linkage of heterogeneous knowledge representations. We therefore introduce a conceptual model of dialogue interaction based on multiple knowledge resources for in-store shopping situations and empirically test its utility with end-users.

**Keywords:** heterogeneous knowledge resources, dynamic linkage, dialogue interaction, ontology, empirical study.

## 1 Introduction

What if you could find a product that directly matches your personal preferences by posing a simple question on your mobile device? Today, customers in bricks-and-mortar stores often lack access to helpful product information from the Web, such as product manuals, user and professional reviews or feature comparisons. On the other hand, those customers that have in-store Web access by their mobile device may suffer from information overload due to the sheer quantity of product information available. We therefore investigate how to enable customer's access to comprehensive, helpful product information in physical shopping environments. In this context, we further study the filtering and intuitive presentation of such product information in form of dialogue interaction between customer and product. The use of product-centered dialogue systems in physical shopping environments enables an improved filtering and presentation of relevant product information [1] and thus satisfy the communication needs of customers as intuitive as possible. To realize such natural language communication between customers and physical products, dialogue systems and comprehensive knowledge representations are necessary [2]. But, knowledge resources around products suffer from heterogeneous nature and diverse semantics, e.g., product descriptions by several manufacturers, user and professional reviews on the web, explanations of product features by diverse providers, pricing and bundling information of specific portals. Here, we assume that product-centered

P.F. Patel-Schneider et al.(Eds.): ISWC 2010, Part II, LNCS 6497, pp. 145–160, 2010.

dialogue interaction requires bundling of these product-related knowledge resources and their dynamic linkage at a specific "hot spot" for instance a web service. In this paper, we therefore introduce a conceptual model of dynamic linkage of product-related knowledge resources within dialogue interaction in in-store shopping situations. By means of the resulting dialogue system prototype, we evaluate the utility of dialogue interaction based on heterogeneous knowledge resources from an end-user perspective. The focus of this contribution lies on the bundling of diverse heterogeneous knowledge resources, e.g., digital product descriptions, at a specific "hot spot" that enables not only standardized access to different types of information but also the dynamic linkage of these resources within in-store dialogue interaction. The rest of this paper is organized as follows. We will discuss related work in Sec. 2. Afterwards, our motivation will be illustrated by an example. In Sec. 4, the approach of bundling diverse heterogeneous knowledge resources is described with a focus on digital product descriptions. We then present our model of dynamic linkage of product-related knowledge resources within dialogue interaction for in-store shopping situations (Sec. 5). Finally, we show and evaluate an implementation of the model from an end-user perspective in Sec. 6, summarize our results and provide an outlook on future work (Sec. 7).

## 2    Related Work

In our work, dialogue interaction between customers and physical products in in-store shopping environments refers to the application of dialogue systems in physical environments and the usage of ontologies as knowledge representations. Dialogue systems provide the opportunity to interact with a system similar to human-human communication [3]. They can be divided into two basic types: dialogue grammars and frames as well as plan-based and collaborative systems. Dialogue grammars identify and represent surface patterns of dialogue or speech acts. Frame-based approaches extend grammars regarding their flexibility. Plan-based and collaborative systems assume that humans communicate to achieve goals and thus, focus on intentional structures [4]. When designing Natural Language Processing (NLP) modules, there are two extremes: full natural language processing or fixed linguistic question templates. A sweet spot between the two extremes is to constrain natural language in order to create a formal, user-friendly query language [5] or a controlled language for posing questions [6]. There are diverse examples for current dialogue systems, for instance SmartKom - a multimodal dialogue system that combines speech, gesture and mimics input [7] as well as DELFOS, an dialogue manager system that enables the integration of OWL ontologies as external knowledge resources for dialogue systems [8]. The combination of NLP and ontologies facilitates the development of novel dialogue systems that use ontologies as a core knowledge component regarding linguistic and non-linguistic knowledge representations. In our case, product information as part of the non-linguistic knowledge base plays an important role. The effective handling of this heterogeneous product knowledge distributed among various steps in the product lifecycle has become essential [9]. Meanwhile, there are several ontology-based developments that address the description of products. For instance, Product Design Ontology (AIM@SHAPE project) focuses on the formalization of knowledge concerning processes, tools and shapes during product development whereas other ontologies as GoodRelations [10] and SearchMonkey Product (by Yahoo!) are used to annotate digital products and service offerings on the web. An ontology within NLP constitutes the conceptualized description of the domain of the dialogue system [11]. We assume that NLP benefits from the appliance of semantic knowledge representations as well as semantic technologies, e.g., SWRL, in general. In the current work, dialogue interaction requires the generation of situation-specific questions and answers on run-time. Multiple pieces of information have to be combined while answering questions that are not anticipated at the time of system construction [12]. The important role of the combination of language technologies, ontology engineering and machine learning is also described by Buitelaar et al. [13]. The semantic web technologies and standards will be used for the specification of web-based, standardized language resources. However, building ontologies in the first place requires experienced knowledge engineers [12]. The linkage of multiple ontology-based knowledge resources pose a challenge concerning the combination of knowledge of different resources and the answering of queries by considering multiple resources [14]. On the other hand, the effort of building up the knowledge base of the dialogue system can be decreased or sourced out. Furthermore, the coverage of dialogue systems is extended by integrating knowledge of multiple resources as it will be described in the current work [15,14].

## 3 Motivation

The motivation for dialogue interaction in in-store shopping situations shall be described by a futuristic example of a sales talk. A customer enters a drugstore because she searches for a whitening toothpaste. She carries a smart phone that is also used to request additional information about products. She wants to take a look at the toothpastes that are right for her. She scans the barcode of a toothpaste with her smart phone to identify the product and poses the following question: "Which whitening toothpastes are available?" The toothpaste answers via mobile device: "There are three whitening toothpastes available: StarLight, Smile and WhiteSky. You can find it on the second floor. If you prefer, somebody will get it for you." Customer: "Yes, please." Toothpaste: "Are you interested in a video clip that explains the application of toothpaste StarLight?" Customer: "Why not!" A corresponding clip is shown on a display nearby. Customer: "Very good. Which mouthwashes fit to this toothpaste?" Toothpaste: "The best options are these two." Both are shown on her mobile device. Toothpaste: "If you buy this toothpaste and one of these mouthwashes you will get a 5% discount." Customer: "That's a good deal." The example has shown complex relationships between customers, manufacturers, products and product-related knowledge. We will depict these relationships in the next section.

# 4  Handling Clouds of Product-Centered Knowledge Resources

In shopping environments, product-related knowledge is retrieved from different sources that use different semantics. Primarily, we speak about digital product descriptions that represent the informational basis to realize a customer-product dialogue. Currently, physical products are described non-standardized or standardized in terms of static databases (e.g., STEP ISO 10303 [16]) or XML structures (e.g., BMEcat: bmecat.org). Furthermore, product-centered knowledge resources also cover comparisons of products as well as their features. On the one hand, this information is provided by single manufacturer web sites exclusively considering their own products, e.g., Apple, Dell. On the other hand, consumer portals, e.g. Ciao! (ciao.co.uk), allocates comparisons of products manufactured by different companies. Further product-centered knowledge types are definitions or explanations of product features. Regarding the complexity of physical products, some features require explanations to enable customers to make confident purchase decisions. Currently, such explanations are available via diverse websites of manufacturers, search engines such as Google or online encyclopedias such as Wikipedia. In addition, the example in Sec. 3 shows the integration of pricing and product bundling information within the customer-product dialogue. Knowledge about matching products is provided by manufacturers exclusively concerning their own assortment as well as by shopping portals, e.g., Amazon, based on collaborative filtering mechanisms according to the principle "customers who bought this item also bought [...]". Pricing information on the web rarely exceed the scope of comparisons between retailing portals, for instance provided by ConsumerSearch (consumersearch.com). However, dynamic prices or discount bargains against dynamic parameters like customer type, current situation or inventory are not available. Finally, the aforementioned example presents a natural language dialogue between customer and product. Such dialogue interaction requires question-answering structures represented by linguistic knowledge resources. Currently, such linguistic knowledge resources are not freely accessible on the web. This short overview points out that there is a cloud of product-centered knowledge resources on the web, all of different semantics and formats. How can these "cloudy" information structures be integrated into a purposeful customer-product dialogue? In the following subsections, we will elaborate our approach for bundling such heterogeneous knowledge resources at a specific "hot spot" that enables standardized access to these different types of information. At this point, we focus on the handling of different formats of digital product descriptions as they represent essential product knowledge of each physical product. Finally, digital product descriptions build the basis for the calculation of dynamic prices [17] as well as product bundling results [18].

## 4.1  Digital Product Information in Physical Environments

As mentioned before, current physical products are mainly described in terms of static databases (e.g., eCl@ss: eclass-online.com) or XML structures (e.g.,

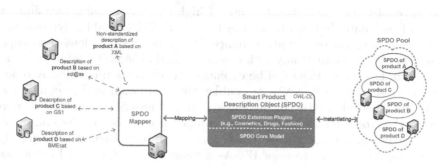

**Fig. 1.** Mapping of product descriptions of different formats into SPDO structure

BMEcat). Modeling enterprises or business processes is generally sophisticated (e.g., Business Process Modeling Notation: bpmn.org), but the description of products rarely exceeds the scope of classification. Furthermore, the effort in maintenance and extension of such product descriptions is high. So, product information is often incomplete or out-of-date. As product descriptions of diverse physical products in shopping environments serve as basis for our customer-product dialogue, we have to handle the different formats of these descriptions (cf. Fig.1). We developed a mapping module that enables the automatic mapping of diverse XML-based product description formats into a standardized semantic product description structure - *Smart Product Description Object (SPDO)* - and vice versa. The web-based SPDO mapping interface is able to retrieve initial product description data from diverse servers, e.g., repositories of manufacturers. Afterwards, each product description is mapped automatically onto the core model of the SPDO structure. In case of imprecise relations between the concepts of the initial product description and SPDO structure, the web-based mapper interface allows the manual arrangement of concepts by the user. SPDO represents semantic and dynamic product information. It is a product ontology for physical products in in-store shopping environments [19]. As shown in Fig. 1, SPDO consists of a core model[1] and SPDO specific extension plugins both formalized in OWL-DL. The further covers prototypical aspects of consumer products which is all domain-independent information of the product itself, e.g., name, color, price, manufacturer etc. Domain-specific conceptualizations, more precisely product information regarding specific product domains, can be imported into SPDO as extension plugins, for instance cosmetics[2], drugs or fashion plugins. While mapping the original description of a physical product to SPDO structure, an instance of SPDO is generated. That is, each product description is transformed into one instantiated SPDO file formalized in OWL-DL (SPDO Pool)(cf. Fig.1). When we consider our in-store shopping environment, we can say that each physical product is then described by one SPDO instance. At the beginning of this subsection, we mentioned that product descriptions are

---

[1] http://im.dm.hs-furtwangen.de/ontologies/spdo/2010b/SPDO.owl
[2] http://im.dm.hs-furtwangen.de/ontologies/spdo/2010b/SPDO_Cosmetics.owl

often incomplete or out-of-date because of high effort in maintenance and extension of such data. In order to address this issue, SPDO enables dynamic and automatic data extension and maintenance. Underspecified relations or concepts can be completed automatically with reasoning mechanisms. Therefore, product information based on SPDO will be extended by application of rules. A repository of web-based rules (SWRL[3]) enables the generation and integration of statements about alternative or matching products into SPDO instances, more precisely the processing of specific rules combines product descriptions automatically in real-time according to parameters that can be configured dynamically, for instance "color A fits to color B". As a result, an advanced and automatic processing in terms of updates and extensions is possible that forces the emergence of dynamically changing product networks.

## 4.2    Bundling the Product-Centered Knowledge Cloud

The network of SPDOs is one part of the product-centered knowledge cloud that needs to be processed to realize an intuitive representation of comprehensive product information in form of dialogue interaction. Fig. 2 shows our approach for bundling heterogeneous knowledge resources at a "hot spot" to enable standardized access to different types of product-centered knowledge. Besides SPDO Pool, we assume that following heterogeneous knowledge resources have to be bundled at a "hot spot" (cf. top layer in Fig. 2):

- Product-centered knowledge requested by *external services* is used to extend SPDOs, e.g., integration of explanations of specific product features or product reviews. Furthermore, external services like accessible thesauri or search engines are requested for linguistic information to extend the lexicon of the linguistic knowledge representation, e.g., adequate articles, plural forms of nouns etc.
- Currently, *linguistic knowledge representations* that match our needs regarding dialogue interaction in in-store shopping environments are not freely accessible on the web. We developed a light-weight linguistic knowledge representation formalized in OWL-DL that represents the structural backbone of dialogue interaction between customers and physical products. This knowledge resource will be elaborated in detail in the next section.
- The repository of *web-based rules* covers rules regarding matching or alternative products as well as rules concerning dynamic bundle prices and discounts. Rules are applied on SPDO Pool and provide results that are integrated in dialogue interaction.
- Sales and inventory figures of manufacturers or retailers represent large *economic data* resources that should be processed within a dialogue between customer and product. So, retailers are able to offer slow sellers in the context of matching products or allow dynamic discounts dependent on sales figures.

---

[3] Semantic Web Rule Language - http://www.w3.org/Submission/SWRL/

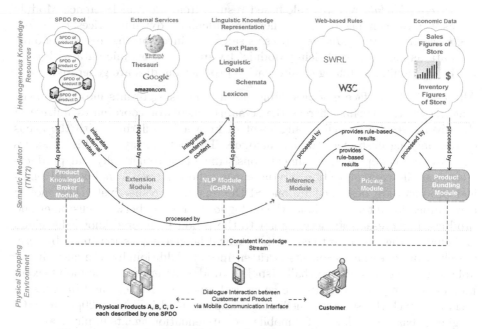

**Fig. 2.** Approach of bundling product-centered knowledge resources

These heterogeneous knowledge resources are bundled by a semantic mediator (cf. middle layer of Fig. 2) that represents the "hot spot" for generating a consistent knowledge stream based on resources of different semantics and formats. The semantic mediator consists of several modules that are responsible for requesting and processing diverse knowledge resources as well as allocating preliminary results. We discern internal (marked light grey) and interface modules (marked dark grey) (cf. Fig. 2). Internal modules generate preliminary results and forward these results to interface modules which directly contribute to the consistent knowledge stream. The semantic mediator contains two internal modules: *Extension Module* and *Inference Module*. The Extension Module requests knowledge from external services, e.g. definitions of product features via Wikipedia, and integrates this content into SPDOs or the linguistic knowledge representation. The second internal module - the Inference Module processes rules of the web-based rule repository and offers rule-based results to the Pricing and Product Bundling Module. Furthermore, Fig. 2 shows the following interface modules:

– *Product Knowledge Broker Module* processes product information stored in the SPDO Pool and induct this knowledge into the consistent knowledge stream.
– *NLP Module* processes the linguistic knowledge representation and allocates linguistic question-answer structures to realize dialogue interaction.

- *Pricing Module* receives rule-based results from internal Inference Module as well as economic data and therefore calculates dynamic prices.
- *Product Bundling Module* processes sales and inventory figures of economic data as much as the Pricing Module. By merging economic data with rule-based results, matching or alternative product bundles are generated.

All described interface modules induct their processing results into the consistent knowledge stream whereas the semantic mediator coordinates linkage of the diverse results as well as alignment of knowledge offerings with the needs of the ongoing customer-product dialogue. The bottom layer of Fig. 2 shows the physical shopping environment consisting of customer, physical products and a mobile communication interface, e.g., the customer's mobile. All physical products are described by one SPDO instance, respectively. The mobile communication interface enables the dialogue interaction between customer and product. It represents an access point to the product-centered knowledge cloud and enables the customer to construct natural language questions term-by-term via choosing questions segments (written mode). Additionally, the customer is able to pose her question verbally (spoken mode). Afterwards, the mobile communication interface presents the generated answers with text and images. On technical level, the semantic mediator is implemented in TNT2 [20] - an OSGi (osgi.org) based middleware for mobile recommendation agents in physical environments. The NLP Module is realized by the Conversational Recommendation Agent (CoRA) [21] that constitutes an OSGi plugin of TNT2. Furthermore, the client of the mobile communication interface was exemplarily developed on an Android (android.com) based mobile phone.

## 5    Linkage of Heterogeneous Knowledge Resources for Dialogue Interaction

Now, we know how to bundle heterogeneous knowledge resources based on a semantic mediator. But how can we dynamically merge these knowledge resources for the generation of answers of a dialogue interaction system? We want to generate situation-specific questions and answers for a dialogue at run-time that means questions and answers depend on concrete physical products in physical shopping situations with a specific context. These generation processes rely on the dynamic integration of heterogeneous knowledge resources. In detail, for answering queries by the user multiple external information resources are considered automatically. The integration of these external resources decrease the effort of building up knowledge representations for dialogue systems. In Fig. 3, we present a model of dynamic linkage of heterogeneous knowledge resources for dialogue interaction. In this contribution, we focus on the linkage of product information from SPDO Pool with linguistic structures of the linguistic knowledge representation, because the NLP and Product Knowledge Broker module constitute the conceptual basis for dialogue interaction. The right part of Fig. 3 shows the conceptual design of the linguistic knowledge representation that represents the linguistic structures of the NLP approach [21]. The representation

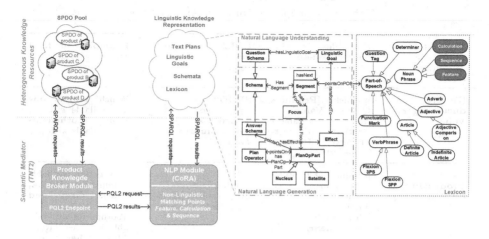

**Fig. 3.** Model of linkage of heterogeneous knowledge resources within dialogue inter-action

was modeled based on an analysis of a German speech corpus of sales conversa-tions and consulting talks concerning consumer electronics at a trade fair. After having transcribed and analyzed the corpus, we derived question structures of these purchase conversations. The linguistic representation can be subdivided into three parts: Natural Language Understanding (NLU), Natural Language Generation (NLG) and lexicon. NLU covers generic schemata of questions that are posed within the shopping domain. A semantic tree of questions is spanned based on question segments that are filled by words or phrases of the lexicon. An example of question schemata is listed as linguistic skeleton in Tab. 1. Similar to NLU, NLG represents generic linguistic structures of text plans that consists of answer schemata, which respond to the aforementioned questions. Examples of answer schemata are listed as linguistic skeletons in Tab. 1. Segments of ques-tion and answer schemata are filled by the German lexicon whose items base on the Penn Treebank Tagset [22]. As shown in Fig. 3, the lexicon contains three specific sub concepts of noun phrase: *Calculation, Feature* and *Sequence.* These concepts represent *Non-Linguistic Matching Points* that operate as gates for linking product-domain-specific information with the linguistic skeleton of questions or answers. Calculation, Feature and Sequence contain several sub-types that represent question and answer segments with specific functions (cf. Tab. 2). While processing the generic linguistic structure (cf. Tab. 1, Col. 2), the NLP Module detects noun phrases of type Calculation, Feature or Sequence and triggers the Product Knowledge Broker Module that allocates a *PQL2* endpoint (cf. Fig. 3). A PQL2 request concerning the detected subtype of Non-Linguistic Matching Point, e.g., "ProductCategoryValue" is sent to the PQL2 endpoint by the NLP Module (cf. Tab. 1, Col. 3). PQL2 is a high-level semantic query language that allows to request pools of multiple ontologies via a light-weight

**Table 1.** Examples of Question-Answer Flows

| Type of Non-Linguistic Matching Point | Examples of Linguistic Skeleton | PQL2 endpoint query | Examples of final output |
|---|---|---|---|
| (Feature) ProductCategoryValue | \<Which\> [ProductCategoryValue] \<are available\> \<with\> [ProductProperty] [ProductPropertyValue]? | SPDO* \<pql2:product-category\> \<pql2:plural\> ?value | Which **bodylotions** are available with [...] Which **toothpastes** are available with [...] |
| (Sequence) ProductPropertyValueSEQ | [ProductCategoryValue] \<is available in\> [NumberOf] [ProductProperty]: [ProductPropertyValueSEQ] | SPDO \<pql2:product-property[property]\> \<pql2:seq\> ?value | Jackets are available in four colors: **White, Black, Grey and Blue.** Toothpastes are available in three flavors: **Mint, Cherry and Orange.** |
| (Calculation) BundlePrice | [ProductValueSEQ] \<cost\> \<as a bundle only\> [BundlePrice]. | SPDO* \<pql2:bundle-price(\<pql2:product-value[product-value]\>\<pql2:seq\>)\> ?value | iPod nano and iPad cost as a bundle only **850 USD**. Shoes Sunshine and t-shirt Summer cost as a bundle only **55 USD**. |

**Table 2.** Subtypes of Non-Linguistic Matching Points

| Feature | Sequence | Calculation |
|---|---|---|
| ProductCategoryValue | ProductPropertySEQ | BundlePrice |
| ProductCategoryValuePlural | ProductPropertyValueSEQ | Discount |
| ProductProperty | ProductValueSEQ | NumberOf |
| ProductPropertyPlural | ProductCategorySEQ | PriceAverage |
| ProductPropertyValue | ProductProperty - PropertyValueSEQ | PriceThreshold |
| ProductValue | | |

endpoint with simple queries that will be internally transformed in an SPARQL[4] request. Below, an exemplary PQL2 query of Tab. 1 is elaborated:

```
SPDO <pql2:product-property[property]> <pql2:seq> ?value
```

This PQL2 query represents the request of a sequence (\<pql2:seq\>) of values of a specific property of one product (SPDO), e.g., \<pql2:product-property[color]\>. In contrast, the term SPDO* expresses that the whole pool of SPDOs is requested (cf. Tab. 1). In summary, PQL2 offers the following features:

– PQL2 analyzes the ontological structure of the ontologies in target concerning their concepts and relation. Then, PQL2 offers specific requesting

---

[4] http://www.w3.org/TR/rdf-sparql-query/

items according to the ontological structure, e.g., <pql2:product-category> <pql2:plural>. This means that external modules that use PQL2 need no knowledge about the constitution of the ontological knowledge base they request.
- PQL2 allows to request data of multiple ontologies, e.g., SPDO Pool, via a single simplified PQL2 request (cf. Fig. 3).
- PQL2 enables the integration of semantic statements into the ontological structures, e.g., based on information of external services or rule-based results.
- With PQL2 other modules of the semantic mediator can be requested such as Pricing or Product Bundling modules.

After receiving the PQL2 request by the NLP Module, the Product Knowledge Broker Module transforms the request into one or several SPARQL queries to request the SPDO Pool or further modules. The results are sent to the NLP Module that inserts them into the linguistic skeleton to generate the final output (cf. Tab. 1, Col. 4). The linkage of the linguistic knowledge representation and SPDO pool via PQL2 queries is shown by means of an example consisting of question and answer in Fig. 4. The red labeled words and phrases are filled from on PQL2 queries (the right part of Fig. 4). Imagine, the user wants to know which fragrances are available for the product *Sunshine Bodylotion*. While composing the desired question, she selects the question segment "fragrances" from a list of product properties (*ProductProperySEQ*) that are available for the product.

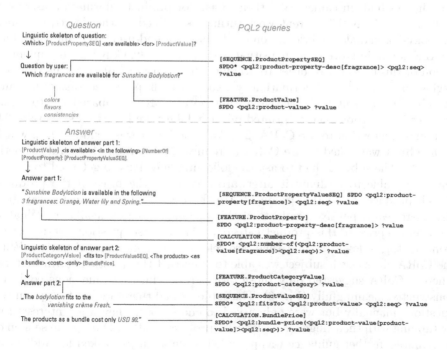

**Fig. 4.** Example of Question-Answering regarding PQL2 requests

In this context, further possible properties would be Color, Flavor or Consistency. The right part of the figure shows the PQL2 queries that are processed to enrich the linguistic skeleton. The question "Which fragrances are available for Sunshine Bodylotion?" is responded by an answer consisting of two parts. First, "Sunshine Bodylotion is available in the following 3 fragrances: Orange, Water lily and Spring." The Sequence (*ProductPropertyValueSEQ*) of fragrance values as well as the name of the property "fragrance" (*ProductProperty*) and the calculation of the number of fragrances available (*NumberOf*) is generated by a PQL2 request. The second part of the answer presents a sequence of matching products (*ProductValueSEQ*) gained via a PQL2 request that triggers the Product Bundling Module whereas the corresponding bundle price (*BundlePrice*) is filled with results of the Pricing Module.

## 6    Utility of Ontology-Based Dialogue Interaction

Having described the model for ontology-based dialogue interaction above, we now test its utility for in-store shopping situations from an end-user perspective. For this purpose, we implemented a *Conversational Recommendation Agent (CoRA)* that is derived from the proposed model and provides a communicative interface between consumers and physical products at the point of sale. Technically, CoRA is an OSGi plugin of TNT2. The CoRA client is implemented on a mobile phone. It allows consumers to identify a product by barcode via the phone's built-in camera and then to ask for product information as shown in Fig. 5 to 7. In the current work, utility is defined as the degree to which ontology-based dialogue interaction is adequate for end-users to request product information in in-store shopping situations. Accordingly, CoRA is an implementation of this concept. In order to evaluate the utility of CoRA, we use the following constructs from information systems research: perceived ease of use and perceived enjoyment [23], relative advantage of CoRA compared (1) to static product information such as printed product labels, and (2) to a sales talk [24] and finally, intention to use CoRA [25]. An experiment was conducted, in which each subject was asked to use CoRA to request information of several cosmetic products. The subjects had to ask the following questions to get used to CoRA and to be able to evaluate it afterwards: What is the price of the product? Which products fit to this product? Are there alternative products available? Are there less expensive products of this product category available? What is the average price of this product category? All of these questions were derived from in-store sales talks and thus, are relevant in a shopping situation. With the CoRA client, each subject was able to construct the questions term-by-term whereas CoRA suggested only those terms from which the questions could be constructed meaningfully (Fig. 6). The subjects had therefore not to type in the questions manually but were only asked to chose terms they were interested in by tapping with their finger (e.g., Which - products - fit...). During the session of 30 minutes, further guidance was provided when a subject asked for additional help with CoRA. Then, in the second part of the experiment, the subjects were

 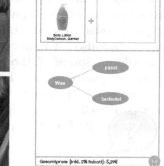 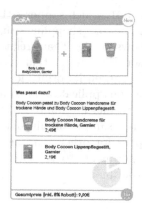

**Fig. 5.** Subject with CoRA in front of a product shelf

**Fig. 6.** Step-by-step composition of a question

**Fig. 7.** Presentation of the answer

asked to rate questionnaire items with regard to the theoretical constructs described above. Consistent with prior research, we adopted 7-point Likert scales that range from strongly disagree (1) to strongly agree (7).

All in all, 19 female and 37 male subjects studying at a business university participated in the experiment. Their age ranged from 20 to 24 (n=34), 25 to 29 (n=13). The seven remaining subjects were above 30. We employed one-sample t-tests with a neutral test value of 4 to indicate whether the results are significantly positive or negative resulting in high or low utility scores for CoRA. The descriptive statistics and the results of the one-sample t-tests are shown in Tab. 3. All multi-item research constructs were reliable as Cronbach's Alpha lies above the recommended value of .70 [26]. The one-sample t-tests indicate that almost all constructs were perceived positive at the highest level of significance

**Table 3.** Descriptive statistics and results of the one-sample t-test for the 54 participants; Note: SD = standard deviation

| Construct | Items | Alpha | Mean | SD | p-value | Interpretation |
|---|---|---|---|---|---|---|
| Perceived ease of use of CoRA | 3 | .761 | 5.70 | 0.87 | < .001 | CoRA was easy to use |
| Perceived enjoyment of CoRA | 3 | .838 | 5.31 | 1.09 | < .001 | CoRA has made fun during usage |
| Perceived relative advantage of CoRA when compared to static information | 3 | .748 | 4.59 | 1.26 | < .01 | CoRA was perceived better relative to static product information |
| Perceived relative advantage of CoRA when compared to a sales talk | 3 | .802 | 4.03 | 1.25 | > .05 | CoRA was neither perceived better nor worse than a sales talk |
| Intention to use CoRA | 1 | n/a | 5.59 | 1.39 | < .001 | The participants would intend to use CoRA |

at .001, which supports the utility of CoRA for product information acquisition in in-store shopping situations. Only when compared to a sales talk, CoRA shows no significant relative advantage but also no significant disadvantage. We therefore assume that CoRA is comparable to a sales talk, which does not only strengthen its utility for consumers but also for retailers that may offer such a mobile application in addition to sales personnel. Although this experiment and its results are limited to the domain of cosmetics and are based on a relatively small sample obtained from a university, the results are promising and may apply for other consumer products as well. We therefore will conduct further experiments to validate the positive results for other product domains in field experiments.

## 7  Conclusion and Future Work

Dialogue interaction between customers and products encompasses the capability for improved filtering and presentation of relevant product information in in-store shopping situations. Thus, information needs of customers can be addressed more intuitive. In order to realize such interaction, dialogue systems and comprehensive knowledge representations are necessary such as product information, linguistic representations and user reviews. Because of their heterogeneous nature and diverse semantics, these product-related knowledge resources have to be bundled and linked to enable standardized information access. In this article, we have introduced a conceptual model of dynamic linkage of product-related knowledge resources for dialogue interaction in in-store shopping situations. It was shown how multiple heterogeneous knowledge resources are bundled by a *Semantic Mediator* that enables standardized access to different types of product-centered knowledge resources. Especially, the mapping of product descriptions of diverse formats, e.g., BMEcat, into a semantic product description structure is elaborated. This standardized ontological product description of physical products is dynamically updated and extended via SWRL. Afterwards, we have described our model of dynamic linkage of these bundled knowledge resources, i.e. the linkage of an ontological linguistic knowledge base of the dialogue system with a pool of product descriptions. We apply ontologies as non-linguistic and linguistic core knowledge components of our dialogue system. The semantic knowledge is merged automatically with external non-semantic contents. This is enabled by a high level semantic query language that maps ontological data into Java structures and thus allows object-oriented querying of semantic data. With a prototype of the dialogue system, we have then shown the utility of in-store dialogue interaction based on heterogeneous knowledge resources by an end-user study. In our future work, we will focus on three issues: (1) extension of the linkage model with more product-related knowledge resources, (2) enhancement and standardization of PQL2 functionality, and (3) augmentation of our linguistic resource by linking up further external services such as dbpedia.org or zemanta.com.

# Acknowledgement

We would like to thank Eva Blomqvist for her comprehensive advices concerning SPDO. This paper resulted from project 'Interactive Knowledge Stack (IKS)' (FP7 231527) co-funded by the European Commission.

# References

1. Gurevych, I., Mühlhäuser, M.: Natural language processing for ambient intelligence. Künstliche Intelligenz/Special Issue: Ambient Intelligence und Künstliche Intelligenz (2), 10–16 (2007)
2. Sabou, M., Kantorovitch, J., Nikolov, A., Tokmakoff, A., Zhou, X., Motta, E.: Position paper on realizing smart products: Challenges for semantic web technologies. In: The 2nd International Workshop on Semantic Sensor Networks, collocated with ISWC 2009 (2009)
3. Bel-Enguix, G., Dediu, A.-H., Jimenez-Lopez, M.: A dialogue-based system for man-machine interaction. In: Conf. on Human System Interactions, pp. 141–146 (25-27, 2008)
4. Catizone, R., Wilks, Y., Worgan, S., Turunen, M.: Some background on dialogue management and conversational speech for dialogue systems. In: Wilks, Y., Catizone, R. (eds.) Computer, Speech and Language (2010) (special issue on dialogue)
5. Warren, H.D.D., Pereira, C.N.F.: An efficient easily adaptable system for interpreting natural language queries. Computational Linguistics (8), 110–122 (1982)
6. Clark, P., Chaw, S.Y., Barker, K., Chaudhri, V., Harrison, P., Fan, J., John, B., Porter, B., Spaulding, A., Thompson, J., Yeh, P.: Capturing and answering questions posed to a knowledge-based system. In: K-CAP 2007: Proc. of the 4th International Conf. on Knowledge Capture, pp. 63–70. ACM, New York (2007)
7. Alexandersson, J., Becker, T., Pfleger, N.: Overlay: The basic operation for discourse processing. In: Wahlster, W. (ed.) SmartKom: Foundations of Multimodal Dialogue Systems, pp. 255–267 (2006)
8. Perez, G., Amores, G., Manchon, P., Gonzalez, O.G.Y.J., Julietta, G.I.: Integrating owl ontologies with a dialogue manager. Technical report, CiteSeerX - Scientific Literature Digital Library and Search Engine (2006)
9. Chen, Y.J., Chen, Y.M., Chu, H.C.: Development of a mechanism for ontology-based product lifecycle knowledge integration. Expert Syst. Appl. 36(2), 2759–2779 (2009)
10. Hepp, M.: Goodrelations: An ontology for describing products and services offers on the web. In: Gangemi, A., Euzenat, J. (eds.) EKAW 2008. LNCS (LNAI), vol. 5268, pp. 329–346. Springer, Heidelberg (2008)
11. Ou, S., Pekar, V., Orasan, C., Spurk, C., Matteo, N.: Development and alignment of a domain-specific ontology for question answering. In: European Language Resources Association (ed.) Proc. of the Sixth International Language Resources and Evaluation (LREC 2008), Marrakech, Morocco (2008)
12. Clark, P., Thompson, J., Porter, B.: A knowledge-based approach to question-answering. In: Proc. AAAI 1999 Fall Symposium on Question-Answering Systems, pp. 43–51 (1999)
13. Buitelaar, P., Declerck, T., Calzolari, N., Lenci, A.: Language resources and the semantic web. In: Proc. of the ELSNET/ENABLER Workshop (2003)

14. Lopez, V., Uren, V., Motta, E., Pasin, M.: AquaLog: an ontology-driven question answering system for organizational semantic intranets. Web Semantics 5(2), 72–105 (2008)
15. Hayashi, Y., Declerck, T., Buitelaar, P., Monachini, M.: Ontologies for a global language infrastructure. In: Webster, J., Ide, N., Fang, A.C. (eds.) Proc. of the 1st International Conf. on Global Interoperability for Language Resources (ICGL 2008), Hong Kong, China, pp. 105–112 (2008)
16. Anderl, R., Trippner, D.: Step standard for the exchange of product model data. Technical report, STEP (2000)
17. Kowatsch, T., Maass, W.: Towards a framework for knowledge-based pricing services improving operational agility in the retail industry. In: D'Andrea, V., Gangadharan, G.R., Iannella, R., Weiss, M. (eds.) CEUR Workshop Proc., vol. 530 (2009)
18. Kowatsch, T., Maass, W., Filler, A., Janzen, S.: Knowledge-based bundling of smart products on a mobile recomendation agent. In: ICMB 2008: Proc. of the 7th International Conf. on Mobile Business, Washington, DC, USA, pp. 181–190. IEEE Computer Society, Los Alamitos (2008)
19. Maass, W., Janzen, S.: A pattern-based ontology building method for ambient environments. In: Blomqvist, E., Sandkuhl, K., Scharffe, F., Svatek, V. (eds.) Proc. of the Workshop on Ontology Patterns (WOP 2009), collocated with ISWC 2009, Washington D.C., vol. 516, CEUR Workshop Proc. (2009)
20. Maass, W., Filler, A.: Towards an infrastructure for semantically annotated physical products. In: Gesellschaft für Informatik e. V (ed.) Conf. Proc. Informatik 2006 (2006)
21. Janzen, S., Maass, W.: Ontology-based natural language processing for in-store shopping situations. In: Proc. of Third IEEE International Conf. on Semantic Computing (ICSC 2009), pp. 361–366. IEEE Computer Society, Los Alamitos (2009)
22. Marcus, M.P., Marcinkiewicz, M.A., Santorini, B.: Building a large annotated corpus of english: the penn treebank. Comput. Linguist. 19(2), 313–330 (1993)
23. Kamis, A., Koufaris, M., Stern, T.: Using an attribute-based decision support system for user-customized products online: An experimental investigation. MIS Quarterly 32(1), 159–177 (2008)
24. Moore, G., Benbasat, I.: Development of an instrument to measure the perceptions of adopting an information technology innovation. Information Systems Research 2, 173–191 (1991)
25. Davis, F.D.: Perceived usefulness, perceived ease of use, and user acceptance of information technology. MIS Quarterly 13(3), 319–339 (1989)
26. Nunnally, J.C.: Psychometric Theory. McGraw-Hill, New York (1967)

# ISReal: An Open Platform for Semantic-Based 3D Simulations in the 3D Internet

Patrick Kapahnke, Pascal Liedtke, Stefan Nesbigall,
Stefan Warwas, and Matthias Klusch*

German Research Center for Artificial Intelligence, Saarbrücken, Germany
`firstname.surname@dfki.de`

**Abstract.** We present the first open and cross-disciplinary 3D Internet research platform, called ISReal, for intelligent 3D simulation of realities. Its core innovation is the comprehensively integrated application of semantic Web technologies, semantic services, intelligent agents, verification and 3D graphics for this purpose. In this paper, we focus on the interplay between its components for semantic XML3D scene query processing and semantic 3D animation service handling, as well as the semantic-based perception and action planning with coupled semantic service composition by agent-controlled avatars in a virtual world. We demonstrate the use of the implemented platform for semantic-based 3D simulations in a small virtual world example with an intelligent user avatar and discuss results of the platform performance evaluation.

## 1 Introduction

In the Internet of today, navigation and display of content mostly remains two-dimensional. On the other hand, the proliferation of advanced 3D graphics for multi-player online games, affordable networked high-definition display and augmented reality devices let Internet users increasingly become accustomed to and expect high-quality 3D imagery and immersive online experience. The 3D Internet (3DI) is the set of 3D virtual and mixed reality worlds in the Internet that users can immersively experience, use and share with others for various applications [25,5,2,31].

As of today, the 3DI offers, for example, various alternative worlds like SecondLife (2L)[1], questville, Croquet and WorldOfWarcraft, and mirror worlds like Twinity[2]. Applications include socializing and business collaboration in 3D meeting spaces, the 3D exploration of virtual cities, the participation in cross-media edutainment events like concerts and lectures, the trading of real and virtual assets, the functional 3D simulation of production lines and architecture at design time, as well as advanced visual 3D information search by using 3D Web

---

* The work presented in this paper has been partially funded by the German Ministry for Education and Research (BMB+F) under project grant 01IWO8005 (ISReal).
[1] http://secondlife.com
[2] http://www.twinity.com

P.F. Patel-Schneider et al.(Eds.): ISWC 2010, Part II, LNCS 6497, pp. 161–176, 2010.

browsers such as SpaceTime and ExitReality. In such virtual worlds, the user is usually represented by and driving the behavior of an avatar as her digital alter-ego.

Major challenges of the 3DI are (a) the more realistic, standard-based 3D graphical display in 3D Web browsers, and (b) the making of user avatars behave more intelligent in their 3D environment. For example, the intelligence of most avatars in virtual worlds today is either restricted to direct execution of non-verbal user commands, or rather simple event-rule-based but resource-optimized means of AI planning with massive volumes of action scripts in online games. Besides, in most cases, avatars are not even capable of understanding the semantics of their perceived 3D environment due to the lack of standard-based semantic annotations of 3D scenes and reasoning upon them or do not exploit 3D scene semantics for intelligent action planning in a virtual world they are involved in.

To address these challenges, we developed the first open, cross-disciplinary 3DI research platform, called ISReal, that integrates semantic Web, semantic services, agents and 3D graphics for intelligent 3D simulation of realities. In this paper, we describe the innovative interplay between its components with focus on semantic 3D scene annotation and query processing, and the semantic-based action planning of intelligent agent-controlled avatars together with a discussion of our experimental performance evaluation of the platform in a simple virtual 3D world. To the best of our knowledge, there is no other such integrated 3DI platform available yet.[3]

The remainder of the paper is structured as follows. Section 2 provides an overview of the ISReal platform while sections 3 and 4 describe the global semantics and intelligent agents for semantic-based 3D simulation. Section 5 demonstrates the use of the platform for a simple use case, followed by performance evaluation results and comments on related work in Sections 7 and 8.

## 2   ISReal Platform: Overview

**Virtual world descriptions in XML3D.** The ISReal platform can be used to develop and simulate virtual worlds in XML3D[4] which is a 3D graphics-oriented extension of HTML4. A virtual world scene is graphically described in form of a single XML3D scene graph that includes all objects of the 3D scene to be displayed as its nodes. In contrast to X3D[5], XML3D scene descriptions can be directly embedded into a standard HTML page such that every scene object becomes part of and accessible in the standard HTML-DOM (Document Object Model) by any XML3D-compliant Web browser capable of rendering the scene without any specific viewer plug-in required. Graphical changes in the virtual

---

[3] Major barriers of an 3DI uptake by people today refer to its potential physio-cognitive, social and economic impacts on individual users of virtual worlds which discussion is outside the scope of this paper.

[4] http://www.xml3d.org

[5] http://www.web3d.org/x3d/specifications/

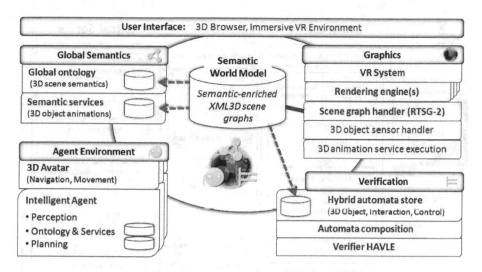

**Fig. 1.** ISReal platform components

world during its simulation such as user interactions with the scene and 3D object animation in the browser correspond to changes of its XML3D scene graph in the Web page of the virtual world scene which is loaded and processed by the ISReal client. In the following, we give an overview of the platform components and its communication architecture for virtual 3D world simulations.

**Platform Components.** The ISReal platform consists of five groups of components that are the user interface, the global semantics, 3D graphics, intelligent agents and verification environment (see Fig. 1). The graphics environment maintains the given set of XML3D scene graphs of virtual worlds by its internal RTSG-2 (real-time scene graph) system [24] and renders them by a pluged-in 3D rendering engine for high-quality 3D display such as our world-fastest ray-tracer RTFact [9] at run time. For immersive 3D interaction with simulated virtual worlds, it additionally provides an open, immersive VR (virtual reality) system. The global semantics environment (GSE) is responsible for managing global scene ontologies each of which describing the semantics of a virtual world in its application domain as well as the execution handling of globally registered semantic services which groundings have an effect on these ontologies such as the change of the position of some object in the scene by the respective 3D animation in the graphics environment (cf. Section 3). The verification environment manages and composes hybrid automata that describe spatial and temporal properties of scene objects and their interactions and verifies them against given safety requirements at design time; for reasons of space, we omit a description of this platform component. The semantic world model of the platform is the set of semantically annotated 3D scene graphs with references to the global semantics and the verification component. The agent environment manages the avatar-controlling intelligent agents capable of scene perception, local scene ontology management and semantic-based action planning to accomplish its tasks given

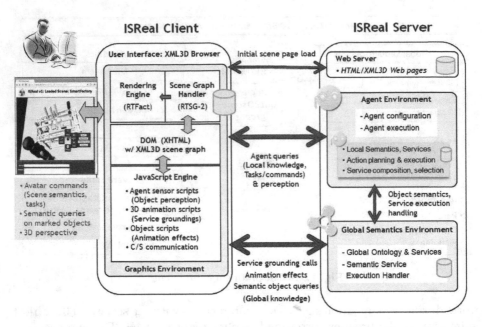

**Fig. 2.** ISReal v1.1 communication architecture for virtual 3D world simulations

by the user or other agents (cf. Section 4). Finally, the user can interact with a 3D simulated virtual world scene by alternative means of 3D Web-based or immersive 3D virtual reality system-based user interface of the platform. The interface is either of both XML3D-compliant versions of Google Chrome and Mozilla Firefox browsers or an immersive VR environment based on the open VR system Lightning (which we connected with multi-touch display, space mouse, tracking system and iPhone as 3D input devices). A user can (non-verbally) query the semantics of marked single objects in the simulated scene with or without her avatar, and to command her avatar to answer complex semantic queries and to pursue given tasks in the scene.

**Communication Architecture for Virtual 3D World Simulation.** The client-server-based communication architecture of the ISReal platform 1.1 for single-user virtual 3D world simulation is shown in Figure 2.

The ISReal client is exclusively responsible for maintaining and rendering the complete virtual world scene with its embedded 3D graphics environment, and communicates with the ISReal server hosting all other components (and a Web server, in case of XML3D browser as ISReal client) for intelligent simulation. Asynchronous and bidirectional client-server communication is impemented by use of the WebSockets API[6]. Once the initial world scene page in HTML/XML3D is loaded by the ISReal client from the ISReal server, the client connects to server-sided components, triggers the scene-relevant configuration of

---

[6] http://dev.w3.org/html5/websockets/

semantics and agents at the server, and is responsible for user-interaction-based updates and rendering of the XML3D scene graph (Web-based or immersive 3D).[7] The open ISReal platform in its current version 1.1 has been fully implemented in Java and JavaScript.

# 3   ISReal Global Semantics

Semantic-based 3D simulation of virtual worlds is a key feature of the ISReal platform. In this section, we describe the semantic annotation of 3D scene objects and the global semantics environment of the platform in more detail.

## 3.1   Semantic 3D Scene Object Annotation

The semantic world model of the platform is the set of all semantically annotated XML3D scene graphs for simulated virtual worlds. Any 3D scene object in a virtual world is represented as a node of the XML3D scene graph that graphically describes this world. The semantics of a 3D scene object can be described by annotating its XML3D scene graph node by use of standard RDFa[8] with links to (a) the uniquely assigned object in a given global scene ontology described in (the OWL-Horst fragment of) standard OWL2 that represents the conceptual and assertional knowledge about the scene and application domain, (b) semantic services in OWL-S that describe the operational functionality of the scene object and are grounded in respective 3D animation scripts, and (c) hybrid automata that describe object properties with respect to continuous time and space in FOL linear arithmetics.

Figure 3 shows an example of semantic annotation of a virtual worlds scene object, that is a door connecting room A with room B. The representation of this object in the XML3D scene graph refers to a node labeled "doorAB" that includes its graphical description and semantic annotation. The first case refers to the 3D geometry (mesh) data required for rendering the scene object "doorAB" as defined in its respective subnode. The semantic annotation of the "doorAB" node is in RDFa with references to (a) an uniquely assigned object "doorAB" which semantics is defined in a given global scene ontology, (b) a set of semantic services describing the opening and closing of "doorAB" each of which grounded with an appropriate 3D animation script to be executed by the graphics environment, and (c) a hybrid automaton describing the temporal-spatial property that "doorAB" can be opened and closed with angular speed of 10 degrees per second, which is not possible to encode and reason upon in OWL2. Both the given global ontology and semantic object services are maintained in the global semantics environment of the ISReal platform.

---

[7] We are working on a multi-user/server architecture where the ISReal server maintains the global scene graph and provides multiple clients with only update instructions of how to change and render their local views on the scene based on user interaction events.

[8] The same principle of semantic annotation can be applied to X3D scene graphs as well. For a discussion of the benefits of XML3D over X3D, we refer to [27].

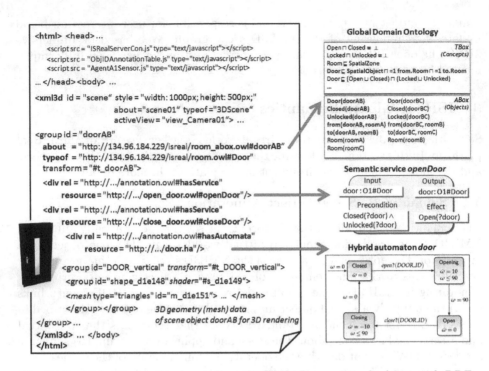

**Fig. 3.** Example of semantic annotation of a XML3D scene graph object with RDFa

## 3.2 Global Semantics Environment

**Architecture.** The global semantics environment (GSE) consists of two components as shown in Figure 4, that are the global ontology management system (OMS) and the semantic service handler (SemSH). The OMS maintains a given set of global ontologies each of which describing the conceptual (TBox) and factual (ABox, fact base) knowledge about one simulated virtual world in OWL2. It handles the processing of different types of semantic queries issued by the user or agents against the global ontology of the actually simulated virtual world[9]. We assume that the TBox of the global ontology, in contrast to its ABox, does not change during simulation. The selected global scene ontology is materialized in, updated and queried through a selected RDF store of the OMS as usual. Other semantic queries (which answering is not possible by triple stores) are routed by the OMS query decider to the appropriate semantic reasoner(s) depending on its type or indicated by the user. The SemSH maintains the global semantic service repository that is assumed to contain all services in OWL-S which are related to the global scene ontology in terms of having either a precondition to be checked against its fact base, a grounding that may update the fact base as an effect, or both.

---

[9] In the following, we focus on the global ontology of one virtual world.

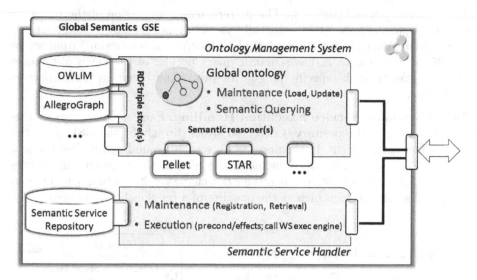

**Fig. 4.** Architecture of the global semantics environment (GSE)

**Implementation.** The implemented GSE has two architectural key features. First, its OMS has an open plug-in (API) architecture for using any RDF/S store and semantic reasoner as appropriate and is realized with the LarKC platform[10]. The OMS query decider routes semantic queries to OMS plug-ins available for the RDF triple stores SwiftOLIM (with RDF materialization of OWL2 under OWL-Horst semantics) and AllegroGraph, the semantic OWL-DL reasoner Pellet[11] with internal Jena RDF store, and the RDF relational reasoner STAR[14]. Second, semantic query answering and service handling by the GSE is upon request only, in particular, the GSE does not actively communicate semantic updates of the global ontology to other components; this avoids communication bottleneck and supports the paradigm of perception-based knowledge for BDI agents (cf. Section 4).

**Semantic 3D Scene Query Processing.** As a result of its open plug-in architecture, the types of semantic queries the OMS is capable of answering depends on the respective functionality of its plug-ins for triple stores and semantic reasoners. For example, the OMS can (a) efficiently answer object (and OWL-Horst concept) queries with its RDF store SwiftOWLIM using SPARQL, (b) more complex (OWL2-DL) concept queries with Pellet using SPARQL-DL, and (c) relational object queries with STAR. For example, a relational object query like "How are scene objects doorAB, doorBC and roomC related ?" is processed by STAR by reduction to the corresponding NP-hard Steiner-Tree problem for the RDF graph of the materialized global ontology followed by the polynomial computation of an approximated solution in $O(nlogn)$ in terms of minimal RDF

---

[10] http://www.larkc.eu/resources/
[11] http://clarkparsia.com/pellet

object property-based path [14]. The pattern-based conversion of the result by our STAR-plugin of the OMS eventually yields a more human-readable answer (rather than just a list of RDF triples) like "doorAB leads to roomB from where doorBC leads to roomC." The semantic query decider of the OMS distributes semantic queries to the specific plug-in for processing based on the respective query type.

**Global Semantic Service Execution Handling.** Each semantic service registered at the global repository is executed either directly by the GSE or by the 3D graphics environment. In the first case, a service grounding updates the fact base without any 3D animation, while in the second case a grounding triggers both the 3D animation of an object and the change of its factual semantics in the global scene ontology such as the opening of a previously closed door.

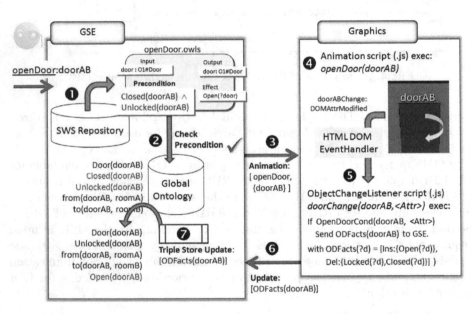

**Fig. 5.** Example of semantic 3D animation service execution handling by the GSE and 3D graphics

Figure 5 illustrates the execution of semantic services with precondition and grounding in animation that effects the global ontology. In this example, the SemSH of the GSE receives the call of a semantic service "openDoor" for opening the scene object "doorAB" by some avatar (or its user). After retrieving the OWL-S description from its repository, it checks whether the service precondition (in SWRL) holds in the fact base of the OMS with a respective SPARQL ASK query. If successful, the SemSH triggers the execution of the service grounding, that is the 3D animation script for opening the door "doorAB" in the scene by the graphics environment in the ISReal client. This animation may result in a change of the attribute values of the (HTML-DOM) XML3D scene graph object "doorAB" which is observed by an object script doorChange() through its registration at the HTML-DOM event handler for this object. In case the change has been encoded in the object script by the scene developer to correspond with a change of the animated object semantics in the global fact base, the script sends a semantic update query with respective insert-delete lists of object facts to the GSE for updating the global fact base by the OMS.

# 4   ISReal User Agent

In ISReal, the intelligent behavior of any avatar apart from direct user commands is determined by an intelligent agent it is uniquely associated with in the considered virtual world. The avatar represents the appearance of its user as her alter-ego but also its agent in the virtual world, and as such it is described as just another scene object in the XML3D scene graph of this world. The idea is that the user does not distinguish between her avatar and the intelligent agent that is driving its intelligent behavior; only in this sense, the terms "avatar" and "agent" can be used interchangeably. But how to design such an intelligent agent that is capable of understanding the semantics of the simulated scene it is involved in and to perform semantic-based action planning?

## 4.1   Architecture

For the development of ISReal agents, we adopted the reactive BDI (belief-desire-intention) architecture [22] that is known to be particularly appropriate for fast perception, deliberation and action in dynamically changing environments such as virtual 3D worlds. In very brief, the BDI agent is equipped with a plan library of domain-dependent and -independent plan patterns and a BDI planner to satisfy given goals by reactive action planning from second principles and execution. Based on the perception of its environment, that is the simulated 3D scene it is involved in, and given task to pursue, an ISReal agent selects an appropriate BDI plan pattern of operators, instatiates it with variable bindings in its local fact base into actions which execution in turn affect the perceived and locally updated state of the scene. Figure 6 outlines the architecture of an ISReal agent which consists of (a) an uniquely associated avatar that is running in the graphics environment (b) a semantic perception facility that interacts with an agent sensor running in the graphics environment and the GSE, (c) a BDI plan library and BDI planner, and (d) a local semantics environment (LSE) that maintains its local knowledge about the virtual 3D world it is involved in.

**Local Semantics.** The LSE of an ISReal agent differs from the GSE in several aspects. Though the local TBox is a copy of the global one copied from the OMS of the GSE during scene initialisation at the ISReal server, the local fact base includes only facts about scene objects the agent individually perceives via its sensor (cf. Section 4.1). The local service repository includes semantic services each of which encoding a plan operator from its BDI plan library; only services with an effect on the global fact base of the scene are also registered with the global repository of the GSE. Further, the local SemSH handles the execution of agent services

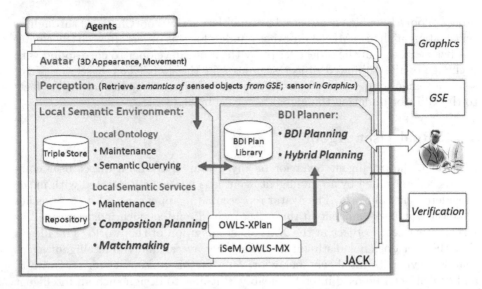

**Fig. 6.** ISReal user agent architecture

(not registered with the GSE) which groundings have no effect on the global fact base. This avoids that agents arbitrarily execute animations without checking the correlated change in the global scene semantics - such as walking through closed doors. Further, the LSE also offers plug-ins for local selection and composition planning of semantic services registered with the local repository.

### 4.2   Semantic Perception and Action Planning

**Semantic Perception of 3D Scene Objects.** An ISReal agent perceives its environment via its uniquely assigned agent sensor that is continuously running in the 3D graphics environment. Each sensor is individually configurable regarding the frequency, resolution and range of object sensing. It sends the set of previously unknown or semantically updated XML3D scene graph objects (including the avatar object for self-perception) in the specified range together with their hashed metadata record (from the scene graph) to the individual agent. The semantics of perceived objects are then requested by the agent from the GSE which returns the set of (a) all object-related facts (corresp. with terminological object abstraction) from the global fact base and (b) all semantic object service descriptions. The subsequent update of the local fact base and registration of object services with the local repository completes the semantic perception. As a result, an ISReal agent only knows about those parts of the scene it perceives such that its local fact base may be inconsistent with the global one hosted by the OMS of the GSE. In particular, individual sensors or local update strategies (by default: immediately) of different agents may lead to different views on the global scene.

**Semantic-Based Action Planning and Query Processing.** In principle, an ISReal agent can answer the same type of semantic queries over its local ontology as the GSE over the global ontology of the simulated world. In addition, it can satisfy action goals like "Go to next room." or declarative goals like "Show me how to produce A with machine X?" by use of its BDI planner and semantic service composition planner over its local ontology and service repository. By default, an ISReal agent is equipped with domain-independent BDI plans for acting in virtual worlds such for basic 3D animations of its avatar, processing of different types of semantic queries in its local ontology, and execution handling of its actions (services) via the GSE or by itself. Other domain-dependent plans have to be added or customized for 3D scene simulation by the 3D scene agent developer.

In case there is no BDI plan pattern in the library that can be instantiated to satisfy a given action goal, the agent tries to solve this problem by action planning from first principles, that is the application of a semantic service composition planner to a given initial state in order to reach a given goal state. The initial state is a local fact base copy, the set of actions are and the goal state is either explicitly given or derived from the BDI plan context conditions at run time. Alternatively, the agent may search first for query objects missing in the local fact base by reactive BDI action planning before subsequent semantic service composition planning can be performed.

**Implementation.** The implemented agent environment is hosted by the ISReal server and consists of one intelligent ISReal agent and avatar by default, a model-driven BDI agent development tool, ISReal agent configuration tool for avatar assignment to and initialisation of an agent in a given scene, and the BDI agent execution platforms JACK and JADEX for server-sided running of the agent with client-sided 3D simulation of its avatar. The agent plug-ins for semantic service selection and composition are OWLS-MX, iSeM and OWLS-XPlan 2.0[12].

## 5  Use Case Example

The implemented ISReal platform 1.1 has been used to develop and simulate several virtual 3D worlds. In this section, we demonstrate this by means of a simple example for 3D simulation of production lines.

**Small Virtual World "SmartFactory".** The small virtual world "SmartFactory" consists of two rows of three rooms each (rooms A to C, D to F) where seven doors connect adjacent rooms, and one automatic ampoulle filling station X located in room F. The station is capable of filling different types of pills into a RFID-tagged cup that is placed on a transport wagon circulating between different pill production stations on demand. The filling state of the cup is read via RFID sensors of designated control points while filling tasks are saved on its RFID chip. There is one default user avatar "Nancy" initially placed in room A. Figure 7 shows the layout and a screenshot of the user interface in the XML3D Chrome browser with the user avatar in front of the filling station.

---

[12] http://www.dfki.de/-klusch/i2s/html/software.html

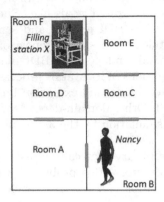

**Fig. 7.** Small virtual world SmartFactory: Layout and Web user interface snapshot

**Implementation.** We graphically modelled this world including the avatar Nancy by using 3DSMAX and stored the designated HTML scene page with embedded XML3D description of the initial scene graph at the ISReal Web server. The global ontology for this world was developed by using Protege and then used for semi-automated semantic annotation of objects. The ontology defines 89 concepts, 38 properties with initially 57 facts about 38 semantically annotated scene objects (7 doors, 1 avatar, 1 station with 29 parts) while the global repository includes 29 object services in OWL-S. The ISReal agent for the single user avatar Nancy is modelled as a novice which knows nothing about the world objects: Its local fact base is empty and no object services are registered with the local repository yet.

**Integrated Example of Intelligent Simulation in a Training Scenario.** Once loaded into the XML3D browser client, the user can command her avatar to explore the scene and to demonstrate the functionality of the filling station on request. Consider the simple training scenario in which the user asks her avatar "show me how to produce a pill XYZ with the filling station X?". To satisfy the declarative goal by semantic object service composition planning, the agent first searches for the unknown query (goal state) objects pill XYZ, station X by reactive BDI action planning with in-room navigation, semantic perception of objects and respectively incremental update of the local fact base during its search. The subsequently generated service composition plan with OWLS-XPlan over the local ontology and 11 object services in the local repository is then reported to the user in form of text as an answer and then executed by corresponding 3D object service animations in the XML3D browser for demonstration.

## 6    Performance Evaluation

In this section, we discuss results of our preliminary performance evaluation of the ISReal platform 1.1 with focus on the global semantics and agent

environment for the use case. The tests were performed with an average re-
sourced notebook (Intel Quad Core Q9400, 2.66GHz, 8GB RAM) [13].

**Global semantics environment.** The results of our performance evaluation
of the OMS plug-ins (SwiftOLIM, AllegroGraph, Pellet) for LUBM benchmark
essentially are in compliance with those reported by others elsewhere [18,16] and
at the RDF Store Benchmarking site[14]. For our small use case, the triple store
contained 1087 (747 explicit) triples. Figure 8 summarizes the thruput of the
GSE (number of operations per simulation time). It shows, in particular, that
within 1 second of simulation in the use case, the GSE can perform 15 updates
and 75 queries over its store (precondition checks, fact retrieval; cf. Section 3).
The STAR reasoner moderately scales up to 35k triples with 35s average query
response time (AQRT).

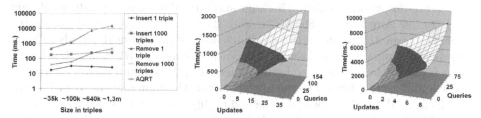

**Fig. 8.** Avg ontology update time for use case (1k triples) and LUBM(1, 5, 10) [Left];
AQRT in relation to updates for use case [Middle] and LUBM(35k) [Right]

**Semantic perception by agent for use case.** The agent needs avg. 80ms
(320ms) to semantically perceive a scene object without (with) any annotated
service. Semantic perception time is the period from received sensor perception
event until completed update of the LSE with ontology and repository (cf. Section 4.2). Object fact retrieval from the GSE takes 6.5ms, 74ms for local fact
base update and 240ms for registering object services in the local repository.

**Semantic-based action planning by agent for use case.** Preparation of the
planning domain for offline semantic service composition planning using OWLS-
XPlan takes 400ms and plan generation without execution (3D animations) takes
5s. The plan execution (service grounding in animation) appears fast with about
530ms: Precondition check per service by the GSE in 6.5ms, 3D animation by
the graphics environment in 300ms and update of the global fact base in 27ms.

**Discussion.** Semantic-based 3D simulation with ISReal appears reasonable for
small virtual worlds with medium-sized global scene ontology, few hundreds of
annotated 3D scene objects and up to a few dozens of agents. For example, the
simulation time of 30 agents concurrently trying to open 30 doors would take

---

[13] Using the alternative immersive VR user interface of ISReal requires substantially
more resources for reasonable 3D simulation performance (e.g. 44 Cores with 16GB
RAM for about 15 frames per second).

[14] http://esw.w3.org/RdfStoreBenchmarking

only 1.5 seconds: Checking of 30 preconditions by the GSE (195ms), concurrent animation of door opening (500ms) and 30 updates of global fact base (810ms). The GSE slows down the overall 3D simulation for one agent with 33.5ms and for 30 agents with 1 second. The reactive BDI action planning appeared extremely fast (avg 25ms per plan operation check) but the main slow down of semantic-based planning within simulation is caused by semantic service composition with 5 seconds.

## 7    Related Work

It appears common sense that the use of semantics can greatly improve the management and retrieval of 3D content [11,19,21,10] as it has been impressively demonstrated for various practical applications in different domains such as arts, bioinformatics, gaming, cultural heritage and virtual museums, partly in relevant projects like Aim@Shape, FocusK3D and 3DVisa[15]. To the best of our knowledge, the open 3DI platform ISReal significantly differs from this body of work in general: Semantic annotation of 3D scenes in standard RDFa and OWL2 enables users and avatars alike to better understand the semantics of simulated 3D objects and their relations in the virtual world; the representation of 3D scene graphs in XML3D allow for all-in-one and highly realistic 3D scene rendering by any XML3D browser with our real-time raytracer RTFact; any ISReal avatar is potentially capable of behaving more intelligent than other types of avatars in virtual worlds available today thanks to the capabilities of its associated intelligent agent for semantic reasoning, semantic-based action planning and service composition.

In particular, many approaches to integrate semantic Web with virtual 3D worlds [20,12] put a strong emphasis on exploiting semantic 3D content annotation in RDF, RDFS or proprietary formats for semantic object search and querying in specific application context such as the virtual 3D furniture shop in [8,15] or the virtual 3D museum tour guide in [4] - but without semantic-based action planning by user avatars like in ISReal. On the other hand, related work on virtual agents such as in [1,3] and STEVE (SOAR Training Expert for Virtual Environments)[23] focus rather on multi-modal user-agent interaction in immersive VR environments and use of AI planning by agents with pre-coded planning domain knowledge and plan patterns - but without any semantic-based 3D scene querying or service composition planning from first principles like in ISReal.

## 8    Conclusions

We presented the first open 3DI research platform for semantic-based 3D simulations in virtual worlds that uses semantic Web, semantic services, intelligent agents and 3D graphics. ISReal user avatars are coupled with intelligent agents

---

[15] http://3dvisa.cch.kcl.ac.uk/project86.html

that understand the semantics of their annotated 3D environment and perform semantic-based action planning to satisfy goals (queries) of their users. Such intelligent 3D simulations with the implemented platform are reasonably fast for small virtual worlds with medium-sized global ontology, small number of annotated 3D scene objects and up to a few dozens of agents.

However, scalability of semantic 3D scene query processing and semantic service composition planning remains an issue for intelligent simulation of time-critical applications of large virtual worlds with potentially thousands of semantically annotated scene objects and hundreds of agents. Our ongoing work for ISReal 2.0 is on multi-agent planning scenarios, multi-user applications, and scalable semantic query processing respecting relevant work [26,28] and research results from LarKC and SEALS[16]. The implemented ISReal platform 1.1 together with the SmartFactory use case will be released under GPL license.

# References

1. Abaci, T., Ciger, J., Thalmann, D.: Planning with Smart Objects. In: WSCG SHORT Papers Proceedings, UNION Agency-Science Press (2005)
2. Alpcan, T., Bauckhage, C., Kotsovinos, E.: Towards 3D Internet: Why, What, and How? In: IEEE Intl. Conference on Cyberworlds. LNCS, vol. 2174. Springer, Heidelberg (2007)
3. Anastassakis, G., Ritchings, T., Panayiotopoulous, T.: Multi-agent Systems as Intelligent Virtual Environments (2001)
4. Chou, S.-C., Hsieh, W.-T., Gandon, F.L., Sadeh, N.M.: Semantic Web Technologies for Context-Aware Museum Tour Guide Applications. In: Proc. 19th Intl. IEEE Conf. on Advanced Information Networking and Applications (2005)
5. Daras, P., Alvarez, F.: A Future Perspective on the 3D Media Internet. Towards the Future Internet - A European Research Perspective. IOS Press, Amsterdam (2009)
6. de Silva, L., Sardina, S., Padgham, L.: First Principles Planning in BDI Systems. In: Proc. 8th Intl. Conf. on Autonomous Agents and Multiagent Systems (2009)
7. Davies, N., Mehdi, Q.: BDI for Intelligent Agents in Computer Games. In: Proc. 8th Intl. Conf. on Computer Games (2006)
8. De Troyer, O., Kleinermann, F., Mansouri, H., Pellens, B., Bille, W., Fomenko, V.: Developing semantic VR-shops for e-Commerce. In: Magoulas, G.D., Lepouras, G., Vassilakis, C. (eds.) Special Issue of Virtual Reality: Virtual Reality in the e-Society. Springer, Heidelberg (2006)
9. Georgiev, I., Rubinstein, D., Hoffmann, H., Slusallek, P.: Real Time Ray Tracing on Many-Core-Hardware. In: Proc. 5th INTUITION Conf. on Virtual Reality (2008)
10. Geroimenko, V., Chen, C.: Visualizing the Semantic Web: XML-based Internet and Information Visualization. Springer, Heidelberg (2005)
11. Ibanez-Martinez, J., Mata, D.: Virtual Environments and Semantics. European Journal for the Informatics Professional 7(2) (2006)
12. Ibanez-Martinez, J., Mata, D.: A Basic Semantic Common Level for Virtual Environments. Virtual Reality 5(3) (2006)
13. Kallmann, M.: Object Interaction in Real-Time Virtual Environments. PhD thesis, École Polytechnique Fédérale de Lausanne (2001)

---

[16] www.seals-project.eu

14. Kasneci, G., Ramanath, M., Sozio, M., Suchanek, F.M., Weikum, G.: STAR: Steiner Tree Approximation in Relationship-Graphs. In: Proc. 25th IEEE Intl. Conf. on Data Engineering (2009)
15. Kleinermann, F., et al.: Adding Semantic Annotations, Navigation Paths and Tour Guides to Existing Virtual Environments. In: Wyeld, T.G., Kenderdine, S., Docherty, M. (eds.) VSMM 2007. LNCS, vol. 4820, pp. 100–111. Springer, Heidelberg (2008)
16. Kiryakov, A.: Measurable Targets for Scalable Reasoning. EU FP7 Project LarKC, Deliverable D5.5.1 (2008), www.larkc.eu/deliverables/
17. Kleinermann, F.: Adding Semantic Annotations, Navigation Paths and Tour Guides for Existing Virtual Environments. In: Wyeld, T.G., Kenderdine, S., Docherty, M. (eds.) VSMM 2007. LNCS, vol. 4820, pp. 100–111. Springer, Heidelberg (2008)
18. Luther, M., Liebig, T., Böhm, S., Noppens, O.: Who the heck is the father of bob? In: Aroyo, L., Traverso, P., Ciravegna, F., Cimiano, P., Heath, T., Hyvönen, E., Mizoguchi, R., Oren, E., Sabou, M., Simperl, E. (eds.) ESWC 2009. LNCS, vol. 5554, pp. 66–80. Springer, Heidelberg (2009)
19. Ontology Schema (RDFS) for a subset of the X3D language, http://cs.swan.ac.uk/-csowen/SVGtoX3D/examples/X3DOntologyRDFS.htm
20. Pittarello, F., De Faveri, A.: Semantic Description of 3D Environments: A Proposal Based on Web Standarts. In: Proc. Intl. Web-3D Conf., ACM Press, New York (2006)
21. Polowinski, J.: SemVisHistory. Visualisierung von Semantic-Web-Geschichtsdaten in 3D (2007), netzspannung.org/database/401953/de
22. Rao, A.S., Georgeff, M.P.: BDI Agents: From Theory to Practice. In: Proc. Intl. Conf. on Multi-Agent Systems (ICMAS). AAAI Press, Menlo Park (1995)
23. Rickel, J., Johnson, W.L.: Task-oriented collaboration with embodied agents in virtual worlds. In: Embodied Conversational Agents. MIT Press, Cambridge (2001)
24. Rubinstein, D., Georgiev, I., Schug, B., Slussalek, P.: RTSG: Ray Tracing for X3D via a Flexible Rendering Framework. In: Proc. 14th Intl. Conf. on Web3D Technology. ACM Press, New York (2009)
25. Smart, J., Cascio, J., Paffendorf, J.: Metaverse Roadmap - Pathways to the 3D Web (2007), www.metaverseroadmap.org/MetaverseRoadmapOverview.pdf
26. Soma, R.: Parallel Inferencing for OWL Knowledge Bases. In: Proc. 37th Intl. Conf. Parallel Processing. IEEE, Los Alamitos (2008)
27. Sons, K., Klein, F., Rubinstein, D., Byelozyorov, S., Slusallek, P.: XML3D - Interactive 3D Graphics for the Web. In: Proceedings of Web3D 2010 Conference (2010)
28. Urbani, J., Kotoulas, S., Oren, E., van Harmelen, F.: Scalable Distributed Reasoning using MapReduce. In: Bernstein, A., Karger, D.R., Heath, T., Feigenbaum, L., Maynard, D., Motta, E., Thirunarayan, K. (eds.) ISWC 2009. LNCS, vol. 5823, pp. 634–649. Springer, Heidelberg (2009)
29. Walczak, A., Braubach, L., Pokahr, A., Lamersdorf, W.: Augmenting BDI Agents with Deliberative Planning Techniques. In: Proc. 5th Intl. Workshop on Programming Multiagent Systems (2006)
30. Warwas, S., Hahn, C.: The DSML4MAS Development Environment. In: Proc. 8th Int. Conf. on Autonomous Agents and Multiagent Systems (2009)
31. Zahariadis, T., Daras, P., Laso-Ballesteros, I.: Towards Future 3D Media Internet. In: NEM Summit 2008, St. Malo (2008), http://www.ist-sea.eu/Dissemination/SEA_FIA.pdf

# ORE - A Tool for Repairing
# and Enriching Knowledge Bases

Jens Lehmann and Lorenz Bühmann

AKSW research group, University of Leipzig, Germany
lastname@informatik.uni-leipzig.de

**Abstract.** While the number and size of Semantic Web knowledge bases increases, their maintenance and quality assurance are still difficult. In this article, we present ORE, a tool for repairing and enriching OWL ontologies. State-of-the-art methods in ontology debugging and supervised machine learning form the basis of ORE and are adapted or extended so as to work well in practice. ORE supports the detection of a variety of ontology modelling problems and guides the user through the process of resolving them. Furthermore, the tool allows to extend an ontology through (semi-)automatic supervised learning. A wizard-like process helps the user to resolve potential issues after axioms are added.

## 1 Introduction

Over the past years, the number and size of knowledge bases in the Semantic Web has increased significantly, which can be observed in various ontology repositories and the LOD cloud[1]. One of the remaining major challenges is, however, the maintenance of those knowledge bases and the use of expressive language features of the standard web ontology language OWL.

The goal of the ORE (Ontology Repair and Enrichment) tool[2] is to provide guidance for knowledge engineers who want to detect problems in their knowledge base and repair them. ORE also provides suggestions for extending a knowledge base by using supervised machine learning on the instance data in the knowledge base. ORE takes the web aspect of the Semantic Web into account by supporting large Web of Data knowledge bases like OpenCyc and DBpedia.

The main contributions of the article are as follows:

- provision of a free tool for repairing and extending ontologies
- implementation and combination of state-of-the-art inconsistency detection, ranking, and repair methods
- use of supervised learning for extending an ontology
- support for very large knowledge bases available as Linked Data or via SPARQL endpoints
- application tests of ORE on real ontologies

---

[1] http://linkeddata.org

[2] See http://dl-learner.org/wiki/ORE and download at
http://sourceforge.net/projects/dl-learner/files/

P.F. Patel-Schneider et al.(Eds.): ISWC 2010, Part II, LNCS 6497, pp. 177–193, 2010.

The article is structured as follows: In Section 2, we cover the necessary foundations in the involved research disciplines such as description logics (DLs), ontology debugging, and learning in OWL. Section 3 describes how ontology debugging methods were implemented and adapted in ORE. Similarly, Section 4 shows how an existing framework for ontology learning was incorporated. In Section 5, we describe the structure of the ORE user interface. The evaluation of both, the repair and enrichment part, is given in Section 6. Related work is presented in Section 7 followed by our final conclusions in Section 8.

## 2    Preliminaries

We give a brief introduction into DLs and OWL as the underlying formalism, recapitulate the state of the art in ontology debugging and give the definition of the class learning problem in ontologies.

### 2.1    Description Logics and OWL

DLs are usually decidable fragments of first order logic and have a variable-free syntax. The standard ontology language OWL 2 is based on the DL $\mathcal{SROIQ}$. We briefly introduce it and refer to [12] for details.

In $\mathcal{SROIQ}$, three sets are used as the base for modelling: *individual* names $N_I$, *concept* names $N_C$ (called classes in OWL), and *role* names (object properties) $N_R$. By convention, we will use $A, B$ (possibly with subscripts) for concept names, $r$ for role names, $a$ for individuals, and $C, D$ for complex concepts. Using those basic sets, we can inductively build complex concepts using the following constructors:

$$A \mid \top \mid \bot \mid \{a\} \mid C \sqcap D \mid C \sqcup D$$
$$\mid \exists r.\texttt{Self} \mid \exists r.C \mid \forall r.C \mid \leq n\, r.c \mid \geq n\, r.C$$

For instance, $\texttt{Man} \sqcap \exists \texttt{hasChild}.\texttt{Female}$ is a complex concept describing a man who has a daughter. A *DL knowledge base* consists of a set of *axioms*. The signature of a knowledge base (an axiom $\alpha$) is the set $\mathbf{S}$ ($Sig(\alpha)$) of atomic concepts, atomic roles and individuals that occur in the knowledge base (in $\alpha$). We will only mention two kinds of axioms explicitly: Axioms of the form $C \sqsubseteq D$ are called *general inclusion axioms*. An axiom of the form $C \equiv D$ is called *equivalence axiom*. In the special case that $C$ is a concept name, we call the axiom a *definition*.

Apart from *explicit* knowledge, we can deduce *implicit* knowledge from a knowledge base. *Inference/reasoning algorithms* extract such implicit knowledge. Typical reasoning tasks are:

- instance check $\mathcal{K} \models C(a)$? (Does $a$ belong to $C$?)
- retrieval $R_\mathcal{K}(C)$? (Determine all instances of $C$.)
- subsumption $C \sqsubseteq_\mathcal{K} D$? (Is $C$ more specific than $D$?)
- inconsistency $\mathcal{K} \models$ false? (Does $\mathcal{K}$ contain contradictions?)
- satisfiability $C \equiv_\mathcal{K} \bot$? (Can $C$ have an instance?)
- incoherence $\exists C\ (C \equiv_\mathcal{K} \bot)$? (Does $K$ contain an unsatisfiable class?)

Throughout the paper, we use the words ontology and knowledge base as well as complex concept and class expression synonymously.

## 2.2  Ontology Debugging

Finding and understanding undesired entailments such as unsatisfiable classes or inconsistency can be a difficult or impossible task without tool support. Even in ontologies with a small number of logical axioms, there can be several, non-trivial causes for an entailment. Therefore, interest in finding explanations for such entailments has increased in recent years. One of the most usual kinds of explanations are *justifications* [15]. A justification for an entailment is a minimal subset of axioms with respect to a given ontology, that is sufficient for the entailment to hold. More formally, let $\mathcal{O}$ be a given ontology with $\mathcal{O} \models \eta$, then $\mathcal{J}$ is a justification for $\eta$ if $\mathcal{J} \models \eta$, and for all $\mathcal{J}' \subset \mathcal{J}$, $\mathcal{J}' \not\models \eta$. In the meantime, there is support for the detection of potentially overlapping justifications in tools like Protégé[3] and Swoop[4]. Justifications allow the user to focus on a small subset of the ontology for fixing a problem. However, even such a subset can be complex, which has spurred interest in computing *fine-grained* justifications [11] (in contrast to *regular* justifications). In particular, *laconic justifications* are those where the axioms do not contain superfluous parts and are as weak as possible. A subset of laconic justifications are *precise justifications*, which split larger axioms into several smaller axioms allowing minimally invasive repair.

A possible approach to increase the efficiency of computing justifications is module extraction [6]. Let $\mathcal{O}$ be an ontology and $\mathcal{O}' \subseteq \mathcal{O}$ a subset of axioms of $\mathcal{O}$. $\mathcal{O}'$ is a module for an axiom $\alpha$ with respect to $\mathcal{O}$ if: $\mathcal{O}' \models \alpha$ iff $\mathcal{O} \models \alpha$. $\mathcal{O}'$ is a module for a signature **S** if for every axiom $\alpha$ with $Sig(\alpha) \subseteq$ **S**, we have that $\mathcal{O}'$ is a module for $\alpha$ with respect to $\mathcal{O}$. Intuitively, a module is an ontology fragment, which contains all relevant information in the ontology with respect to a given signature. One possibility to extract such a module is syntactic locality [6]. [30] showed that such *locality-based modules* contain all justifications with respect to an entailment and can provide order-of-magnitude performance improvements.

## 2.3  The Class Learning Problem

The process of learning in logics, i.e. trying to find high level explanations for given data, is also called *inductive reasoning* as opposed to the deductive reasoning tasks we have introduced. The main difference is that in deductive reasoning it is formally shown whether a statement follows from a knowledge base, whereas in inductive learning we invent new statements. Learning problems, which are similar to the one we will analyse, have been investigated in *Inductive Logic Programming* [27] and, in fact, the method presented here can be used to solve a variety of machine learning tasks apart from ontology engineering.

The considered supervised ontology learning problem is an adaption of the problem in *Inductive Logic Programming*. We learn a formal description of a class $A$ from inferred instances in the ontology. Let a class name $A \in N_C$ and an ontology $\mathcal{O}$ be given. We define the *class learning problem* as finding a class expression $C$ such that $R_{\mathcal{O}}(C) = R_{\mathcal{O}}(A)$, i.e. $C$ covers exactly all instances of $A$.

---

[3] http://protege.stanford.edu
[4] http://www.mindswap.org/2004/SWOOP/

Clearly, the learned concept $C$ is a description of (the instances of) $A$. Such a concept is a candidate for adding an axiom of the form $A \equiv C$ or $A \sqsubseteq C$ to the knowledge base $\mathcal{K}$. This is used in the enrichment step in ORE as we will later describe. In the case that $A$ is described already via axioms of the form $A \sqsubseteq C$ or $A \equiv C$, those can be either modified, i.e. specialised/generalised, or relearned from scratch by learning algorithms.

Machine learning algorithms usually prefer those solutions of a learning problem, which are likely to classify unknown individuals well. For instance, using nominals (owl:oneOf) to define the class $A$ above as the set of its current instances is a correct solution of the learning problem, but would classify all individuals, which are added to the knowledge base later as not being instance of $A$. In many cases, the learning problem is not perfectly solvable apart from the trivial solution using nominals. In this case, approximations can be given by ML algorithms. It is important to note that a knowledge engineer usually makes the final decision on whether to add one of the suggested axioms, i.e. candidate concepts are presented to the knowledge engineer, who can then select and possibly refine one of them.

## 3  Ontology Repair

For a single entailment, e.g. an unsatisfiable class, there can be many justifications. Moreover, in real ontologies, there can be several unsatisfiable classes or several reasons for inconsistency. While the approach described in Section 2.2 works well for small ontologies, it is not feasible if a high number of justifications or large justifications have to be computed. Due to the relations between entities in an ontology, several problems can be intertwined and are difficult to separate. We briefly describe how we handle these problems in ORE.

*Root Unsatisfiability.* For the latter problem mentioned above, an approach [18] is to separate between root and derived unsatisfiable classes. A derived unsatisfiable class has a justification, which is a proper super set of a justification of another unsatisfiable class. Intuitively, their unsatisfiability may depend on other unsatisfiable classes in the ontology, so it can be beneficial to fix those root problems first. There are two different approaches for determining such classes: The first approach is to compute all justifications for each unsatisfiable class and then apply the definition. The second approach relies on a structural analysis of axioms and heuristics. Since the first approach is computationally too expensive for larger ontologies, we use the second strategy as default in ORE. The implemented approach is sound, but incomplete, i.e. not all class dependencies are found, but the found ones are correct. To increase the proportion of found dependencies, the TBox is modified in a way which preserves the subsumption hierarchy to a large extent. It was shown in [18] that this allows to draw further entailments and improve the pure syntactical analysis.

*Axiom Relevance.* Given a justification, the problem needs to be resolved by the user, which involves the deletion or modification of axioms in it. To assist the user, ranking methods, which highlight the most probable causes for problems, are important. Common methods (see [16] for details) are frequency (How often does the axiom appear in justifications?), syntactic relevance (How deeply rooted is an axiom in the ontology?)

and semantic relevance (How many entailments are lost or added?[5]). ORE supports all metrics and a weighted aggregation of them. For computing semantic relevance, ORE uses the incremental classification feature of Pellet, which uses locality-based modules. Therefore, only the relevant parts of the ontology are reclassified when determining the effect of changes.

*Consequences of Repair Step.* Repairing a problem involves editing or deleting an axiom. Deletion has the technical advantage that it does not lead to further entailments due to the monotonicity of DLs. However, desired entailments may be lost. In contrast, editing axioms allows to make small changes, but it may lead to new entailments, including inconsistencies. To support the user, ORE provides fine-grained justifications, which only contain relevant parts of axioms and, therefore, have minimal impact on deletion. Furthermore, ORE allows to preview new or lost entailments. The user can then decide to preserve them, if desired.

*Workflow.* The general workflow of the ontology repair process is depicted in Figure 1. First, all inconsistencies are resolved. Secondly, unsatisfiable classes are handled by computing root unsatisfiable classes, as well as regular and laconic justifications, and different ranking metrics.

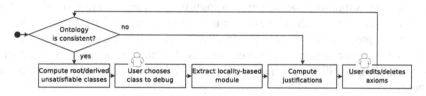

**Fig. 1.** Workflow for debugging an ontology in ORE

*Web of Data and Scalability.* In order to apply ORE to existing very large knowledge bases in the Web of Data, the tool supports using SPARQL endpoints instead of local OWL files as input knowledge bases. To perform reasoning on those knowledge bases, ORE implements an incremental load procedure inspired by [9].

Using SPARQL queries, the knowledge base is loaded in small chunks. In the first step, ORE determines the size of the knowledge base by determining the number of all types of OWL 2 axioms. In the main part of the algorithm, a priority based loading procedure is used. This means that axioms that are empirically more likely to cause inconsistencies in the sense that they are often part of justifications have a higher priority. In general, schema axioms have a higher loading priority than instance data. Before loading parts of the instance data, the algorithm performs sanity checks on the data, i.e. performs a set of simple SPARQL queries, which probe for inconsistent axiom sets. These cases include individuals, which are instances of disjoint classes, properties which are used on instances incompatible with their domain, etc. The algorithm can also be configured to fetch additional information via Linked Data such that consistency

---

[5] Since the number of entailed axioms can be infinite, we restrict ourselves to a subset of axioms as suggested in [16].

of a knowledge base in combination with knowledge from another knowledge base can be tested.

The algorithm converges towards loading the whole knowledge base into the reasoner, but can also be configured to stop automatically after the schema part and sample instances, based on ABox summarisation techniques, of all classes have been loaded. This is done to prevent a too high load on SPARQL endpoints and the fact that most knowledge bases cannot be loaded into standard OWL reasoners on typical hardware available. At the moment, the algorithm uses the incremental reasoning feature available in Pellet such that it is not required to reload the reasoner each time a chunk of data has been received from the SPARQL endpoint.

The general idea behind this component of ORE is to apply state-of-the-art reasoning methods on a larger scale than was possible previously. We show this by applying ORE on OpenCyc and DBpedia in Section 6.3. To the best of our knowledge, none of the existing tools can compute justifications for inconsistencies on those large knowledge bases. This part of ORE aims at stronger support for the "web aspect" of the Semantic Web and the high popularity of Web of Data initiative.

## 4   Ontology Enrichment

Currently, ORE supports enriching an ontology with axioms of the form $A \equiv C$ and $A \sqsubseteq C$. For suggesting such an axiom, we use the DL-Learner framework to solve the class learning problem described in Section 2.3. In particular, we use the CELOE algorithm in DL-Learner, which is optimised for class learning in an ontology engineering scenario. It is a specialisation of the OCEL algorithm [24], which was shown to be very competitive.

The main task of ORE is to provide an interface to the algorithm and handle the consequences of adding a suggested axiom. In this section, we will focus on the latter problem. The learning algorithm can produce false positives as well as false negatives, which can lead to different consequences. In the following, assume $\mathcal{O}$ to be an ontology and $A$ the class for which a definition $A \equiv C$ was learned. Let $n$ be a false positive, i.e. $\mathcal{O} \not\models A(n)$ and $\mathcal{O} \models C(n)$. We denote the set of justifications for $\mathcal{O} \models \eta$ with $\mathcal{J}_\eta$. ORE would offer the following options in this case:

1. assign $n$ to class $A$
2. completely delete $n$ in $\mathcal{O}$
3. modify assertions about $n$ such that $\mathcal{O} \not\models C(n)$: In a first step, ORE uses several reasoner requests to determine the part $C'$ of $C$, which is responsible for classifying $n$ as instance of $C$. The algorithm recursively traverses conjunctions and disjunctions until it detects one of the class constructors below.
   - $C' = B\ (B \in N_C)$: Remove the assignment of $n$ to $B$, i.e. delete at least one axiom in each justification $J \in \mathcal{J}_{B(n)}$
   - $C' = \forall r.D$: Add at least one axiom of the form $r(n, a)$ where $a$ is not an instance of $D$
   - $C' = \exists r.D$:
     (a) Remove all axioms of the form $r(n, a)$, where $a$ is an instance of $D$

(b) Remove all axioms of the form $r(n, a)$

- $C' =\leq mr.D$: Add axioms of the form $r(n, a)$, in which $a$ is instance of $D$, until their number is greater than $m$
- $C' =\geq mr.D$: Remove axioms of the form $r(n, a)$, where $a$ is instance of $D$, until their number is smaller than $m$

The steps above are an excerpt of the provided functionality of ORE. False negatives are treated in a similar fashion. The strategy is adapted in case of learning superclass axioms ($A \sqsubseteq C$). Those steps, where axioms are added, can naturally lead to inconsistencies. In such a case, a warning is displayed. If the user chooses to execute the action, the ORE wizard can return to the inconsistency resolution step described in Section 2.2.

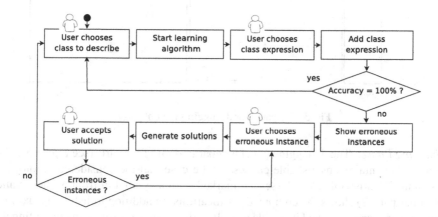

**Fig. 2.** Workflow for enriching an ontology in ORE

*Workflow.* The workflow for the enrichment process is shown in Figure 2. First, the user selects a class for which he wants to learn a description. Alternatively, ORE can loop over all classes and provides particularly interesting suggestions to the user. ORE calls the CELOE learning algorithm and presents the 10 best suggestions to the user. If the user decides to accept a suggestion and if there are false positives or negatives, possible repair solutions are provided.

## 5   User Interface

In ORE, we decided to use a wizard-based user interface approach. This allows a user to navigate through the dialogues step-by-step, while the dependencies between different steps are factored in automatically. This enables the user to perform the repair and enrichment process with only a few clicks and a low learning curve. Changes can be rolled back if necessary. The design of ORE ensures that it can be embedded in ontology editors. Below, we describe the most important parts of the ORE wizard.

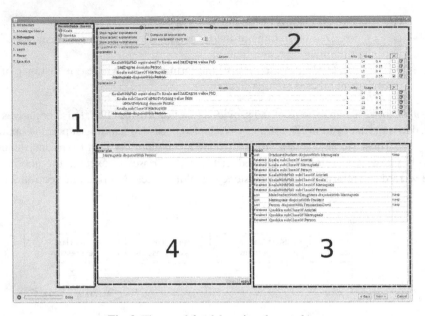

**Fig. 3.** The panel for debugging the ontology

*Debugging Phase.* The debugging panel is separated into four parts (see Figure 3): The left part (1) contains unsatisfiable classes for the case that the considered ontology is consistent. Unsatisfiable root classes are marked with a symbol in front of their name. The upper part (2) shows the computed justifications. In addition to listing the axioms, several metrics are displayed in a table as well as the actions for removing or editing the axiom. The axioms are displayed in Manchester OWL Syntax[6]. To increase readability, key words are emphasised and the axioms are indented. Configuration options allow to set the maximum number of explanations, which should be displayed, and their type (regular/laconic). In (3), lost or added entailments, as a consequence of the selected modifications, are displayed. This part of the user interface allows to preserve those entailments, if desired. Part (4) of the debugging panel lists the axioms, which will be added or removed. Each action can be undone. When a user is satisfied with the changes made, they can execute the created repair plan, which results in the actual modification of the underlying ontology.

*Enrichment Phase.* For the enrichment phase, the panel is separated into three parts (see Figure 4). The right part (2) allows to start or stop the underlying machine learning algorithm, the configuration of it, and the selection whether equivalent or superclass axioms should be learned. In part (1), the learned expressions are displayed. For each class expression, a heuristic accuracy value provided by the underlying algorithm is displayed. When a class expression is selected, an illustration of its coverage is shown in part (3). The illustration is generated by analysing the instances covered by the class expression and comparing it to the instance of the current named class.

---

[6] http://www.w3.org/2007/OWL/wiki/ManchesterSyntax

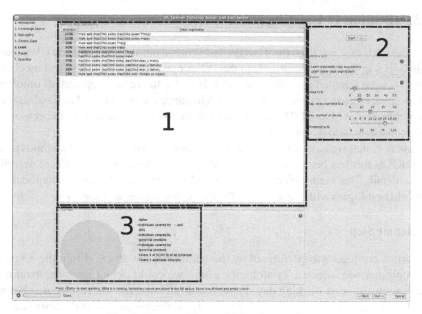

**Fig. 4.** The panel for enriching the ontology

*Repair of Individuals.* Enriching the ontology can have consequences on the classification of individuals in the ontology. For repairing unwanted consequences, a dialogue (see Figure 5) is displayed, which is separated in three parts. The upper part (1) shows the class expression itself. As briefly described in Section 4, the parts of the expression, which cause the problem, are highlighted. Clicking on such a part of an expression, opens a menu, which provides repair suggestions. The middle part (2) displays information about the individual, which is currently repaired. This allows the ontology engineer to observe relevant information at a glance. The lower part (3) lists the repair decisions made and provides an undo method.

**Fig. 5.** The panel for repairing an erroneous instance

# 6    Application to Existing Knowledge Bases

To test the ORE tool, we used the TONES and Protégé ontology repositories. We loaded all ontologies in those repositories into the Pellet reasoner. Inconsistent and incoherent ontologies were selected as evaluation candidates for the repair step and all ontologies which contain at least 5 classes with at least 3 instances, were selected as candidates for the enrichment step. Out of 216 ontologies which could be loaded into the reasoner, 3 were inconsistent, and 32 were incoherent.

Please note that we have not performed an extensive evaluation of all methods underlying ORE as this has been done in the cited articles, where the methods are described in more detail. The main objective was to find out whether the tool is applicable to real-world ontologies with respect to usability, performance, and stability.

## 6.1    Repair Step

This part of our tests was performed by the authors of the article. From the 35 candidate ontologies, we selected 7 ontologies where we could obtain an understanding of the domain within one working day. These ontologies and the test results are shown in Table 1. We used ORE to resolve all occurring problems and, overall, resolved 1 inconsistency and 135 unsatisfiable classes. Generally, the ontologies could be processed without problems and the performance for computing justifications was sufficient. The maximum time required per justification was one second.

**Table 1.** Repair of ontologies from the Protégé and TONES repositories

| ontology | resolved inconsistency | #resolved unsatisfiable classes | #removed axioms | #added axioms | #changed axioms |
|---|---|---|---|---|---|
| http://protege.cim3.net/file/pub/ontologies/camera/**camera.owl** | yes | - | 0 | 0 | 2 |
| http://protege.cim3.net/file/pub/ontologies/koala/**koala.owl** | - | 3 | 3 | 0 | 0 |
| http://reliant.teknowledge.com/DAML/**Economy.owl** | - | 51 | 11 | 5 | 0 |
| http://www.cs.man.ac.uk/ horridgm/ontologies/complexity/**UnsatCook.owl** | - | 8 | 1 | 0 | 0 |
| http://www.co-ode.org/ontologies/pizza/2007/02/12/**pizza.owl** | - | 2 | 3 | 0 | 0 |
| http://www.mindswap.org/ontologies/debugging/**University.owl** | - | 9 | 3 | 1 | 2 |
| http://reliant.teknowledge.com/DAML/**Transportation.owl** | - | 62 | 15 | 28 | 0 |

## 6.2    Enrichment Step

The test of the enrichment step was done by two researchers, who made themselves familiar with the domain of the test ontologies. We are aware that an ideal evaluation procedure would require OWL knowledge engineers from the respective domains, e.g. different areas within biology, medicine, finance, and geography. Considering the budget limitations, however, we believe that our method is sufficient to be able to meet

**Table 2.** Test results on several ontologies. On average, suggestions by the ML algorithm were accepted in 60% of all cases.

| ontology | #logical axioms | #suggestion lists | accept (1) in % | reject (2) in % | fail (3) in % | selected position on suggestion list (incl. std. deviation) | #hidden inconsistencies | #additional instances |
|---|---|---|---|---|---|---|---|---|
| http://www.mindswap.org/ontologies/**SC.owl** | 20081 | 12 | 79 | 21 | 0 | 2.2±2.1 | 0 | 1771 |
| http://www.fadyart.com/ontologies/data/**Finance.owl** | 16057 | 50 | 52 | 48 | 0 | 3.6±2.6 | 0 | 1162 |
| http://www.biopax.org/release/**biopax-level2.owl** | 12381 | 34 | 78 | 22 | 0 | 2.7±2.2 | 1 | 803 |
| http://i2geo.net/ontologies/dev/**GeoSkills.owl** | 8803 | 180 | 56 | 44 | 0 | 1.6±1.2 | 1 | 295 |
| http://reliant.teknowledge.com/DAML/**Economy.owl** | 1625 | 22 | 74 | 26 | 0 | 1.5±0.9 | 0 | 77 |
| http://www.acl.icnet.uk/ mw/**MDM0.73.owl** | 884 | 77 | 56 | 44 | 0 | 3.7±2.6 | 1 | 82 |
| http://www.co-ode.org/ontologies/.../**eukariotic.owl** | 38 | 8 | 91 | 9 | 0 | 2.5±1.2 | 0 | 7 |

our basic test objectives for the first releases of ORE. Each researcher worked independently and had to make 383 decisions, as described below. The time required to make those decisions was 40 working hours per researcher.

From those ontologies obtained in the pre-selection step, described at the beginning of this section, we picked ontologies, which vary in size and complexity. We wanted to determine whether 1.) the underlying adapted learning algorithm is useful in practice, i.e. is able to make sensible suggestions, 2.) to which extent additional information can be inferred when enriching ontologies with suggestions by the learning algorithm (described as *hidden inconsistencies* and *additional instances* below).

We ran ORE in an evaluation mode, which works as follows: For each class $A$, the learning method generates at most ten suggestions with the best ones on top of the list. This is done for learning superclasses ($A \sqsubseteq C$) and equivalent classes ($A \equiv C$) separately. If the accuracy of the best suggestion exceeds a defined threshold, we suggest them to the knowledge engineer. The knowledge engineer then has three options to choose from: 1. pick one of the suggestions by entering its number (accept), 2. declare that there is no sensible suggestion for $A$ in his opinion (reject), or 3. declare that there is a sensible suggestion, but the algorithm failed to find it (fail). If the knowledge engineer decides to pick a suggestion, we query whether adding it leads to an inconsistent ontology. We call this case the discovery of a *hidden inconsistency*, since it was present before, but can now be formally detected and treated. We also measure whether adding the suggestion increases the number of inferred instances of $A$. Being able to infer *additional instances* of $A$, therefore, provides added value (see also the notion of *induction rate* as defined in [5]).

We used the default settings of 5% noise and an execution time of 10 seconds for the algorithm. The evaluation machine was a notebook with a 2 GHz CPU and 3 GB RAM. Table 2 shows the evaluation results.

*Objective 1:* We can observe that the researchers picked option 1 (accept) most of the time, i.e. in many cases the algorithm provided meaningful suggestions. This allows us to answer the first evaluation objective positively. The researchers never declared that the algorithm failed on finding a potential solution. The 7th column shows that many selected expressions are amongst the top 5 (out of 10) in the suggestion list, i.e. providing 10 suggestions appears to be a reasonable choice.

*Objective 2:* In 3 cases a hidden inconsistency was detected. Both researchers independently coincided on those decisions. The last column shows that in all ontologies additional instances could be inferred for the classes to describe if the new axiom would be added to the ontology after the learning process. Overall, being able to infer additional instances was very common and hidden inconsistencies could sometimes be detected.

### 6.3    Very Large Knowledge Bases

We applied ORE to two very large knowledge bases: DBpedia [21] (live version [10]) and OpenCyc. DBpedia is a knowledge base extracted from Wikipedia in a joint effort of the University of Leipzig, the Free University of Berlin and the company OpenLink. It contains descriptions of over 3.4 million entities out of which 1.5 million are classified in the DBpedia ontology. Overall, the DBpedia knowledge base consists of more than one billion triples with more than 250 million triples in the English language edition. OpenCyc is a part of the Cyc artificial intelligence project started in 1984, which provides a huge knowledge base of common sense knowledge. In its current OWL version, it contains more than 50 thousand classes, 20 thousand properties, 350 thousand invididuals. OpenCyc has a sophisticated and large schema, while DBpedia has a smaller schema part and a huge amount of instance data.

*Application to DBpedia.* Most reasoning on DBpedia focuses on very light-weight reasoning techniques, which are usually employed within triple stores like OpenLink Virtuoso. Standard OWL reasoners are not able to load or reason within DBpedia. However, the incremental approach sketched in Section 3 allows to apply standard reasoners to DBpedia, detect inconsistencies and compute justifications with only moderate hardware requirements. Two justifications in Manchester Syntax are shown below[7]:

*Example 1 (Incorrect Property Range in DBpedia).*
Individual: dbr:Purify_%28album%29 Facts: dbo:artist dbr:Axis_of_Advance
Individual: dbr:Axis_of_Advance Types: dbo:Organisation
Class: dbo:Organisation DisjointWith dbo:Person
ObjectProperty: dbo:artist Range: dbo:Person

ORE found an assertion that "Axis of Advance" created the album "Purify". DBpedia states that the range of the "artist" property is a person, hence "Axis of Advance" must be a person. However, it is an organisation (a music band) and organisations and persons are disjoint, so we get a contradiction. In this example, the problem can be resolved by generalising the range of "artist", i.e. not requiring it to be a person.

*Example 2 (DBpedia Incompatible with External Ontology).*
Individual: dbr:WKWS Facts: geo:long -81.76833343505859

---

[7] Used prefixes: dbr = http://dbpedia.org/resource/, dbo = http://dbpedia.org/ontology/, geo = http://www.w3.org/2003/01/geo/wgs84_pos#

Types: dbo:Organisation
DataProperty: geo:long Domain: geo:SpatialThing
Class: dbo:Organisation DisjointWith: geo:SpatialThing

In this example, the longitude property is used on an organisation, which is a contradiction, because an organisation is itself not a spatial entity. The interesting aspect of this example is that information from an external knowledge base (W3C Geo) is fetched via Linked Data, which is an optional feature of ORE. The inconsistency only arises in combination with this external knowledge.

*Application to OpenCyc.* OpenCyc is very large, but still loadable in standard OWL reasoners. However, only few reasoners can detect that it is not consistent. In our experiments, only HermiT 1.2.3 was able to do this given sufficient memory. Nevertheless, computing actual justifications is still not possible when considering the whole knowledge base and could only be achieved using the incremental priority-based load procedure in ORE. Below is an inconsistency detected by ORE in "label view", i.e. the value of rdfs:label is shown instead of the URIs:

*Example 3 (Class Hierarchy Problems in OpenCyc).*
Individual: 'PopulatedPlace' Types: 'ArtifactualFeatureType', 'ExistingStuffType'
Class: 'ExistingObjectType' DisjointWith: 'ExistingStuffType'
Class: 'ArtifactualFeatureType' SubClassOf: 'ExistingObjectType'

The example shows a problem in OpenCyc, where an individual is assigned to classes, which can be inferred to be disjoint via the class hierarchy. (Note that "PopulatedPlace" is used as individual and class in OpenCyc, which is allowed in OWL2.)

## 7 Related Work

The growing interest in Semantic Technologies has led to an increasing number of ontologies, which has, in turn, spurred interest in techniques for ontology creation and maintenance. In [28] and [8], methods for the detection and repair of inconsistencies in frequently changing ontologies were developed. [29] discusses a method for axiom pinpointing, i.e. the detection of axioms responsible for logical errors. A non proof-theoretic method is used in OntoClean [7]. By adding meta-properties (rigidity, identity, dependency) to each class, problems in the knowledge base taxonomy could be identified by using rules. Classes could then be moved in the hierarchy or additional ones can be added. OntoClean supports resolving taxonomy errors, but was not designed for detecting logical errors.

The work on the enrichment part of ORE goes back to early work on supervised learning in DLs, e.g. [3], which used so-called least common subsumers to solve the learning problem (a modified variant of the problem defined in this article). Later, [2] invented a refinement operator for $\mathcal{ALER}$ and proposed to solve the problem by using a top-down approach. [4,13] combine both techniques and implement them in the YINYANG tool. However, those algorithms tend to produce very long and hard-to-understand class expressions, which are often not appropriate in an ontology enrichment context. Therefore, ORE is based on DL-Learner [20], which allows to select between a variety of learning algorithms [22,23,19,24]. Amongst them, CELOE is particularly optimised for learning easy to understand expressions. DL-FOIL [5] is a similar approach mixing upward and downward refinement. Other approaches focus on learning in hybrid language settings [26].

In (semi-)automatic ontology engineering, formal concept analysis [1] and relational exploration [32] have been used for completing knowledge bases. [33] focuses on learning disjointness between classes in an ontology to allow for more powerful reasoning and consistency checking. Naturally, there is also a lot of research work on ontology learning from text. The most closely related approach in this area is [31], in which OWL DL axioms are obtained by analysing sentences which have definitorial character.

There are a number of related tools for ontology repair:

**Swoop**[8][17] is a Java-based ontology editor using web browser concepts. It can compute justifications for the unsatisfiability of classes and offers a repair mode. The fine-grained justification computation algorithm is, however, incomplete. Swoop can also compute justifications for an inconsistent ontology, but does not offer a repair mode like ORE in this case. It does not extract locality-based modules, which leads to lower performance for large ontologies.

**RaDON**[9][14] is a plugin for the NeOn toolkit. It offers a number of techniques for working with inconsistent or incoherent ontologies. It can compute justifications and, similarly to Swoop, offers a repair mode. RaDON also allows to reason with inconsistent ontologies and can handle sets of ontologies (ontology networks). Compared to ORE, there is no feature to compute fine-grained justifications, and the user gets no informations about the impact of repair.

**Pellint**[10][25] is a Lint-based tool, which searches for common patterns which lead to potential reasoning performance problems. In future work, we plan to integrate support for detecting and repairing reasoning performance problems in ORE.

**PION and DION**[11] have been developed in the SEKT project to deal with inconsistencies. PION is an inconsistency tolerant reasoner, i.e. it can, unlike standard reasoners, return meaningful query answers in inconsistent ontologies. To achieve this, a four-valued paraconsistent logic is used. DION offers the possibility to compute justifications, but cannot repair inconsistent or incoherent ontologies.

**Explanation Workbench**[12] is a Protégé plugin for reasoner requests like class unsatisfiability or inferred subsumption relations. It can compute regular and laconic justifications [11], which contain only those axioms which are relevant for answering the particular reasoner request. This allows to make minimal changes to resolve potential problems. We adapted its layout for the ORE debugging interface. Unlike ORE, the current version of Explanation Workbench does not allow to remove axioms in laconic justifications.

---

[8] SWOOP: http://www.mindswap.org/2004/SWOOP/
[9] RaDON: http://radon.ontoware.org/demo-codr.htm
[10] PellInt: http://pellet.owldl.com/pellint
[11] PION: http://wasp.cs.vu.nl/sekt/pion/
   DION: http://wasp.cs.vu.nl/sekt/dion/
[12] http://owl.cs.manchester.ac.uk/explanation/

Most of those tools were designed to detect logical errors or ignore them (PION). PellInt is an exception because it detects problems relevant for reasoning performance. The ORE tool unites several techniques present in those tools and combines them with the DL-Learner framework to enable suggestions for enrichment. It also enhances other tools by providing support for working on SPARQL endpoints and Linked Data.

## 8 Conclusions and Future Work

We have presented a freely available tool for ontology repair and enrichment. It integrates state-of-the-art methods from ontology debugging and supervised learning in OWL in an intuitive, wizard-like interface. It combines the advantages of other tools and provides new functionality like the enrichment part of the tool. An evaluation on real ontologies has shown the need for a repair and enrichment tool and, in particular, the benefits of ORE.

In future work, we aim at integrating support for further modelling problems apart from inconsistencies and unsatisfiable classes. Those problems will be ordered by severity reaching from logical problems to suggested changes for improving reasoner performance. We plan to improve the enrichment part by suggesting other types of axioms, e.g. disjointness. We also plan to evaluate and optimise the SPARQL/Linked Data component of ORE. Possibly, we will provide an alternative web interface and appropriate hardware infrastructure for ORE such that it can be used for online analysis of Web of Data knowledge bases. In addition to those features, a constant evaluation of the underlying methods will be performed to improve the foundations of ORE.

## References

1. Baader, F., Ganter, B., Sattler, U., Sertkaya, B.: Completing description logic knowledge bases using formal concept analysis. In: IJCAI 2007. AAAI Press, Menlo Park (2007)
2. Badea, L., Nienhuys-Cheng, S.-H.: A refinement operator for description logics. In: Cussens, J., Frisch, A.M. (eds.) ILP 2000. LNCS (LNAI), vol. 1866, pp. 40–59. Springer, Heidelberg (2000)
3. Cohen, W.W., Hirsh, H.: Learning the CLASSIC description logic: Theoretical and experimental results. In: KR 1994, pp. 121–133. Morgan Kaufmann, San Francisco (1994)
4. Esposito, F., Fanizzi, N., Iannone, L., Palmisano, I., Semeraro, G.: Knowledge-intensive induction of terminologies from metadata. In: McIlraith, S.A., Plexousakis, D., van Harmelen, F. (eds.) ISWC 2004. LNCS, vol. 3298, pp. 441–455. Springer, Heidelberg (2004)
5. Fanizzi, N., d'Amato, C., Esposito, F.: DL-FOIL concept learning in description logics. In: Železný, F., Lavrač, N. (eds.) ILP 2008. LNCS (LNAI), vol. 5194, pp. 107–121. Springer, Heidelberg (2008)
6. Grau, B.C., Horrocks, I., Kazakov, Y., Sattler, U.: Modular reuse of ontologies: Theory and practice. J. Artif. Intell. Res (JAIR) 31, 273–318 (2008)
7. Guarino, N., Welty, C.A.: An overview of ontoclean. In: Staab, S., Studer, R. (eds.) Handbook on Ontologies, pp. 151–172. Springer, Heidelberg (2004)
8. Haase, P., van Harmelen, F., Huang, Z., Stuckenschmidt, H., Sure, Y.: A framework for handling inconsistency in changing ontologies. In: Gil, Y., Motta, E., Benjamins, V.R., Musen, M.A. (eds.) ISWC 2005. LNCS, vol. 3729, pp. 353–367. Springer, Heidelberg (2005)

9. Hellmann, S., Lehmann, J., Auer, S.: Learning of OWL class descriptions on very large knowledge bases. Int. Journal on Semantic Web and Information Systems 5(2), 25–48 (2009)

10. Hellmann, S., Stadler, C., Lehmann, J., Auer, S.: Dbpedia live extraction. In: Meersman, R., Dillon, T., Herrero, P. (eds.) OTM 2009. LNCS, vol. 5871, pp. 1209–1223. Springer, Heidelberg (2009)

11. Horridge, M., Parsia, B., Sattler, U.: Laconic and precise justifications in OWL. In: Sheth, A.P., Staab, S., Dean, M., Paolucci, M., Maynard, D., Finin, T., Thirunarayan, K. (eds.) ISWC 2008. LNCS, vol. 5318, pp. 323–338. Springer, Heidelberg (2008)

12. Horrocks, I., Kutz, O., Sattler, U.: The even more irresistible SROIQ. In: KR 2006, pp. 57–67. AAAI Press, Menlo Park (2006)

13. Iannone, L., Palmisano, I., Fanizzi, N.: An algorithm based on counterfactuals for concept learning in the semantic web. Applied Intelligence 26(2), 139–159 (2007)

14. Ji, Q., Haase, P., Qi, G., Hitzler, P., Stadtmüller, S.: Radon - repair and diagnosis in ontology networks. In: Aroyo, L., Traverso, P., Ciravegna, F., Cimiano, P., Heath, T., Hyvönen, E., Mizoguchi, R., Oren, E., Sabou, M., Simperl, E. (eds.) ESWC 2009. LNCS, vol. 5554, pp. 863–867. Springer, Heidelberg (2009)

15. Kalyanpur, A., Parsia, B., Horridge, M., Sirin, E.: Finding all justifications of OWL DL entailments. In: Aberer, K., Choi, K.-S., Noy, N., Allemang, D., Lee, K.-I., Nixon, L.J.B., Golbeck, J., Mika, P., Maynard, D., Mizoguchi, R., Schreiber, G., Cudré-Mauroux, P. (eds.) ASWC 2007 and ISWC 2007. LNCS, vol. 4825, pp. 267–280. Springer, Heidelberg (2007)

16. Kalyanpur, A., Parsia, B., Sirin, E., Grau, B.C.: Repairing unsatisfiable concepts in owl ontologies. In: Sure, Y., Domingue, J. (eds.) ESWC 2006. LNCS, vol. 4011, pp. 170–184. Springer, Heidelberg (2006)

17. Kalyanpur, A., Parsia, B., Sirin, E., Grau, B.C., Hendler, J.: Swoop: A web ontology editing browser. Journal of Web Semantics 4(2), 144–153 (2006)

18. Kalyanpur, A., Parsia, B., Sirin, E., Hendler, J.: Debugging unsatisfiable classes in OWL ontologies. Journal of Web Semantics 3(4), 268–293 (2005)

19. Lehmann, J.: Hybrid learning of ontology classes. In: Perner, P. (ed.) MLDM 2007. LNCS (LNAI), vol. 4571, pp. 883–898. Springer, Heidelberg (2007)

20. Lehmann, J.: DL-Learner: learning concepts in description logics. Journal of Machine Learning Research (JMLR) 10, 2639–2642 (2009)

21. Lehmann, J., Bizer, C., Kobilarov, G., Auer, S., Becker, C., Cyganiak, R., Hellmann, S.: DBpedia - a crystallization point for the web of data. Journal of Web Semantics 7(3), 154–165 (2009)

22. Lehmann, J., Hitzler, P.: Foundations of refinement operators for description logics. In: Blockeel, H., Ramon, J., Shavlik, J., Tadepalli, P. (eds.) ILP 2007. LNCS (LNAI), vol. 4894, pp. 161–174. Springer, Heidelberg (2008)

23. Lehmann, J., Hitzler, P.: A refinement operator based learning algorithm for the ALC description logic. In: Blockeel, H., Ramon, J., Shavlik, J., Tadepalli, P. (eds.) ILP 2007. LNCS (LNAI), vol. 4894, pp. 147–160. Springer, Heidelberg (2008)

24. Lehmann, J., Hitzler, P.: Concept learning in description logics using refinement operators. Machine Learning Journal 78(1-2), 203–250 (2010)

25. Lin, H., Sirin, E.: Pellet - a performance lint tool for pellet. In: OWLED 2008. CEUR Workshop Proceedings, vol. 432, CEUR-WS.org (2008)

26. Lisi, F.A.: Building rules on top of ontologies for the semantic web with inductive logic programming. TPLP 8(3), 271–300 (2008)

27. Nienhuys-Cheng, S.-H., de Wolf, R. (eds.): Foundations of Inductive Logic Programming. LNCS, vol. 1228. Springer, Heidelberg (1997)

28. Plessers, P., De Troyer, O.: Resolving inconsistencies in evolving ontologies. In: Sure, Y., Domingue, J. (eds.) ESWC 2006. LNCS, vol. 4011, pp. 200–214. Springer, Heidelberg (2006)
29. Schlobach, S., Cornet, R.: Non-standard reasoning services for the debugging of description logic terminologies. In: IJCAI 2003, pp. 355–362. Morgan Kaufmann, San Francisco (2003)
30. Suntisrivaraporn, B., Qi, G., Ji, Q., Haase, P.: A modularization-based approach to finding all justifications for OWL DL entailments. In: Domingue, J., Anutariya, C. (eds.) ASWC 2008. LNCS, vol. 5367, pp. 1–15. Springer, Heidelberg (2008)
31. Völker, J., Hitzler, P., Cimiano, P.: Acquisition of OWL DL axioms from lexical resources. In: Franconi, E., Kifer, M., May, W. (eds.) ESWC 2007. LNCS, vol. 4519, pp. 670–685. Springer, Heidelberg (2007)
32. Völker, J., Rudolph, S.: Fostering web intelligence by semi-automatic OWL ontology refinement. In: Web Intelligence, pp. 454–460. IEEE, Los Alamitos (2008)
33. Völker, J., Vrandecic, D., Sure, Y., Hotho, A.: Learning disjointness. In: Franconi, E., Kifer, M., May, W. (eds.) ESWC 2007. LNCS, vol. 4519, pp. 175–189. Springer, Heidelberg (2007)

# Mapping Master: A Flexible Approach for Mapping Spreadsheets to OWL

Martin J. O'Connor[1], Christian Halaschek-Wiener[2], and Mark A. Musen[1]

[1] Stanford Center for Biomedical Informatics Research
Stanford, CA 94305, USA
[2] Clados Management, LLC
San Mateo, CA 94401, USA

**Abstract.** We describe a mapping language for converting data contained in spreadsheets into the Web Ontology Language (OWL). The developed language, called $M^2$, overcomes shortcomings with existing mapping techniques, including their restriction to well-formed spreadsheets reminiscent of a single relational database table and verbose syntax for expressing mapping rules when transforming spreadsheet contents into OWL. The $M^2$ language provides expressive, yet concise mechanisms to create both individual and class axioms when generating OWL ontologies. We additionally present an implementation of the mapping approach, Mapping Master, which is available as a plug-in for the Protégé ontology editor.

## 1  Introduction

One of the hurdles that new and existing users of Semantic Web standards continue to face is converting preexisting, non-Semantic Web encoded information into one of the many Semantic Web languages (e.g., RDF, OWL). In some domains, a large deal of this information is represented in spreadsheets (e.g., financial services), which has motivated both academia [1] and industry [2, 8] to develop a variety of general-purpose spreadsheet mapping techniques to avoid manually encoding spreadsheet content in OWL or writing custom extraction programs.

Existing mapping approaches, however, suffer from a variety of limitations. First, many mapping techniques assume very simple data models within spreadsheets [3]. Typically, it is assumed that each table in a spreadsheet adheres to a relational model where each row in the table describes a different entity and each column describes an attribute for that entity; we refer to this as the 'entity-per-row' assumption. Unfortunately, there are numerous real-world spreadsheets that do not adhere to this simple data model, as many spreadsheet-authoring tools are extremely flexible and do not restrict the manner in which users author tabular structures. Common examples of complex layouts can be found in the financial domain. Here, analysts or companies publish sales forecasts or results, which are typically represented by tables that have products or market segments listed in a column, quarters or years listed in a row, and sales figures specified for each product/market segment and date. An example of this type of spreadsheet is illustrated in Figure 1.

P.F. Patel-Schneider et al.(Eds.): ISWC 2010, Part II, LNCS 6497, pp. 194–208, 2010.

| ◇ | A | B | C | D | E |
|---|---|---|---|---|---|
| 1 | **Revenues—Major Pharmaceutical Products** | | | | |
| 2 | (Millions of Dollars) | | YEAR ENDED DECEMBER 31, | | |
| 3 | PRODUCT | PRIMARY INDICATION | 2008 | 2007 | 2006 |
| 4 | **Infectious and respitory diseases** | | | | |
| 5 | Zyvox | Bacterial infections | 1,115 | 944 | 782 |
| 6 | Vfend | Fungal infections | 743 | 632 | 515 |
| 7 | Zithromax/Zmax | Bacterial infections | 429 | 438 | 638 |
| 8 | Difulcan | Fungal infections | 373 | 415 | 435 |
| 9 | ... | | | | |

**Fig. 1.** Drugs, their primary use, and their sales for a number of years[1]

Recently, there have been efforts to overcome the entity-per-row limitation and to support mappings for arbitrary spreadsheets [1]. However, to the best of our knowledge, these approaches use an RDF triples-based approach to encode mapping rules. They can be effective when mapping spreadsheet content to RDF, but are very cumbersome when encoding content in OWL due to its verbose RDF serialization. To illustrate, let's assume a financial analyst wants to model the information in Figure 1 in OWL. First, assume the analyst models each drug as a class that has OWL property restrictions for the drug's treated disease type and primary indication.[2] Using this representation, the drug Zyvox could be modeled as follows (presented using the Manchester Syntax [4]):

> Class: Zyvox SubClassOf: Drug and treatsDisease some
> 'Infectious and respiratory diseases' and        (Ex. 1)
> forIndication some 'Bacterial infections'

Next, assume the analyst models each sales figure as an OWL class that has OWL property restrictions for the drug, date, and actual amount. Thus, the cell C5 could be modeled using a new OWL class, sales1, as follows:

> Class: sales1 SubClassOf: SalesAmount and forDrug some Zyvox and
> forDate has 2008 and amount has "1,115"        (Ex. 2)

To encode the OWL class axioms in Ex. 1 & 2 in RDF, dozens of triples are required because the RDF serializations for owl:intersectionOf, owl:hasValue and owl:someValuesFrom require multiple triples. Therefore, using currently available mapping techniques, even simple mapping rules can be extremely verbose.

To overcome these limitations, we propose a new declarative OWL-centric mapping language that supports arbitrary spreadsheet-to-OWL mappings. The language also supports syntactic transformations of cell contents, as well as inline OWL axioms involving classes, properties and individuals extracted from cell contents. In the end, the mapping language enables mapping information from complex spreadsheets to OWL using a compact, user-friendly syntax.

---

[1] Source: Pfizer 2008 Financial Report:
http://media.pfizer.com/files/annualreport/2008/financial/
financial2008.pdf

[2] A philosophical discussion regarding whether this information should be modeled as classes or individuals is out of the scope of this paper. However, a class-based representation is consistent with modeling conventions used in widely accepted biomedical ontologies.

## 2  Related Work

A variety of systems have been developed to map spreadsheet content to RDF. The earliest approach include Excel2RDF [6] and Convert2RDF [7]. Both systems provide basic mapping languages to support mappings from entity-per-row spreadsheets to RDF. The later RDF123 approach [3] has a mapping language that allows complex mapping conditions that support less restricted spreadsheet data models but the language still fundamentally assumes entity-per-row storage. The recent XLWrap mapping approach [1] attempts to address this shortcoming. It allows data to be organized in essentially arbitrary ways and provides an expressive mapping language for generating RDF content. However, the resulting mapping language is rather verbose. Other spreadsheet mapping systems include MIT's Simile project and Cambridge Semantic's Anzo for Excel [8], though these systems are primarily based on metadata.

Some systems use an XSLT-based approach to map automatically-generated XML representations of spreadsheets to RDF. However, these approaches can be very cumbersome and are generally useful for only a small range of simple mappings. A related, higher level approach is to use importation tools to generate OWL or RDF tabular representations of spreadsheet data and to then map these tabular representations to domain ontologies using rule or scripting languages. For example, TopBraid Composer's SPARQLMotion [2] provides a range of scripting modules for generating RDF from tabular data imported from spreadsheets. The authors have used a similar approach with a data importation tool called DataMaster [9] that uses SWRL [10] rules to map spreadsheet data to domain-level constructs. While these approaches provide great flexibility, a multitude of rules or mapping scripts can quickly accumulate, which can be difficult to manage and debug.

A general shortcoming of existing mapping systems is that they are RDF-centric and are not designed to directly work with OWL. The only exception known to the authors is ExcelImport [11]. However, this tool assumes simple-entity-per row spreadsheets and provides only a small set of OWL constructs that are specified graphically. It additionally does not support a mapping language.

It is lastly noted that techniques have also been developed for mapping information stored in relational database management systems to RDF and OWL [13]. While possibly applicable to simple spreadsheets adhering to the entity-per-row assumption, such techniques are not suited for mapping semi-structured tables such as that presented in Figure 1.

## 3  Mapping Language

The primary goal of this work is to address the limitations of existing mapping tools by developing a new declarative OWL-centric mapping language. Importantly, this domain-specific language (DSL) should support complex spreadsheets that do not conform to the entity-per-row assumption. To ensure the mapping approach is compatible with the workflow familiar to users of spreadsheet tools, the language must also allow mappings of data spread over multiple sheets. A related requirement is that it should support mapping of data that may be distributed non-uniformly in individual sheets; for example, multiple disconnected tables in a sheet representing the same underlying information. Additionally, it should allow the selective extraction of data from within cells.

Full coverage of all OWL constructs is also a primary goal of the mapping language. In addition to supporting the definition of simple OWL entities such as named classes, properties, and individuals, class expressions and potentially complex necessary and sufficient declarations should be expressible. While an RDF triple-based mapping mechanism can in principle generate arbitrary OWL constructs, such an approach is not always practical because of OWL's complex RDF serialization. This approach would also conflict with the goal of producing a concise language.

Additionally, the language should not only be concise but also simple to learn for users familiar with both OWL and spreadsheet tools. A general usability issue when developing a custom language is providing debugging support for that language. The typical levels of complexity when mapping from spreadsheets to OWL makes this support crucial. In particular, the ability to preview the final result of a mapping expression before executing it can greatly assist in debugging. An important language design goal is thus to support instantaneous preview of mapping results before they are executed and to allow those previews to be updated dynamically when the underlying data are changed.

### 3.1  Core $M^2$ Language

Rather than designing a DSL from scratch, the proposed language is built upon the Manchester Syntax [4], a widely used DSL for declaratively describing OWL ontologies. As illustrated in Ex. 1 & 2, this DSL has concise clauses for defining common OWL entities. It also provides full coverage of all OWL constructs and is familiar to most users of OWL since it is the standard presentation syntax used by the Protégé ontology editing tools [14, 15]. It has a very clean language definition, allowing it to be extended in a principled way. The DSL that we have defined—called $M^2$, or *Mapping Master*—is a superset of the Manchester Syntax, so any valid Manchester Syntax expression is also a valid $M^2$ expression. In the remainder of this section we provide an overview of the $M^2$ language. We refer the interested reader to the $M^2$ wiki [5] for a full description of the language. Additionally, its BNF is available at [17].

**$M^2$ Reference Clause.** $M^2$ extends the Manchester Syntax to allow references to spreadsheet content in expressions. It introduces a new *reference* clause to support these references (see Figure 2). This clause indicates one or more cells in spreadsheet. In the DSL, any clause in a Manchester Syntax expression that indicates an OWL class, OWL property, OWL individual, data type, or data value can be substituted with this reference clause.

References clauses are prefixed with the character '@' and are followed by an Excel-style cell reference. In the standard Excel cell notation, cells extend from A1 in the top left corner of a sheet within a spreadsheet to successively higher columns and rows, with alpha characters referring to columns and numerical values referring to rows[3]. For example, a reference to cell A5 in a spreadsheet is written as follows:

$$@A5$$

The above cell specification indicates that the reference is relative, meaning that if a formula containing the reference is copied to another cell then the row and column components of the reference are updated appropriately. An equivalent absolute reference, again adopting Excel notation, can be written as follows:

$$@\$A\$5$$

```
reference  ::= '@' cell-ref [ '(' [ entity-type ]
                                  { value-encoding }
                                  [ shift-setting ]
                                  { empty-value-handling }
                                  [ default-value ]
                                  [ filter ]
                                  { defining-type } ')' ]
cell-ref  ::= [ a valid Excel sheet reference '!' ] column-ref row-ref
column-ref  ::= '*' | a valid Excel column reference
row-ref  ::= '*' | a valid Excel row reference
entity-type :: = 'Class' | 'ObjectProperty' | 'DataProperty' |
                 'Individual' | xsd-type
xsd-type  ::= 'xsd:int' | 'xsd:float' | ...
value-encoding  ::= ( 'rdfs:label' | 'rdf:ID' ) [ value-specification ]
value-specification  ::= '=' '(' value-item { ',' value-item } ')'
value-item  ::= reference | literal | capture-expression
capture-expression  ::= '[' a valid Java pattern expression ']'
default-value  ::= 'mm:default' '=' '(' value-item { ',' value-item } ')'
empty-value-handling  ::= empty-location-handling | empty-ID-handling |
                          empty-label-handling
empty-location-handling  ::= 'mm:ErrorIfEmptyLocation' | ...
shift-setting  ::= 'mm:NoShift' | 'mm:ShiftLeft' | 'mm:ShiftRight' | ...
filter  ::= 'mm:default' '=' '"' a valid Excel Boolean expression '"'
defining-type  ::= reference | classExpression | dataPropertyExpression |
                   objectPropertyExpression | xsd-type
```

Fig. 2. Partial BNF of $M^2$ reference clause

References can also be preceded by a sheet name. For example, a reference to the same cell in the sheet "Sales Data" can be written:

$$@ì Sales Data"!A\$5$$

In many real-world spreadsheets, users may want to evaluate the same mapping formula over a range of spreadsheet cells. For example, an analyst would likely want to evaluate the mapping expression in Ex. 2 over the cell range C5:E8 of the spreadsheet presented in Figure 1. To avoid repeatedly defining mapping expressions for each cell in such a range, $M^2$ allows the user to define a cell range and then use wildcards, denoted by '*', in place of row and/or column references (defined in Figure 2). Then, when the mapping expression is evaluated, the mapper iterates over the cell range and the wildcards are replaced with the current row and column.

---

[3] A formalization of spreadsheets is omitted here due to their widespread use and adoption.

The reference clause can be used in $M^2$ expressions to define OWL constructs using spreadsheet content. For example, an $M^2$ expression can easily be defined to take the name in cell A5 of the spreadsheet in Figure 1 and declare an OWL named class that is a subclass of an existing Drug class as follows:

*Class: @A5 SubClassOf: Drug*

This expression declares an OWL class named by the contents of cell A5 ('Zyvox' in this case) and asserts that it is a subclass of class Drug. If the class has previously been declared and is not already a subclass of Drug then that relationship will be asserted.

Using this approach, any OWL axiom can be declared using the appropriate Manchester Syntax clause, with references used in these clauses to specify spreadsheet content. For example, a $M^2$ expression to instead declare an individual of type Drug using the contents cell A5 as its name can be written·

*Individual: @A5 Types: Drug*

**$M^2$ Mapping Directives.** The $M^2$ language additionally extends the Manchester syntax with a variety of directives, which facilitate the mapping process and help achieve the goals previously described. In this section, we discuss a variety of these directives and illustrate their use.

In the above drug class declaration example, it is clear that @A5 refers to an OWL class. However, the type cannot always be inferred and ambiguities may arise regarding the type of the entity being referenced. To deal with this case, explicit entity type specifications are provided. Specifically, a reference may be optionally followed by a parenthesis-enclosed *entity type specification* to explicitly declare the type of referenced entity. This specification can indicate that the entity is a named OWL class, an OWL object or data property, or an OWL individual or a data type. The $M^2$ keywords to specify the types are: Class, ObjectProperty, DataProperty, Individual, and any XSD type name (e.g., xsd:int). Using this specification, the above drug declaration, for example, can be written:

*Class: @A5(Class) SubClassOf: Drug*

In many cases, specifying the super class, super property, individual class membership, or the data type of referenced entities is also desired. While these types of relationships can be defined using standard Manchester Syntax expressions, this approach will often entail the use of multiple mapping expressions. To concisely support defining these types of relationships, a reference may optionally be followed by a parenthesis-enclosed list of type names. Using this approach, the above drug declaration, for example, can be written as follows:

*Class: @A5(Drug)*

These type specifications can themselves be cell references and can be nested to arbitrary depths, though excessive use of nesting may make expressions difficult to understand and debug. Super properties, individual class membership, and data types can be specified in the same way.

A variety of name encoding strategies are supported when creating entities from spreadsheet content. The primary strategies are to either use direct URI-based names (equivalent to using rdf:about or rdf:ID clauses in an RDF serialization of OWL) or to use `rdfs:label` annotation values. The default naming encoding uses the `rdfs:label` annotation property. The default may also be changed globally (discussed in Section 5). Using `rdfs:label` encoding, the OWL entity generated from a cell referenced is given an automatically generated (and non meaningful) URI and its `rdfs:label` annotation value is set to the content of the cell.

A *name encoding clause* is provided to explicitly specify a desired encoding. As with entity type specifications, this clause is enclosed by parentheses after the cell reference. The $M^2$ keywords to specify the three types of encoding are `rdf:about`, `rdf:ID`, and `rdfs:label`. Using this clause, a specification of `rdf:ID` encoding for the previous drug example can be written:

*Class: @A5(rdf:ID Drug)*

The default $M^2$ behavior is to directly use the contents of the referenced cell when encoding a name. However, this default can be overridden using an optional *value specification clause*. This clause is indicated by the '=' character immediately after the encoding specification keyword and is followed by a parenthesis-enclosed, comma-separated list of *value specifications*, which are appended to each other. These value specifications can be cell references or values. For example, an expression that extends the earlier reference to specify that the entity created from cell A5 is to use `rdfs:label` name encoding and that the name is to be the value of the cell preceded by the string "Sale:" can be written as follows:

*Class: @A5(rdfs:label=("Sale:", @A5) Drug)*

Value specification references are not restricted to the referenced cell itself and may indicate arbitrary cells. More than one encoding can also be specified for a particular reference so, for example, names and annotation values can be generated for a particular entity using the contents of different cells.

A similar approach can be used to selectively extract values from referenced cells. A *regular expression capture group clause* is provided and can be used in any position in a value specification clause. This clause is contained in a quoted string enclosed by square parenthesis. For example, if cell A5 in the previous example contained the string "Pfizer:Zyvox" but only the text following the ':' character is to be used in the label encoding, an appropriate capture expression could be written as:

*Class: @A5(Drug rdfs:label=("Sale:", [":([a-zA-Z][a-zA-Z0-9]*)"]))*

Note that parentheses around the sub-expressions in a regular expression clause specify capture groups and indicate that the matched strings are to be extracted. In some cases, more than one capture group may be matched for a cell value, in which case they are extracted in the order that they are matched and appended to each other.

A *filter clause* may be used to indicate that cells that do not meet particular criteria should be ignored. This clause is indicated by the keyword `mm:filter` and, like value and type specifications, is enclosed in parentheses after a cell specification. This keyword is followed by the '=' character and a quoted condition, which is specified using Excel-style Boolean condition notation. Using this clause, a variant of the previous expression that skips cells with the value 'Zyvox' can be written:

*Class: @A5(mm:filter="A5<>Zyvox" Drug)*

**M² Missing Value Handling.** To deal with missing cell values, default values can also be specified in references. A *default value clause* is provided to assign these values. This clause is indicated by the keyword mm:default and is followed by a parenthesis-enclosed, comma-separated list of value specifications. For example, the following expression uses this clause to indicate that the value "Unknown" should be used as the created class label if cell A5 is empty:

*Class: @A5(rdfs:label mm:default=("Unknown") Drug)*

Additional behaviors are also supported to deal with missing cell values. M²'s default behavior is to skip an entire expression if it contains any references with empty cells. Four keywords are supplied to modify this behavior. These keywords indicate that: (1) an error should be thrown if a cell value is missing and the mapping process should be stopped (mm:ErrorIfEmptyLocation); (2) expressions containing references with empty cells should be skipped (mm:SkipIfEmptyLocation); (3) expressions containing references with empty cells should generate a warning in addition to being skipped (mm:WarningIfEmptyLocation); and (4) expressions containing such empty cells should be processed (mm:ProcessIfEmptyLocation).

The last option allows processing of spreadsheets that may contain a large amount of missing values. The option indicates that the M² language processor should, if possible, conservatively drop the sub-expression containing the empty reference rather than dropping the entire expression. Consider, for example, the following M² expression declaring an individual from cell A5 of a spreadsheet and associating a property hasAge with it using the value in cell A6:

*Individual: @A5 Facts: hasAge @A6(mm:ProcessIfEmptyLocation)*

Here, using the default skip behavior action, a missing value in cell A5 will cause the expression to be skipped. However, the process directive for the hasAge property value in cell A6 will instead drop only the sub-expression containing it if that cell is empty. So, if cell A5 contains a value and cell A6 is empty, the resulting expression will still declare an individual.

Using a similar approach, more fine grained empty value handling is also supported to specify different empty value handling behaviors for rdf:ID and rdfs:label values. Here, the label directives are mm:ErrorIfEmptyLabel, mm:SkipIfEmptyLabel, mm:WarningIfEmptyLabel, and mm:ProcessIfEmptyLabel with equivalent keywords for RDF identifier handling.

One additional option is provided to deal with empty cell values. This option is targeted to the common case in many spreadsheets where a particular cell is supplied with a value and all empty cells below it are implied to have the same value. In this case, when these empty cells are being processed, their location must be 'shifted' to the location above it containing a value. For example, the following expression uses this keyword to indicate that call A5 does not contain a value for the name of the declared class then the row number must be shifted upwards until a value is found:

*Class: @A5(mm:ShiftUp Drug)*

If no value is found, normal empty value handling processing applied. Similar directives provide for shifting down (mm:ShiftDown), and to allow shifting to the left (mm:ShiftLeft) or to the right (mm:ShiftRight).

## 3.2  $M^2$ Mapping Process

The $M^2$ mapping process takes a source spreadsheet, set of $M^2$ expressions, and target ontology as input, and the mappings are processed in three phases. In first phase, every expression is preprocessed and the relevant content specified by references in these expressions is retrieved from the source spreadsheet. This content, which will either specify a data value or the name of a data type or an OWL entity, is substituted for each reference in an $M^2$ expression to generate a valid Manchester Syntax expression.

The second phase declares all referenced OWL entities that are not already declared in the target ontology. The type specification for each reference is used to generate the appropriate declaration clause. Any super class, super property, individual class membership, or data type specifications in the reference are also declared in this phase.

Once the entities have been declared, the third phase involves sending the final Manchester Syntax expression to a Manchester Syntax processor. This processor populates the target ontology with the OWL axioms specified by the expressions. At the end of phase one, the generated expressions can be checked for syntactic correctness. They can also be previewed at this stage if desired, allowing users to see the final entity names expanded within their enclosing $M^2$ expression.

$M^2$ supports several preprocessing directives to specify configuration options for the mapping process. These directives include the ability to declare both a default namespace for generated entities and to specify prefix-to-namespace mappings. The latter option allows $M^2$ to deal with cells that contain both prefixed and fully qualified URI entity names. An option is also supported to indicate that cell values refer to OWL entities using annotation values. In the default case, these names—be they prefixed, fully qualified, or annotated—are assumed to either refer to existing OWL entities or to named entities that are to be declared during the import process. $M^2$ supports a pair of options to modify this behavior. The first option can be set to indicate that an error should be thrown if a name refers to an existing entity in the target ontology; the second option indicates that an error should be thrown if the name does not refer to an existing entity. A related option deals with the possible ambiguity introduced by the use of annotation value references. It can be set to produce an error if more than one existing OWL entity could be named by the value.

$M^2$ provides an *option specification clause* for each option type. The general form of this option specification clause is a keyword followed by a value. For example, the default name encoding for all mappings can be written:

$$mm{:}DefaultNameEncoding = rdfs{:}label$$

It is noted that OWL axioms generated during the mapping process may cause inconsistencies in the target ontology. Further, since users have full control over $M^2$ expression authoring, the expressions can also generate axioms that are inconsistent with each other. To immediately detect such inconsistencies, an ideal implementation would invoke an incremental OWL reasoner after each expression is executed.

# 4  Implementation

We have developed a parser, editor, and a mapper for the $M^2$ DSL. The parser currently supports core Manchester Syntax OWL entity declarations plus arbitrary class expressions, though full coverage is anticipated soon. Additionally, a development environment has been released as an open source plugin to the Protégé-OWL editor [5]. This development environment includes Java APIs for interacting with $M^2$ from software applications and a graphical user interface (Figure 3).

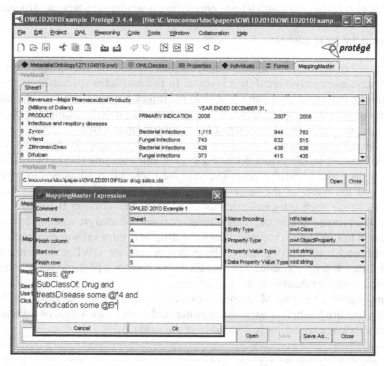

**Fig. 3.** Screen shot of the Mapping Master Protégé plug-in in Protégé-OWL. The top half of the plug-in screen shows the preview screen, which allows users to explore Excel or CSV-based spread sheets. The bottom portion of the plug-in screen shows the configuration control panel, which allows users to set default options for the mapping process and to initiate the mapping and review results. The floating popup shows the $M^2$ expression editor.

The user interface is available as a Protégé-OWL plug-in called Mapping Master and provides an editor for defining, managing and executing $M^2$ expressions. It supports the loading and previewing of spreadsheets defined in both Excel and CSV formats. An interface to interactively specify the array of configuration options supported by $M^2$ is also provided. $M^2$ expressions can also be defined interactively and then executed to map the contents of loaded spreadsheets to a target ontology. The plug-in also includes a persistence mechanism to save and reload these mappings.

## 5   Empirical Evaluation

We evaluated $M^2$ on a variety of third party spreadsheets from several domains. Here, we describe the experiences encountered during these evaluations. One evaluation was performed by the authors on a range of publicly available financial spreadsheets and two were performed in collaboration with other research groups. All three required the generation of OWL ontologies, which ranged from simple ontologies containing basic class and property declarations to ontologies containing definitions of complex necessary and sufficient conditions. The source spreadsheets ranged from entity-per-row layouts to spreadsheets containing irregular non-tabular structures.

### 5.1   Financial Spreadsheets

As discussed in Section 1, financial analysts make extensive use of spreadsheets that often contain extremely varied data layouts. In product sales spreadsheets, for example, a tabular layout of core numerical data is common, which can be associated with surrounding data in complex ways. Using a range of publicly available spreadsheets of this type, we evaluated the ability of $M^2$ to map them to OWL.

The spreadsheet presented in Figure 1 is a typical example. It shows a set of sales figures contained in the grid C5:E8. Each grid cell contains the sales amount for a particular drug in a particular year. The year for each cell is contained in row 3 of its column and the name of the relevant drug in column A of its row. The heading in column A above each drug indicates its category, while column B for each cell row contains the primary use of the named drug. A financial analyst wishing to model this information in OWL must represent both the drug and sales information and associate them with each other. To generate these definitions, an OWL expression must be defined for each drug and sales amount. Ex.1 and 2 show a possible set of expressions that define a single sale and its associated drug.

Figure 4 shows the two $M^2$ mapping expressions for these examples. In these expressions, the drug subclass expression is iterated from rows 5 to 8 of column A and the sales subclass expression is iterated over the grid C5:E8.

```
Class: @A* SubClassOf: Drug and
                      treatsDisease some @A4 and forIndication some @B*
Class: @** SubClassOf: SalesAmount and
                      forDrug some @A* and amount has @**
```

**Fig. 4.** Two $M^2$ expressions to map content of a financial spreadsheet (Figure 1) to OWL expressions (Ex. 1 and 2). In this example, the drug subclass expression is iterated from rows 5 to 8 of column A and the sales subclass expression is iterated over the grid C5:E8.

### 5.2   Ontology for Biomedical Investigations

The Ontology for Biomedical Investigations Consortium (OBI; [12]) is producing an integrated ontology for the description of life science and clinical investigations. The group has developed a spreadsheet-based procedure to allow domain experts to add terms to the OBI ontology that supports complex logical definitions yet is relatively simple to use for non ontology specialists. The procedure is based on editing

| ◇ | A | B | C | D | E | F |
|---|---|---|---|---|---|---|
| 1 | Analyte label | Analyte ID | Evaluant label | Evaluant ID | Unit label | Unit ID |
| 2 | glucose | CHEBI:17234 | blood | FMA:9670 | mmol per liter | UO:0000300 |
| 3 | sodium chloride | CHEBI:26710 | blood plasma | OBI:0100016 | mmol per liter | UO:0000300 |
| 4 | chromium-51 | CHEBI:50076 | cell culture supernatant | OBI:1000023 | ppm | UO:0000169 |
| 5 | glucose | CHEBI:17234 | material_entity | BFO:MaterialEntity | mmol per liter | UO:0000300 |
| 6 | interferon gamma | PRO:000000017 | cell culture supernatant | OBI:1000023 | ug per liter | UO:0000301 |

**Fig. 5.** An example OBI spreadsheet for submitting analyte assay term definition. These definitions include classes defined in several external ontologies, such as the Chemical Entities of Biological Interest (ChEBI) ontology and the Foundational Model of Anatomy (FMA).

definitions in a spreadsheet format, which is subsequently converted to OWL. An example spreadsheet is shown in Figure 5.

It contains several example entities necessary to specify an *analyte assay*. An analyte assay measures the concentration of a molecular entity in a material entity, such as measuring glucose concentration in blood. Each row in the spreadsheet contains the information necessary to define a single assay with specific columns containing its parameters. Each OBI OWL definition of an assay relates the material in which the concentration is measured (the *evaluant*; e.g., blood), the molecular entity that is detected (the *analyte*; e.g., glucose), and the units of the measurement being made (e.g., microgram per liter). For example, the 'Analyte label' column is expected to contain the name of the analyte, and the 'Evaluant ID' column the name of the evaluant. The content of these cells contains the URIs of terms in external ontologies.

Figure 6 shows the $M^2$ expression that generates the analyte assay definition containing necessary and sufficient conditions from this spreadsheet. As can be seen, the contents of columns A, B, D, and F are specified by references in this expression and are incorporated into the definition of each assay. In this expression, a default rdf:ID name encoding is used to resolve the URIs in columns B, D, and F; the analyte assay class uses rdfs:label encoding for its name. This single $M^2$ expression replaces a custom script that was developed by an OBI developer and was written and executed using the Mapping Master Protégé plug-in. The OBI team is currently using this tool to develop a range of additional mappings.

```
Class: @A*(rdfs:label 'analyte assay')
EquivalentTo:
(achieves_planned_objective some 'analyte measurement objective') and
(realizes some ('evaluant role' and (role_of some
                                    @D*(material_entity)))) and
(realizes some ('analyte role' and
                (role_of some ('scattered molecular aggregate' and
                              ('has grain' only
                                    @B*('molecular entity'))))))
SubClassOf:
 has_specified_output some
      ('scalar measurement datum' and
      ('is quality measurement of' some 'molecular concentration') and
      ('has measurement unit label' some
        @F*('measurement unit label')))
```

**Fig. 6.** $M^2$ expressions to map content of OBI spreadsheet to necessary and sufficient definitions of analyte assays in the OBI ontology. Here, the expression is iterated from row 2 to the end of the spreadsheet.

## 5.3   International Disease Classification

The International Statistical Classification of Diseases and Related Health Problems (ICD; [16]) provides a standard classification for diseases and a wide variety of health-related indications. The standard has gone through a variety of revisions with the 11th revision (ICD-11) due in 2015. A key to the development of ICD-11 is a content model designed to support detailed descriptions of the characteristics of each disease category and clear relationships to other terminologies, all of which are to be modeled in OWL. Part of this content model is to be populated with definitions previously developed and encoded in spreadsheets and primarily describe individual disease classifications. A sample of these definitions is presented below in Figure 7.

| | A | B | C | D | E | F | G | H | I | J | K | L | M | N | O | P | Q | R |
|---|---|---|---|---|---|---|---|---|---|---|---|---|---|---|---|---|---|---|
| 1 | ICD category | | | | | | | | | ICD code | Linearizatio | Synonyms | Definition | Exclusions | Inclusions | Note | Comment | Class name |
| 2 | Disorders of the thyroid gland and thyroid hormones system | | | | | | | | | E00-E07 | | | | | | | | E00-E07 |
| 3 | | Congenital hypothyroidism | | | | | | | | | | | Congenital hypothyroidism is the principle cause of preventable intellec | | | | | |
| 4 | | | Congenital hypothyroidism due to iodine deficiency | | | | | | | E00 | | Congenital iodine-deficiency syndrome | | | | | Section trar E00 | |
| 5 | | | | Congenital iodine-deficiency syndrome, neurological type | | | | | | E00.0 | | Endemic hypothyroidism, neurological type | | | | | | E00.0 |
| 6 | | | | Congenital iodine-deficiency syndrome, myxoedematous type | | | | | | E00.1 | | Endemic hypothyroidism, myxoedematous type | | | | | | E00.1 |
| 7 | | | | Congenital iodine-deficiency syndrome, mixed type | | | | | | E00.2 | | Endemic hypothyroidism, mixed type | | | | | | E00.2 |
| 8 | | | | Congenital iodine-deficiency syndrome, unspecified | | | | | | E00.9 | | Congenital iodine-deficiency hypothyroidism NOS | | | | | | E00.9 |
| 9 | | Permanent congenital hypothyroidism | | | | | | | | | | | | | | | | |
| 10 | | | | Permanent congenital hypothyroidism with diffuse goitre | | | | | | E03.0 | | Familial thyroid dyshormonogenesi | | | Congenital hypothyroidism due to I | | | E03.0 |
| 11 | | | | | Primary congenital hypothyroidism due to impaired hormo | | | | | E07.1 | | | | | | | | E07.1 |
| 12 | | | | | | Pendred syndrome | | | | E07.1 | | Conductive Hypothyroi | Pendred syndrome is characterized by the association of cc | | | | | E07.1 |
| 13 | | | | | Hypothyroidism due to peripheral resistance to thyroid hormones | | | | | | | | | | | | | |
| 14 | | | | Permanent congenital hypothyroidism without goitre | | | | | | E03.1 | | | | | | | | E03.1 |
| 15 | | | | | Primary congenital hypothyroidism due to a developmental anomaly | | | | | | | Thyroid dysgenesis | | | | | | E03.1 |
| 16 | | | | | | Thyroid ectopia | | | | E03.1 | | | | | | | | E03.1 |
| 17 | | | | | | Thyroid agenesis | | | | E03.1 | | Athyreosis | | | | | | E03.1 |
| 18 | | | | | | Thyroid hemiagenesis | | | | E03.1 | | | | | | | | E03.1 |
| 19 | | | | | | Thyroid hypoplasia | | | | E03.1 | | | | | | | | E03.1 |
| 20 | | | | | Primary congenital hypothyroidism due to TSH receptor mutations | | | | | | | Resistance to TSH binding or signalling | | | | | | |

**Fig. 7.** Spreadsheet containing ICD disease classes, their position in a disease hierarchy, and data about each disease

Column R contains the disease classes to be constructed. The OWL content model requires that the content of this column is used as the `rdf:ID` of the created class if a value if present for a row; if no value is present, an `rdf:ID` should be automatically generated for the new class. The class's label comes from the closest value to its left in columns A through I. These rows contain a hierarchy of the classes, with the left-most column containing the root classes. As can be seen, most of the cells in this range are empty and should therefore be skipped. Property values for each created class are also added to the class created for column R using columns L-O (which will be OWL Full). Four object properties (`synonym`, `definition`, `exclusion`, `inclusion`) are associated with each generated class (with types `SynonymTerm`, `DefinitionTerm`, `ExclusionTerm`, `InclusionTerm`) with individuals generated for each. Each individual is assigned a property called `label` from the contents of the cell in its column. Again, if these cells contain no values for a particular row, they are skipped for that property (but other properties for that class get any values that are present).

In association with the developers of content model, we developed a collection of $M^2$ expressions to support mapping these spreadsheets to OWL, which replaced a custom mapping script they had previously developed. Figure 8 shows a sample of the resulting $M^2$ expressions. The first expression declares the disease class using the cell content as its `rdf:ID` and sets its `rdfs:label` to the closest value in columns A through I. The second expression declares the subclass pairs for columns A and B. There is one of these expressions for each adjacent column pair. The third expression associates an individual with four property values using columns L through M. The final expression sets the `label` property for one of these individuals. There is one expression for each of the four property value individuals.

```
Class: @R*(rdf:ID rdfs:label=(@I*(mm:ShiftLeft)))

Class: @B*(rdf:ID=(@R*) rdfs:label mm:SkipIfEmptyLabel)
  SubClassOf: @A*(rdf:ID=(@R*) rdfs:label mm:SkipIfEmptyLabel mm:ShiftUp)

Individual: @R*(Class rdf:ID rdfs:label=(@I*(mm:ShiftLeft)))
  Facts: synonym     @L*(mm:SkipIfEmptyLocation SynonymTerm)
         definition @M*(mm:SkipIfEmptyLocation DefinitionTerm )
         exclusion  @N*(mm:SkipIfEmptyLocation ExclusionTerm)
         inclusion  @O*(mm:SkipIfEmptyLocation InclusionTerm)

Individual: @L*(mm:SkipIfEmptyLocation SynonymTerm)
  Facts: label @L*
```

**Fig. 8.** $M^2$ expressions to map the content of a spreadsheet to an ICD content model. These expressions are iterated from row 2 to the final row of the spreadsheet.

### 5.4 Evaluation Results

In summary, real-world use of the $M^2$ language and mapping process has demonstrated that it provides a compact, user-friendly mechanism for mapping complex spreadsheets to OWL. During the evaluation, we found that $M^2$ was expressive enough to capture the desired mappings in all examples. Further, the use cases demonstrated the utility of the language's novel features not supported in previous spreadsheet mapping work.

## 6  Conclusion

Recent approaches for mapping information contained in spreadsheets to OWL suffer from a variety of limitations, including assuming well-formed spreadsheets reminiscent of a single relational database table and verbose syntaxes for expressing mapping rules. In this paper, we have overcome these limitations by developing a mapping language, $M^2$, which is based on an extension of the OWL Manchester Syntax. This mapping language supports arbitrary spreadsheet cell references and provides a compact, user-friendly, OWL-centric approach for expressing mapping rules for arbitrary spreadsheets. The language also supports syntactic transformations of cell contents, as well as inline OWL axioms involving classes, properties and individuals extracted from cell contents. Lastly, we have recently released a free, open source implementation of the approach as a Protégé plug-in called Mapping Master. As described in this paper, this plug-in has been used successfully by several research groups.

Future work includes extending the mapping approach to work directly within Microsoft Excel, which will allow mapping expressions to be authored directly in cells and use native Excel cell references and functions. This will additionally enable standard Excel formula operations, such as copy and paste, for mapping expressions associated with cells, as well as allow interactive previews of $M^2$ expressions in cells using references substituted with cell values. Other potential future work includes supporting user-defined functions in mapping expressions.

## Acknowledgments

This research was supported in part by STTR Award #0750543 from the National Science Foundation and by the National Library of Medicine under grants LM007885 and LM009607. We would also like to thank Peter Moore, Natasha Noy, Samson Tu, and members of the OBI Consortium. An initial, shorter version of this paper appeared in [18].

## References

[1]  Langegger, A., Woss, W.: XLWrap – querying and integrating arbitrary spreadsheets with SPARQL. In: Bernstein, A., Karger, D.R., Heath, T., Feigenbaum, L., Maynard, D., Motta, E., Thirunarayan, K. (eds.) ISWC 2009. LNCS, vol. 5823, pp. 359–374. Springer, Heidelberg (2009)

[2]  TopBraid Composer, http://www.topbraidcomposer.com

[3]  Han, L., Finin, T.W., Parr, C.S., Sachs, J., Joshi, A.: RDF123: From spreadsheets to RDF. In: Sheth, A.P., Staab, S., Dean, M., Paolucci, M., Maynard, D., Finin, T., Thirunarayan, K. (eds.) ISWC 2008. LNCS, vol. 5318, pp. 451–466. Springer, Heidelberg (2008)

[4]  Manchester OWL Syntax, http://www.w3.org/TR/owl2-manchester-syntax/

[5]  Mapping Master,
     http://protege.cim3.net/cgi-bin/wiki.pl?MappingMaster

[6]  Reck, R.P.: Excel2RDF for Microsoft Windows,
     http://www.mindswap.org/~rreck/excel2rdf.shtml

[7]  Grove, M.: Mindswap Convert2RDF Tool,
     http://www.mindswap.org/~mhgoeve/convert/

[8]  Huynh, D., Karger, D., Miller, R.: Exhibit: lightweight structured data publishing. In: Proceedings of the 16th International Conference on World Wide Web (2007)

[9]  O'Connor, M.J., Shankar, R.D., Tu, S.W., Nyulas, C.I., Das, A.K.: Developing a Web-Based Application using OWL and SWRL. In: AAAI Spring Symposium, Stanford, CA, USA (2008)

[10]  SWRL Submission, http://www.w3.org/Submission/SWRL

[11]  ExcelImport, http://code.google.com/p/co-ode-owl-plugins/wiki/ExcelImport

[12]  OBI Consortium, http://obi-ontology.org/page/Consortium

[13]  Bizer, C.: D2R MAP - A Database to RDF Mapping Language. In: 12th International World Wide Web Conference, Budapest, Hungary (2003)

[14]  Gennari, J., Musen, M., Fergerson, R., Grosso, W., Crubezy, M., Eriksson, H., Noy, N., Tu, S.: The evolution of Protégé-2000: An environment for knowledge-based systems development. International Journal of Human-Computer Studies 58(1), 89–123 (2003)

[15]  Knublauch, H.: An AI tool for the real world: Knowledge modeling with Protégé. JavaWorld, June 20 (2003)

[16]  World Health Organization. Production of ICD-11: The overall revision process (2007), http://www.who.int/classifications/icd/ICDRevision.pdf

[17]  Mapping Master BNF,
      http://swrl.stanford.edu/MappingMaster/1.0/BNF/MappingMasterParser.html

[18]  O'Connor, M.J., Halaschek-Wiener, C., Musen, M.: M2: a Language for Mapping Spreadsheets to OWL OWL: Experiences and Directions (OWLED). In: Sixth International Workshop, San Francisco, CA (2010)

# dbrec — Music Recommendations Using DBpedia*

Alexandre Passant

Digital Enterprise Research Institute,
National University of Ireland, Galway
alexandre.passant@deri.org

**Abstract.** This paper describes the theoretical background and the implementation of dbrec, a music recommendation system built on top of DBpedia, offering recommendations for more than 39,000 bands and solo artists. We discuss the various challenges and lessons learnt while building it, providing relevant insights for people developing applications consuming Linked Data. Furthermore, we provide a user-centric evaluation of the system, notably by comparing it to last.fm.

**Keywords:** Semantic Web Applications, Linked Data, Recommendation Systems, Semantic Distance, DBpedia.

## 1  Introduction

Since its first steps in 2007, the Linking Open Data (LOD) cloud has grown considerably, as shown in Fig. 1[1]. However, besides recent initiatives outreaching how to build applications using it [5] [7], there is still room for more end-user applications (*i.e.* not semantic search engines nor APIs) that *consume* Linked Data. While we can argue that the data itself is the most valuable component, building innovative applications would lead to a virtuous circle enriching the value of this global network, by analogy with Metcalfe's law [11].

In this paper, we describe dbrec — http://dbrec.net —, a music recommendation system based on Linked Data (in particular on DBpedia) offering recommendations for more that 39,000 bands and solo artists. In addition, a core component of dbrec is its explanation feature, provided as a side effect of using Linked Data for computing the recommendations. We provide a user-centric evaluation of the system in order to identify how it compares to existing systems, in particular with last.fm[2], and how users rate its novel recommendations. Furthermore, besides presenting the theoretical background and the architecture of the system, we also discuss some lessons learnt when building it, in terms of data quality, architecture considerations as well as query patterns

---

* The work presented in this paper has been funded by Science Foundation Ireland under Grant No. SFI/08/CE/I1380 (Líon-2).

[1] Based on [2] and http://richard.cyganiak.de/2007/10/lod/
[2] http://last.fm

P.F. Patel-Schneider et al.(Eds.): ISWC 2010, Part II, LNCS 6497, pp. 209–224, 2010.

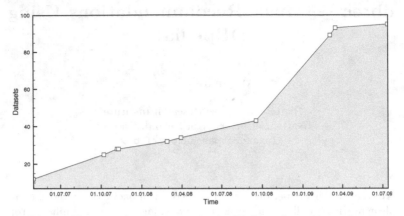

**Fig. 1.** The growth of datasets in the Linking Open Data cloud

and scalability. Thus, our aim is to provide a set of insights and best practices that can be re-used when building end-user applications that consume Linked Data.

The rest of the paper is organised as follows. In Section 2, we briefly describe the *LDSD* algorithm — *Linked Data Semantic Distance* —, used as a basis of our recommendation engine. In addition, we detail its related ontology, used to represent these distances and their explanations in RDF. In Section 3, we discuss the dbrec architecture, explaining how the previous algorithm has been applied to DBpedia to compute recommendations for more than 39,000 resources. Section 4 describes the evaluation of the system, including comparison with last.fm, evaluation of the novel recommendations and of the system as a whole. Then, in Section 5, we discuss three particular lessons learnt when building dbrec, which are however relevant in the broader Linked Data context: (1) data quality; (2) architecture considerations; and (3) SPARQL query patterns and scalability. We then discuss related work in Section 6 before concluding the paper with an overview of future challenges for dbrec.

## 2    Linked Data Semantic Distance

### 2.1    Motivation

Our main motivation was to identify how semantic distance [21] measures could be applied to resources published on the Web as Linked Data [1]. Specifically, and while semantic distance have been studied over time in various contexts [21] [3] [6], our goal was to identify and to apply such measures by considering some of the main characteristics of Linked Data:

- relying only on *links* — *i.e.* not taking into account literal values and their linguistic proximity;

- relying only on *instance data* — *i.e.* not taking into account ontologies used to describe resources, since LOD is more oriented towards publishing instance data than using formal ontologies;
- considering *dereferencable URIs* — so that distances can be computed simply by accessing URIs and retrieving corresponding RDF data.

Our aim was then to identify the usefulness of the Linked Data principles for computing semantic distance between particular resources.

## 2.2   A Conceptual Model for Linked Data

While Linked Data is generally introduced using its four publishing principles [1] — which make sense from a programmatic point of view — there is a need to ground it into a theoretical framework to define algorithms using it. We thus provide the following definition of a Linked Data dataset, whether it is centralised or distributed on the Web — and we can then consider $LOD = \bigcup_i G_i$.

**Definition 1.** *A dataset following the Linked Data principles is a graph G such as $G = (R, L, I)$ in which $R = \{r_1, r_2, ..., r_n\}$ is a set of resources — identified by their URI —, $L = \{l_1, l_2, ..., l_n\}$ is a set of typed links — identified by their URI — and $I = \{i_1, i_2, ..., i_n\}$ is a set of instances of these links between resources, such as $i_i = \langle l_j, r_a, r_b \rangle$.*

This definition voluntary excludes literals, as we focused only on the URI-linking aspect of Linked Data, as discussed in [1]: *"The simplest way to make linked data is to use, in one file, a URI which points into another"*.

## 2.3   LDSD — Linked Data Semantic Distance

Based on this definition, we defined a *Linked Data Semantic Distance* (LDSD) measure to compute the distance between two resources published as Linked Data[3], normalised in the $[0, 1]$ interval. So far, our measure considers only resources linked either directly or through a third resource, and recursive patterns such as SimRank [15] may be used in the future. Since LDSD — and some of its initial variants — has already been discussed in [19], we will not present it in too many details. At a glance, for two resources $r_a$ and $r_b$, LDSD identifies four dimensions (direct and indirect links, both incoming and outcoming) to compute their distance, using the following definitions.

**Definition 2.** *$C_d$ is a function that computes the number of direct and distinct links between resources in a graph G. $C_d(l_i, r_a, r_b)$ equals 1 if there is an instance of $l_i$ from resource $r_a$ to resource $r_b$, 0 if not. By extension $C_d$ can be used to compute (1) the total number of direct and distinct links from $r_a$ to $r_b$*

---

[3] Note that we use the term *distance* while the measure may actually not be symmetric.

$(C_d(n, r_a, r_b))$ as well as (2) the total number of distinct instances of the link $l_i$ from $r_a$ to any node $(C_d(l_i, r_a, n))$.

**Definition 3.** $C_{io}$ and $C_{ii}$ are functions that compute the number of indirect and distinct *links*, *both* outcoming and incoming, *between resources in a graph* $G$. $C_{io}(l_i, r_a, r_b)$ equals 1 if there is a resource $n$ that satisfy both $\langle l_i, r_a, n \rangle$ and $\langle l_i, r_b, n \rangle$, 0 if not. $C_{ii}(l_i, r_a, r_b)$ equals 1 if there is a resource $n$ that satisfy both $\langle l_i, n, r_a \rangle$ and $\langle l_i, n, r_b \rangle$, 0 if not. By extension $C_{io}$ and $C_{ii}$ can be used to compute (1) the total number of indirect and distinct links between $r_a$ and $r_b$ $(C_{io}(n, r_a, r_b)$ and $C_{ii}(n, r_a, r_b)$, respectively outcoming and incoming) as well as (2) the total number of resources $n$ linked indirectly to $r_a$ via $l_i$ $(C_{io}(l_i, r_a, n)$ and $C_{ii}(l_i, r_a, n)$, respectively outcoming and incoming)

$$LDSD(r_a, r_b) = \frac{1}{1 + \sum_i \frac{C_d(l_i, r_a, r_b)}{1 + log(C_d(l_i, r_a, n))} + \sum_i \frac{C_d(l_i, r_b, r_a)}{1 + log(C_d(l_i, r_b, n))} + \sum_i \frac{C_{ii}(l_i, r_a, r_b)}{1 + log(C_{ii}(l_i, r_a, n))} + \sum_i \frac{C_{io}(l_i, r_a, r_b)}{1 + log(C_{io}(l_i, r_a, n))}}$$

**Fig. 2.** The *LDSD* measure

### 2.4  The LDSD Ontology

In addition to the measure itself, and since we focus on a Linked Data approach, our aim was to provide the output of such measures also available on the Web as Linked Data. We thus designed a lightweight LDSD ontology[4], accompanying the previous measure and containing two main classes:

- `ldsd:Distance`, in order to represent the distance between two resources (using `ldsd:from` and `ldsd:to`) and its value (`ldsd:value`[5])
- `ldsd:Explanation` (and four subclasses: `ldsd:DirectIn`, `ldsd:DirectOut`, `ldsd:IndirectIn` and `ldsd:IndirectOut`), in order to store the links and the property-value pairs (`ldsd:property` and `ldsd:node`) used to measure the distance, and how much similar links appear in the dataset (`ldsd:total`).

Here lies one of the first advantages of using Linked Data to compute semantic distance. The links that are traversed by the algorithm are all typed, and this is consequently easy to know how the distance has been computed, as we will show when presenting dbrec's user-interface (Section 3.4). As an example, the following snippet of code (Listing 1.1) represents that Elvis Presley is at a distance of 0.09 from Johnny Cash, because (among others) both have the same value for their `rdf:type` property ( `http://dbpedia.org/class/yago/SunRecordsArtists`), shared only by 19 artists in the `http://dbpedia.org` dataset.

---

[4]  Available at `http://dbrec.net/ldsd/ns#`
[5]  `rdf:value` was not used due to its lack of formalism —
    `http://www.w3.org/TR/rdf-schema/#ch_value`

```
@prefix ldsd: <http://dbrec.net/ldsd/ns#> .
<http://dbrec.net/distance/774a32aa-dede-11de-84a3
   -0011251e3563> a ldsd:Distance ;
   ldsd:from <http://dbpedia.org/resource/Johnny_Cash> ;
   ldsd:to <http://dbpedia.org/resource/Elvis_Presley> ;
   ldsd:value "0.0977874534544" .
<http://dbrec.net/distance/774a32aa-dede-11de-84a3
   -0011251e3563> ldsd:explain [
   a ldsd:IndirectOut ;
   ldsd:property <http://www.w3.org/1999/02/22-rdf-syntax
      -ns#type> ;
   ldsd:node <http://dbpedia.org/class/yago/
      SunRecordsArtists> ;
   ldsd:total "19" ] .
```

**Listing 1.1.** Representing distance between Johnny Cash and Elvis Presley.

## 3   The dbrec Recommendation System

### 3.1   System Architecture

Based on our previous findings, we implemented a music recommendation system in order to demonstrate the usability of the *LDSD* measure for an end-user application. To do so, we computed semantic distance for all artists referenced in DBpedia. While it does not involve cross-datasets recommendations, which are possible using our algorithm, it however offers two main advantages. First, there are more than 39,000 artists available in DBpedia for which recommendations can be built. Second, DBpedia also provides pictures and description of artists that can be used to build the system's user interface.

In order to build the system, we followed four steps (Fig. 3): (1) identify the relevant subset from DBpedia; (2) reduce the dataset for query optimisation; (3) compute distances using the *LDSD* algorithm and represent them using its ontology; (4) build a user-interface for browsing recommendations.

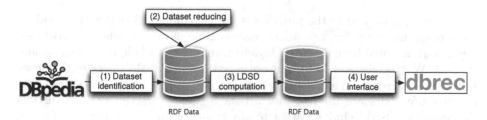

**Fig. 3.** The dbrec architecture

## 3.2   Identifying the Relevant Dataset from DBpedia

While the LDSD algorithm can be simply translated to SPARQL queries and applied to any public endpoint, this approach has some drawbacks. Indeed, DBpedia's public endpoint is limited to a certain number of answers per query, so each query must be split in sub-queries, and results must then be recomposed.

Consequently, we setup our own replica of the dataset to compute the recommendations locally. Instead of relying on a complete DBpedia dump, and as we aim at building music recommendations only, we limited ourselves to all instances of `dbpedia:MusicalArtist` and `dbpedia:Band` from DBpedia. In addition, according to the *LDSD* algorithm, we needed both incoming and outcoming links for each artist. Fortunately, each data file in DBpedia (retrieved when dereferencing the resource URI with the proper HTTP header) provides this information, for both incoming and outcoming links. This also means that the distance could be measured live, by dereferencing URIs of relevant resources, while it would obviously be more time consuming.

The original dataset, including more than 39,000 resources, included 3,004,351 triples. We then cleaned it to get a smaller and more accurate dataset, for two main reasons. On the one hand, we wanted to remove datatype properties, as they are not relevant for our experiment[6]. Removing them lead to a dataset containing 2,247,019 triples, thus reducing the original one from about 25.2% — implying that 1/4 of DBpedia assertions, in our dataset, involve literals. On the other hand, we identified lots of redundancy and inconsistencies in our DBpedia subset[7]. Especially, many links between resources are defined redundantly as `http://dbpedia.org/ontology/xxx` and at the same time as `http://dbpedia.org/property/xxx`. We then removed duplicates, leading to 1,675,711 triples, *i.e.* only 55.7% of the original dataset.

We also analysed the dataset to identify how artists are related to each other (by direct links) in DBpedia (Fig. 4). We observed that 21,211 of them (more than 50%) are not linked to any artists, and 9,555 are linked to three of them, the maximum being 14 links from one artist to 14 others. Then, by using indirect links for computing semantic distance with *LDSD*, we are able to provide recommendations for these 21,211 isolated artists.

## 3.3   Cleaning and Reducing the Dataset

While being optimised in the previous step, the computation time was still far from optimum. Even for a recommendation time of 40 seconds per artist (see Section 5), it would have taken 15 days to compute the whole recommendations dataset. We then focused on further optimisations not at the query-engine level, but at the dataset level, analysing it more deeply, and we identified that: (1) 188 distinct properties are used to link artists together directly; (2) 578 distinct properties are used to link an artist to any resource (including artists) ; (3) 767

---

[6] We agree that using and comparing literals may help in the distance measurement, but our focus was to consider only a link-based approach.

[7] We relied on DBpedia 3.3.

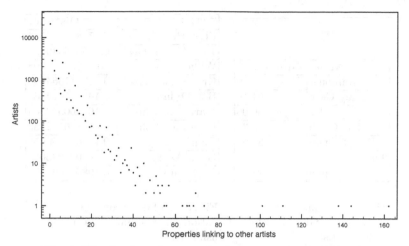

**Fig. 4.** Distribution of properties between artists in DBpedia

distinct properties are used to link any resource (including artists) to an artist. We then focused on data curation: (1) on the one hand to remove properties and property-values that are useless for computing the LDSD measures, and (2) on the other hand, to solve some data quality issues in DBpedia.

From the 188 properties linking two artists, we identified that 18 were used as links between artists while it was not their main purpose[8], such as the property `dbprop:notableInstruments` — used to link an artist to its instrument(s) — or `dbprop:nationalAnthem` — linking a country to its anthem. Moreover, we identified 35 properties that were wrongly defined — while however used two times of less —, such as `http://dbpedia.org/property/extra18` and `http://dbpedia.org/property/klfsgProperty`. Then, from the 578 properties used to link artists to resources, 183 were used only one time and were consequently useless for our recommendations, since it imply there is no more than one artist using on. In addition, 36 of these properties were wrongly defined. Furthermore, we identified 11 useless property-value combinations to compute our recommendations, by being too generic such as `rdf:type foaf:Agent`. Finally, from the 767 properties used to link any resource to an artist, 336 were removed as used only to link to a single artist, and 115 were wrongly written.

We then cleaned-up the dataset and reduced it to a total of 1,073,077 triples. We eventually ran LDSD on this dataset. The computation time took a total of 9,797 minutes[9], and resulted in 50,753,494 new triples describing the recommendations (and the explanations) modelled using the previous LDSD ontology. The time to compute the recommendation for a single resource obviously depended on the artist and related properties, as one can see in Fig. 5[10]. In addition,

---

[8] At least from what their general usage on DBpedia can tell, since they do not have any domain or range.

[9] On a 2 x AMD Opteron 250 with 4GB memory running Ubuntu 8.10/x86_64.

[10] Average time of 5 consecutive runs.

| Artist | Time (sec.) |
|---:|---:|
| Ramones | 25.20 |
| Johnny Cash | 61.16 |
| U2 | 50.06 |
| The Clash | 43.34 |
| Bar Religion | 34.98 |
| The Aggrolites | 7.35 |
| Janis Joplin | 23.12 |

| Artist | Distance |
|---|---|
| Elvis Presley | 0.0977874534544 |
| June Carter Cash | 0.105646049225 |
| Willie Nelson | 0.13221654708 |
| Kris Kristofferson | 0.140717564665 |
| Bob Dylan | 0.146635674481 |
| Marty Robbins | 0.167300943904 |
| Rosanne Cash | 0.17826142135 |
| Charlie McCoy | 0.183656756953 |
| Gene Autry | 0.191014026051 |
| Carl Smith | 0.198003626307 |

**Fig. 5.** Computation time of recommendations for various artists

**Fig. 6.** 10 first recommendations for Johnny Cash using *LDSD*

Fig. 6 displays the result of the computation (distance only, 10 first results) for dbpedia:Johnny_Cash.

## 3.4   User-Interface

Thanks to the use of Linked Open Data, building the user-interface was quite straightforward. As each artist and band is identified by a reference URI, abstracts and pictures can be obtained by simply dereferencing it. We then build a front-end providing recommendation (ranked by distance) for any of the 39,000

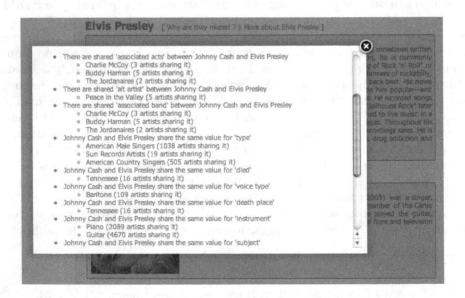

**Fig. 7.** Example of dbrec explanations

artists and bands of our dataset, including related pictures and abstracts. Recommendation pages are rendered *via* SPARQL queries ran over the computed LDSD data, and pages also provide links to YouTube videos, Twitter messages and last.fm profiles in order to enhance the browsing experience and let users listen to related songs. In addition, in order to let developers build third-party applications on top of dbrec, recommendations are also available as RDFa using the LDSD ontology.

Explanation are provided on demand, through a *"Why are they related?"* link that opens a pop-up launching another query to retrieve the explanations. These explanations are provided using human-readable labels of the property and their values, as seen in Fig. 7, explaining the recommendation of Elvis Presley for one user browsing the page about Johnny Cash.

# 4  Evaluation

## 4.1  Context of the Evaluation

In order to evaluate dbrec, we focused on standard user evaluations protocols for recommender systems, both off-line an on-line [12]. We interviewed 10 participants: 2 women and 8 men, ages ranging from 24 to 34. Interviews were conducted face to face (besides one that has to be done by phone) and last between 35 and 55 minutes. Before the evaluation, we asked users to submit a list of 10 to 15 bands they listen to and appreciate, from which we randomly selected 10 bands (ensuring that all belong to the dbrec dataset).

Then, the interviews involved two main steps. First, an *off-line evaluation*: for five (randomly chosen) bands from the previous list, we provided users with two sets of ten recommendations each. One was generated from dbrec, the other one from last.fm. Users were not aware of this and were just given the two lists randomly, simply telling them they came from different systems. We asked interviewees to rate to each recommendation (from 1 — poor — to 5 — excellent —) or to mention if that was an unknown recommendation. Note that we asked them to rank the relevance of the recommendations, not if they like that particular band or artist.

Then, we conducted an *on-line evaluation* for the five remaining bands. Users sat in front of the system and were asked to browse the recommendation list and to rate the first 10 recommendations. However, as opposed to the first part, when a band was unknown, users could read the description of each recommended artist and check the explanations provided by dbrec. We also told users that checking the explanations was not mandatory, as we wanted to observe how often they use it or not[11].

## 4.2  dbrec versus Last.fm

Regarding the off-line evaluation, the average mark for dbrec recommendations was $3.37(\pm 1.19)$ (and $3.44(\pm 1.25)$) when combined with the results from the

---

[11] For the phone interview, we asked the user to tell us if he was using them, since we were not able to setup a screen-sharing teleconference.

Table 1. Precision of recommendations: dbrec *versus* last.fm

|        | dbrec (off-line only) | dbrec (off-line and on-line) | last.fm |
|--------|-----------------------|------------------------------|---------|
| t=2    | 92.05                 | 90.59                        | 98.32   |
| t=3    | 76.63                 | 77.72                        | 87.91   |
| t=4    | 49.06                 | 51.23                        | 58.05   |
| t=5    | 20.09                 | 25                           | 25.165  |

on-line part), while the average mark for last.fm was $3.69(\pm1.01)$. We also evaluated the precision of both recommendations, considering the number of relevant items provided in both lists. To do so, we considered different threshold in the multi-point scale used to evaluate the recommendations. Table 1 shows the different values for both dbrec and last.fm ($t=x$ means that we consider a recommendation as being relevant it it is ranked $x$ or higher). In spite of a slight advantage for last.fm, dbrec achieves a reasonable score, especially considering that it does not use any collaborative filtering approach, and relies only on links between resources. Measuring the recall of recommendation was however not possible, as it would have implied users to know and check all bands of the dbrec dataset.

### 4.3   Evaluating Novel Recommendations

An interesting outcome of the evaluation was that many recommendations were unknown to users: 62% for dbrec (59.6% when combining off-line and on-line parts) and 40.4% for last.fm. However, as argued by [4]: *"novel recommendations are sometimes necessary in order to improve the users experience and discovery in the recommendation workflow"*. To that end, we used the on-line setup to evaluate quality of the novel recommendations provided by dbrec. For that on-line part, 310 recommendations (on a total of $500^{12}$) were identified as unknown, *i.e.* being novel. Among these 310, 274 have been evaluated. One user justified that, without listening to the music and even with the explanations, he was not able to provide any mark for them, while other users were able to do so, judging the recommendations by reading descriptions and explanations. Among these 274 remaining recommendations, the average rate for novel recommendations was $3.05(\pm 1.09)$. In [4], the authors also showed that, based on user-centric evaluation, the average mark for novel recommendations was less than $3^{13}$ and argued *"this probably emphasises the need for adding more context when recommending unknown music. Users might want to understand why a song was recommended"*. We hence believe than the features provided by dbrec, namely the description of each recommendation and most of all its explanation, made users better understand and appreciate the recommendations — and consequently put this average mark higher.

Furthermore, we also evaluated the precision of these recommendations, considering various thresholds as previously (Table 2). We the observed than even

---

[12] 10 *users* × 5 *bands* × 10 *recommendations*.

[13] Respectively $3.03(\pm1.19)$ for Collaborative Filtering, $2.77(\pm1.20)$ for Hybrid and $2.57(\pm1.19)$ for Audio Content-Based recommendations.

**Table 2.** Precision for novel recommendations on dbrec

| Precision | | | |
|-----------|-------|-------|------|
| t=2 | t=3 | t=4 | t=5 |
| 89.42 | 70.80 | 37.59 | 7.3 |

with a threshold of 3 (*i.e.* only good, very good or excellent recommendation), the precision is more than 70%, while still more than 37% considering only very good or excellent ones.

Overall, in terms of recommendations, dbrec achieves respectable performances comparable to last.fm and to other systems. However, instead of relying on collaborative-filtering algorithms (based on proprietary data from million of users), it only requires a set of publicly available open-data. This clearly shows the advantage of the Linking Open Data initiative for building such recommender systems.

### 4.4 Evaluating the UI and the Explanations

Finally, in addition to the recommendations themselves, we asked users to agree (or not) on a set of adjectives describing (1) the system and its user-interface in general, and (2) the explanations in particular. As results show (Table 3), all users positively acknowledged both the system and its explanations. There are however efforts to be made regarding the explanations and their related presentation, still considered as "Too geeky" by six users.

Furthermore, we observed that users relied on explanations for 198 of the 310 unknown recommendations. In addition, they relied on it for 24 of the 190 known recommendations, wanting to understand the reason of the recommendation, as discussed in [16].

**Table 3.** User-feedback on the overall dbrec system

| | User-interface | Explanations |
|-------------|----------------|--------------|
| Enjoyable | 9 | 7 |
| Useful | 9 | 9 |
| Enriching | 8 | 10 |
| Easy to use | 10 | 9 |
| Confusing | 0 | 2 |
| Complicated | 0 | 2 |
| Too geeky | 1 | 6 |

## 5 Lessons Learns and Discussions

### 5.1 Data Quality

A first lesson learnt concerns the data quality within the LOD cloud, which is far from perfect to build applications using it. As exposed in Section 3.3, we had to rely on (manual) curation of the dataset, and identified issues with the underlying data model, such as similar properties defines at both /property and /ontology URLs in DBpedia, or many having neither domain nor range, making difficult to identify inconsistencies. While we focused only on the DBpedia dataset, similar observations have been identified more globally on the Web, implying a need for more data curation in the LOD cloud [13].

### 5.2   Use, but Replicate

Then, while data is openly available on the Web, and while some services provide public SPARQL endpoints (such as DBpedia), local mirroring is required to ensure scalability and efficiency in the development process. For example, due to results restrictions on the public DBpedia endpoint, simply retrieving all bands and artists from DBpedia implies to (1) get the number $n$ of results satisfying that pattern (which furthermore relies on COUNT, not supported by SPARQL 1.0); (2) split the query into $\lceil n/5000 \rceil$ queries using the LIMIT and OFFSET clauses; (3) run the queries and recompose the results, while also taking care of network issues that may break that loop. Then, we had to replicate data in a local store, which conforms to what [10] discussed, by proposing a reference architecture for Semantic Web applications based on empirical analysis of existing services. Such solutions however raise the issue of synchronising datasets between original services and local repositories, but also shows business opportunities for Linking Open Data services providers that could deploy commercial SPARQL capabilities with enhanced quality of service.

### 5.3   SPARQL: Be Quick or Be Neat

Another lesson learnt concerns the use of SPARQL, where we observed that decomposing queries provides much faster answering time than running single queries covering complex paths.

For example, in order to translate $LDSD$ to SPARQL queries, one of our need was to identify, from a resource $r_i$, all resources $r_j$ that are linked to a third resource $r_x$ through the same path as $r_i$ is linked to $r_x$ — that is $\langle l_i, r_i, r_x \rangle, \langle l_j, r_j, r_x \rangle$ — looking for resource sharing a common property-value. In addition, we had to ensure that $r_x$ was also either a band or a solo artists. To do so, we considered three different options:

1. running a single query covering the full pattern, thus retrieving at the same time all the property-value pairs, as well as the corresponding resources;
2. running a first SPARQL query to identify all the property associated to $r_i$, and then identifying all resources sharing a property (plus its value) (*Property-slicing*);
3. running a first SPARQL query to identify all the property-values related to $r_i$, and then identifying all resources sharing that property-value pair (*Complete-slicing*).

As Table 4 shows, while up to 135 queries were needed when we initially needed only one, the computation time was up to 75% shorter when using the

**Table 4.** Comparing strategies to identify indirectly related artists

|  | Direct-SPARQL | | Property-Slicing | | Complete-Slicing | |
|---|---|---|---|---|---|---|
|  | queries | time | queries | time | queries | time |
| Ramones | 1 | 139.97 | 20 | 109.51 | 66 | 37.84 |
| Johnny Cash | 1 | 257.81 | 30 | 152.60 | 135 | 75.35 |
| U2 | 1 | 155.53 | 22 | 122.91 | 70 | 44.03 |
| The Clash | 1 | 146.43 | 20 | 110.84 | 79 | 42.61 |
| Bad Religion | 1 | 104.08 | 23 | 86.49 | 97 | 47.35 |
| The Aggrolites | 1 | 145.92 | 13 | 114.52 | 28 | 28.33 |
| Janis Joplin | 1 | 230.88 | 27 | 151.00 | 98 | 62.81 |

complete-slicing approach[14]. This means that optimisation must be done by the query authors, as writing extensive queries for a complex graph-matching is not yet the best solution regarding scalability. Further work should probably to be done to optimise complex SPARQL query processing and decomposition of patterns [23], so that developers could write single queries instead of relying on decomposition and recomposition of results through external scripts.

# 6   Related Work

In the realm of large-scale Semantic Web based recommender systems, the most know approach is probably the FOAFing-the-music project [18], that uses the distributed social networking capabilities of FOAF to provide music recommendations based on users' and friends' tastes. Focusing on a similar idea of cross-social-networking recommendations, [20] presented some ways to use FOAF, SIOC and MOAT to compute recommendations and also discussed some first steps on using Linked Data to build explanatory recommender systems. More recently, [9] followed a similar idea by developing a first prototype applied to cross-sites collaborative filtering using Linked Data. However, as discussed in introduction, our motivation was to rely on Linked Data from a resource-centric point of view, not considering social aspects but only links between resources. In that context, LODations recently focused on LOD-based music recommendations[15], while using a simpler approach not ranking the recommendations nor combining multiple features automatically. Furthermore, [17] also focused on the use of ontologies for recommender systems.

Regarding ontologies and data modelling, extensive work has been done around the Music Ontology [22]. This also includes MuSim[16] — The Music Similarity Ontology [14] — that could be mapped to our LDSD ontology in the future.

---

[14] While we ran these tests only with our local endpoint, using 4store, we observed that the initial full-query time-outs on the public DBpedia endpoint while other strategies ran properly, albeit the query limit that we mentioned earlier, and similar timing issues.

[15] http://lodations.heroku.com/

[16] http://grasstunes.net/ontology/musim/musim.html

More recently, a Recommendation Ontology has also been proposed[17], and we may also consider alignment with our model and SCOVO — the Statistical Core Vocabulary [8] — to represent statistic information about the explanations of the recommendations.

# 7    Future Work

In terms of future work, we first plan to investigate additional criteria to tune the distance measurement. It could include using the transitivity of genres, defined as skos:Concepts and hierarchically ordered in DBpedia. We may also investigate link propagation and recursivity, as done by SimRank, in order to recommend artists that are more than one node away of the seed one.

Moreover, feature selection is also an issue that needs to be tackled. Indeed, we identified that geolocation properties are often used for recommendation, but not always relevant. This is especially a problem for bands having a poor description in DBpedia, especially non-international ones where often, the only property besides their genre is their location. Then, it makes recommendation based mostly on the genre and location, which is often not relevant enough. We could have imagined excluding or weighting geolocation properties, but the issue is actually more complex. For instance, for a pop-band, being from Washington or San Francisco is probably not relevant. However, for a punk-hardcore one, this makes a lot of sense since the two scenes are radically different, and someone enjoying east-cost punk-hardcore may not listen to west-coast one. However, this would probably require manual classification of such graph patterns, in order to identify their relevance or not in certain contexts.

In addition, while currently limited to DBpedia, we aim at integration further sources of information (Freebase, MusicBrainz, etc.) to compute the recommendations, making the system targeted towards a wider Linking Open Data perspective.

# 8    Conclusion

In this paper, we discussed how semantic distance measures can be applied to Linked Data, and how they can be used to build music recommendation systems. We provided an algorithm to enable such measures on any Linked Data dataset, and an ontology to represent the distances and their explanations.

In addition, we have build dbrec, a recommendation system using DBpedia and providing open and explanatory recommendations for more than 39,000 bands and solo artists. The system was evaluated with use-centric evaluation, both off-line and on-line. We showed how it competes with last.fm, in addition of providing relevant novel recommendations, while relying only on public and open data, and not of listening behaviours of a large user set.

---

[17] http://smiy.sourceforge.net/rec/spec/recommendationontology.html

Finally, more than the distance measurement and the application, we discussed some set of lessons learnt from building the system, in terms of data quality, architectures for Semantic Web applications and optimisation of SPARQL queries. We hope that such lessons could be useful for implementers and provide some useful insights for anyone building applications consuming Linked Data circa'2010.

# References

1. Berners-Lee, T.: Linked Data. Design Issues for the World Wide Web, World Wide Web Consortium (2006), http://www.w3.org/DesignIssues/LinkedData.html
2. Bizer, C., Heath, T., Ayers, D., Raimond, Y.: Interlinking Open Data on the Web. In: Franconi, E., Kifer, M., May, W. (eds.) ESWC 2007. LNCS, vol. 4519. Springer, Heidelberg (2007)
3. Budanitsky, E., Hirst, G.: Semantic distance in wordnet: An experimental, application-oriented evaluation of five measures. In: Proceedings of the NAACL 2001 Workshop on WordNet and Other Lexical Resources (2001)
4. Celma, Ò., Herrera, P.: A new approach to evaluating novel recommendations. In: RecSys 2008: Proceedings of the 2008 ACM Conference on Recommender Systems, pp. 179–186. ACM, New York (2008)
5. Corlosquet, S., Delbru, R., Clark, T., Polleres, A., Decker, S.: Produce and Consume Linked Data with Drupal! In: Bernstein, A., Karger, D.R., Heath, T., Feigenbaum, L., Maynard, D., Motta, E., Thirunarayan, K. (eds.) ISWC 2009. LNCS, vol. 5823, pp. 763–778. Springer, Heidelberg (2009)
6. Euzenat, J., Shvaiko, P.: Ontology Matching. Springer, Berlin (2007)
7. Hausenblas, M.: Exploiting linked data to build web applications. IEEE Internet Computing 13(4), 68–73 (2009)
8. Hausenblas, M., Halb, W., Raimond, Y., Feigenbaum, L., Ayers, D.: SCOVO: Using Statistics on the Web of Data. In: Aroyo, L., Traverso, P., Ciravegna, F., Cimiano, P., Heath, T., Hyvönen, E., Mizoguchi, R., Oren, E., Sabou, M., Simperl, E. (eds.) ESWC 2009. LNCS, vol. 5554, pp. 708–722. Springer, Heidelberg (2009)
9. Heitmann, B., Hayes, C.: Using Linked Data to build open, collaborative recommender systems. In: Linked AI: AAAI Spring Symposium "Linked Data Meets Artificial Intelligence", AIII (2010)
10. Heitmann, B., Kinsella, S., Hayes, C., Decker, S.: Implementing Semantic Web applications: reference architecture and challenges. In: Proceedings of the 5th Workshop on Semantic Web Enabled Software Engineering. CEUR Workshop Proceedings, vol. 524, CEUR-ws.org (2009)
11. Hendler, J.A., Golbeck, J.: Metcalfe's law, Web 2.0, and the Semantic Web. Journal of Web Semantics 6(1), 14–20 (2008)
12. Herlocker, J.L., Konstan, J.A., Terveen, L.G., Riedl, J.T.: Evaluating Collaborative Filtering Recommender Systems. ACM Transactions on Information Systems (TOIS) 22(1), 5–53 (2004)
13. Hogan, A., Harth, A., Passant, A., Decker, S., Polleres, A.: Weaving the Pedantic Web. In: 3rd International Workshop on Linked Data on the Web (LDOW2010) at WWW 2010. CEUR Workshop Proceedings, vol. 628, CEUR-ws.org (2010)
14. Jacobson, K., Raimond, Y., Sandler, M.: An Ecosystem for Transparent Music Similarity in an Open World. In: International Symposium on Music Information Retrieval (2009)

15. Jeh, G., Widom, J.: Simrank: a measure of structural-context similarity. In: KDD 2002: Proceedings of the 8th ACM SIGKDD International Conference on Knowledge Discovery and Data Mining, pp. 538–543. ACM, New York (2002)
16. McSherry, D.: Explanation in Recommender Systems. Artificial Intelligence Review 24(2), 179–197 (2005)
17. Middleton, S.E., Alani, H., De Roure, D.: Exploiting Synergy Between Ontologies and Recommender Systems. CoRR, cs.LG/0204012 (2002)
18. Celma, Ò., Ramirez, M., Herrera, P.: Foafing the music: A music recommendation system based on RSS feeds and user preference. In: Proceedings of the 6th International Conference on Music Information Retrieval, ISMIR (2005)
19. Passant, A.: Measuring Semantic Distance on Linking Data and Using it for Resources Recommendations. In: Linked AI: AAAI Spring Symposium "Linked Data Meets Artificial Intelligence", AIII (2010)
20. Passant, A., Raimond, Y.: Combining Social Music and Semantic Web for Music-related Recommender Systems. In: Proceedings of the First Workshop on Social Data on the Web (SDoW2008). CEUR Workshop Proceedings, vol. 405, CEUR-ws.org (2008)
21. Rada, R., Mili, H., Bicknell, E., Blettner, M.: Development and application of a metric on semantic nets. IEEE Transactions on Systems, Man and Cybernetics 19, 17–30 (1989)
22. Raimond, Y., Abdallah, S., Sandler, M., Giasson, F.: The Music Ontology. In: International Conference on Music Information Retrieval, pp. 417–422 (September 2007)
23. Stocker, M., Seaborne, A., Bernstein, A., Kiefer, C., Reynolds, D.: SPARQL basic graph pattern optimization using selectivity estimation. In: WWW 2008: Proceeding of the 17th International Conference on World Wide Web, pp. 595–604. ACM, New York (2008)

# Knowledge Engineering for Historians on the Example of the *Catalogus Professorum Lipsiensis*

Thomas Riechert[1], Ulf Morgenstern[2], Sören Auer[1],
Sebastian Tramp[1], and Michael Martin[1]

[1] AKSW, Institut für Informatik, Universität Leipzig, Pf 100920, 04009 Leipzig
`lastname@informatik.uni-leipzig.de`
`http://aksw.org`
[2] Historisches Seminar, Universität Leipzig, Pf 100920, 04009 Leipzig
`lastname@uni-leipzig.de`
`http://www.uni-leipzig.de/histsem`

**Abstract.** Although the Internet, as an ubiquitous medium for communication, publication and research, already significantly influenced the way historians work, the capabilities of the Web as a direct medium for collaboration in historic research are not much explored. We report about the application of an adaptive, semantics-based knowledge engineering approach for the development of a prosopographical knowledge base on the Web - the *Catalogus Professorum Lipsiensis*. In order to enable historians to collect, structure and publish prosopographical knowledge an ontology was developed and knowledge engineering facilities based on the semantic data wiki *OntoWiki* were implemented. The resulting knowledge base contains information about more than 14.000 entities and is tightly interlinked with the emerging Web of Data. For access and exploration by other historians a number of access interfaces were developed, such as a visual SPARQL query builder, a relationship finder and a Linked Data interface. The approach is transferable to other prosopographical research projects and historical research in general, thus improving the collaboration in historic research communities and facilitating the reusability of historic research results.

## 1 Introduction

The World Wide Web, as an ubiquitous medium for publication and exchange, already significantly influenced the way historians work: the online availability of catalogs and bibliographies allows to efficiently search for content relevant for a certain investigation; the increasing digitization of works from historical archives and libraries, in addition, enables historians to directly access historical sources remotely. The capabilities of the Web as a medium for collaboration, however, are only starting to be explored. Many, historical questions can only be answered by combining information from different sources, from different researchers and organizations. Also, after original sources are analyzed, the derived information is often much richer, than can be captured by simple keyword indexing. These factors pave the way for the successful application of knowledge engineering techniques in historical research communities.

P.F. Patel-Schneider et al.(Eds.): ISWC 2010, Part II, LNCS 6497, pp. 225–240, 2010.

In this article we report about the application of an adaptive, semantics-based knowledge engineering approach for the development of a prosopographical knowledge base. In prosopographical research, historians analyze common characteristics of historical groups by studying statistically relevant quantities of individual biographies. Untraceable periods of biographies can be determined on the basis of such accomplished analyses in combination with statistically examinations as well as patterns of relationships between individuals and their activities.

In our case, researchers from the historical seminar at Universität Leipzig aimed at creating a prosopographical knowledge base about the life and work of professors in the 600 years history of Universität Leipzig ranging from the year 1409 till 2009 - the *Catalogus Professorum Lipsiensis* (CPL). In order to enable historians to collect, structure and publish this prosopographical knowledge an ontological knowledge model was developed and incrementally refined over a period of three years. The community of historians working on the project was enabled to add information to the knowledge base using an adapted version of the semantic data wiki OntoWiki [1][1]. For the general public, a simplified user interface[2] is dynamically generated based on the content of the knowledge base. For access and exploration of the knowledge base by other historians a number of access interfaces was developed and deployed, such as a graphical SPARQL query builder, a relationship finder and plain RDF and Linked Data interfaces. As a result, a group of 10 historians supported by a much larger group of volunteers and external contributors collected information about 1,300 professors, 10,000 associated periods of life, 400 institutions and many more related entities.

The benefits of the developed knowledge engineering platform for historians are twofold: Firstly, the collaboration between the participating historians has significantly improved: The ontological structuring helped to quickly establish a common understanding of the domain. Collaborators within the project, peers in the historic community as well as the general public were enabled to directly observe the progress, thus facilitating peer-review, feedback and giving direct benefits to the contributors. Secondly, the ontological representation of the knowledge facilitated original historical investigations, such as historical social network analysis, professor appointment analysis (e.g. with regard to the influence of cousin-hood or political influence) or the relation between religion and university. The use of the developed model and knowledge engineering techniques is easily transferable to other prosopographical research projects and with adaptations to the ontology model to other historical research in general. In the long term, the use of collaborative knowledge engineering in historian research communities can facilitate the transition from largely individual-driven research (where one historian investigates a certain research question solitarily) to more community-oriented research (where many participants contribute pieces of information in order to enlighten a larger research question). Also, this will improve the reusability of the results of historic research, since knowledge represented in structured ways can be used for previously not anticipated research questions.

The article is structured as follows: we present the overall technical architecture of the knowledge engineering approach in Section 2. We describe how the collaboration in

---

[1] Online at: http://ontowiki.net

[2] Available at: http://www.uni-leipzig.de/unigeschichte/professorenkatalog/

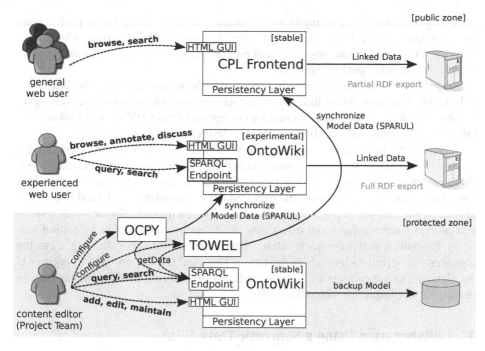

**Fig. 1.** Architectural overview about the project platforms

the historian community was facilitated by the semantic data wiki approach in Section 3. The underlying ontology is presented in Section 4. We elaborate on the knowledge engineering methodology in Section 5. Different exploration and access interfaces are showcased in Section 6. We present some prosopographical use cases in Section 7 and conclude in Section 8 with an outlook on future work.

## 2   Architectural Overview

The system architecture of CPL comprises a combination of different applications, which interact using standardized interfaces as illustrated in Figure 1. The project-set-up provides specialized human user interfaces for different user-groups according to the needs and tasks as well as generic access interfaces for machines including Linked Data and SPARQL endpoints. We divided the architecture into two separated zones (public and protected zone) due to technical constraints and in order to prevent security problems.

The semantic data wiki OntoWiki located in the protected layer [3] uses the *Catalogus Professorum Model* (CPM), which comprises several ontologies and vocabularies for structuring the prosopographical information (cf. Section 4). The project team, consisting of historians supported by knowledge engineers and semantic web experts, is working collaboratively and spatially distributed (e.g. in archives or libraries) to collect, structure and validate information about persons and institutions relevant to this

---

[3] http://professoren.ontowiki.net [restricted access]; OntoWiki-Version 0.85

knowledge domain. The resulting knowledge base is accessible only by the project team and is backed-up daily. Using the two configurable tools OCPY[4] (**O**ntology **C**o**PY**) and TOWEL[5] (**L**ightweight **O**ntology **E**xport **T**ool) the knowledge base is exported in order to make it accessible for the public.

For general web users the catalog is integrated in the public website of the University of Leipzig[6]. Due to technical limitations and security considerations on the web server of the university, a simplified user interface consisting of plain HTML and Linked Data Information is generated nightly from the knowledge base, using TOWEL.

Domain experts, i.e. historians, are able to interact with CPL via an experimental version[7] of OntoWiki, which is presented in Section 3. The version of the catalog available there is synchronized using the tool OCPY, that exports data from the protected OntoWiki installation, transforms the exported data considering any linked knowledge bases and imports the changed data into this experimental installation. This experimental deployment in particular offers new functionality of OntoWiki for testing purposes. In addition to this complete catalog, smaller subsets are provided (e.g. a catalog containing only professors born in the city of Dresden[8]). Users are able to register at the platform, to use community features such as resource commenting and tagging, or sharing SPARQL queries with other users.

## 3   Collaboration Using a Semantic Data Wiki

The core of CPL is OntoWiki - a tool for browsing and collaboratively editing RDF knowledge bases. It differs from other Semantic Wikis insofar as OntoWiki uses RDF as its natural data model instead of Wiki texts. Information in OntoWiki is always represented according to the RDF statement paradigm and can be browsed and edited by means of views. These views are generated automatically by employing the ontology features such as class hierarchies or domain and range restrictions. OntoWiki adheres to the Wiki principles by striving to make the editing of information as simple as possible and by maintaining a comprehensive revision history. This history is also based on the RDF statement paradigm and allows to roll-back prior change-sets. OntoWiki has recently been extended to incorporate a number of Linked Data[9] features, such as exposing all information stored in OntoWiki as Linked Data as well as retrieving background information from the Linked Data Web [5]. Apart from providing a comprehensive user interface, OntoWiki also contains a number of components for the rapid development of Semantic Web applications, such as the RDF API Erfurt[10], methods for authentication, access control, caching and various visualization components.

OntoWikis main interface consist of two types of views as shown in Figure 2.

---

[4] http://catalogus-professorum.org/tools/ocpy/
[5] http://catalogus-professorum.org/tools/towel/
[6] http://www.uni-leipzig.de/unigeschichte/professorenkatalog/
[7] http://catalogus-professorum.org/
[8] http://catalogus-professorum.org/Dresden/
[9] http://linkeddata.org/
[10] http://aksw.org/Projects/Erfurt/

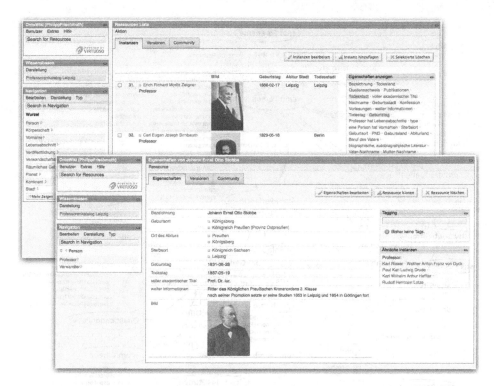

**Fig. 2.** OntoWiki views: (background) A tabular list view, which contains a filtered list of resources highlighting some specific properties of those resources and (foreground) a resource view which allows to tag and comment a specific resource as well as editing all property values.

*List views.* deal with the filtering and managing of resource lists. The user creates a list view by submitting a search keyword or selecting a class in the navigation module[11]. Subsequently, the user can apply multiple filter conditions to the list, which modify the underlying SPARQL query operating on the triple store.

*Resource views.* combine all information stored in OntoWiki about a specific resource. They are rendered by selecting a resource from a list view, requesting a resource directly via bookmark or link from an external page or selecting a resource in any other OntoWiki module. Resource views also allow the user to manipulate the selected resource. Starting from a resource view one can easily add and delete statements, as well as tags or comments.

## 4  The Catalogus Professorum Model

In this section we give an overview on the *Catalogus Professorum Model* (CPM), which is used to structure the prosopographical knowledge base. Although, the conceptual

---

[11] This module is not restricted to display class hierarchies, but allows to navigate through all types of hierarchies (e.g. group, geo-spatial or taxonomic hierarchies).

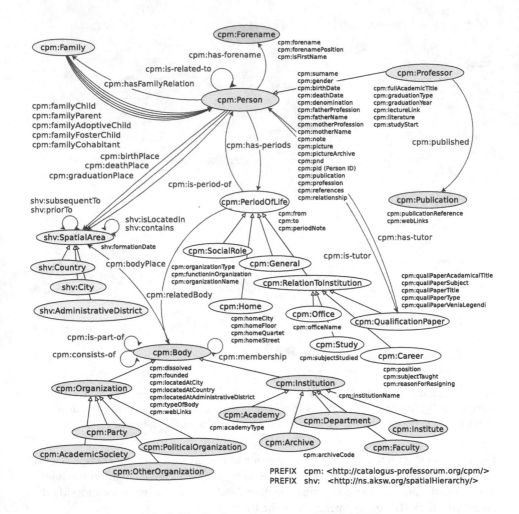

**Fig. 3.** The Catalogus Professorum Model (CPM)

model is initially only used for the Catalogus Professorum Lipsiensis the model was designed to be used for similar prosopographical knowledge bases at other universities. To facilitate reusability, CPM resides in its own namespace[12] and it uses identifiers with localized labels (currently English and German). This will in particular allow a simple integration of information from different sources at a later stage.

In its current version CPM contains 30 classes and 104 properties, 25 of which are object and 79 data properties. An birds-eye view of CPM is depicted in Figure 3. The central concepts in CPM are cpm:Person (with the subclass cpm:Professor), cpm:FamilyRelation, cpm:Body (with subclasses cpm:Organization and cpm:Insitution), cpm:Publication, cpm:PeriodOfLife and cpm:SpatialArea (with subclasses cpm:Country, cpm:City and cpm:AdministrativeDistric).

---

[12] CPM namespace: http://catalogus-professorum.org/cpm/

From a knowledge representation point-of-view, CPM currently uses rather shallow description logic expressivity. The used features are currently limited to subclass-superclass and subproperty-superproperty relationships, domain, range and simple cardinality restrictions, inverse properties and disjointness axioms. Most of the defined properties are defined as either object or datatype properties. CPM does not contain any deep class hierarchies. In the following paragraphs we describe some of the core concepts of CPM in more detail:

*Historic persons and professors.* The core information about persons in a prosopographical catalog comprises their name, information about birth and death, graduation and many more (cf. Figure 3). We explain in Section 5 why existing vocabularies such as FOAF [13], ULAN[14] or VIAF[15] were not reused and how a mapping from properties such as `cpm:birthDate` to these vocabularies will be achieved. A crucial resource for the interlinking of person data in the German speaking region is the Personen Namen Datei (PND) of the German National Library. Projects such as PND/BAECON[16] or LinkedHistory[17] enable the interlinking of existing databases on the basis of this identifier.

*Periods of Life.* For prosopographical research it is paramount to have a fine-grained representation of different periods within the life of a certain person. In order to capture such knowledge we introduced the concept `cpm:PeriodOfLife`, which is associated with a `Person` through the properties `cpm:has-period` and its inverse `cpm:is-period-of`. Different types of periods are distinguished by the subclasses `cpm:SocialRole`, `cpm:Home`, `cpm:RelationToInstitution`, `cpm:Office` (e.g. dean, rector), `cpm:Study`, `cpm:Career`, `cpm:QualificationPaper` (e.g. dissertation, habilitation) or `cpm:General`. Each of these types of periods of life is the domain of a number of properties, which are used to describe a particular instance in more detail, however, all inherit the delimiting properties `cpm:from` and `cpm:to`. Different periods of life of the same person can overlap, e.g. the `cpm:Home` usually overlaps with other periods.

*Bodies.* The bodies class is used to describe organizations (e.g. parties) and institutions (e.g. academies). Persons can be members of more than one body.

*Historic specifics of spatial areas.* CPM uses the independently developed *spatial hierarchy vocabulary* (SHV)[18] to represent spatial information. The core of SHV is the class `shv:SpatialArea` with subclasses `shv:City`, `shv:Country`, `shv:AdminstrativeDistrict`. Using the properties `shv:contains`

---

[13] Friend Of A Friend, http://www.foaf-project.org/
[14] Union List of Artist Names Online,
   http://www.getty.edu/research/conducting_research/vocabularies/ulan/
[15] Virtual International Authority File, http://viaf.org/
[16] http://meta.wikimedia.org/wiki/BEACON
[17] http.//linkedhistory.aksw.org/pnd/
[18] http://ns.aks.org/spatialHierarchy/

and `shv:isLocatedIn` instances belonging to these classes can be arranged in spatial hierarchies of arbitrary depth. SHV is aligned with the World Geodetic System vocabulary WGS84[19]. In order to support the representation of historic administrative divisions, each spatial area can be equipped with properties `shv:formationDate`, `shv:dissolutionDate` as well as `shv:priorTo` and `shv:subsequentTo`. Thus, the historic administrative evolution can be easily represented. Königsberg, the early capital of Prussia, for example, is located in the following historic administrative districts (respective countries): Duchy of Prussia (from 10 April, 1525), the united Duchy of Prussia and Brandenburg (from August 27, 1618), Kingdom of Prussia (from 18 January, 1701), the Free State of Prussia (from 9 November, 1918) until its abolition at 30 January, 1934.

*Representing family relationships.* Family relationships are represented in CPM using the `cpm:Family` class. Instances of the class `cpm:Person` are then related to an instance of the `cpm:Family` class using one of the following properties: `cpm:familyChild`, `cpm:familyAdoptiveChild`, `cpm:familyFosterChild`, `cpm:familyParent`, `cpm:familyCohabitant`. Genealogy is a subfield of prosopographical research. Within this popular area of research some large database we already developed. A popular vocabulary for representing genealogical information is GEDCOM[20], which is also based on the family concept as central information asset. For evolutionary reasons, we still also included the properties `is-related-to` and `cpm:relationship`, which allow to model a family relationship more directly.

## 5   Knowledge Engineering Methodology

In order to describe the knowledge engineering methodology followed in CPL we statistically analyzed a number of usage indicators (cf. Figure 4). Editing activities using OntoWiki in the restricted zone was logged since September 2008. Statistics about accessing data via the user interface of the CPL frontend was logged since its launch in April 2009. This statistics does, however, not include any access information from the linked data interface.

It can be clearly seen, that most of the editing activity were additions of statements. In the learning phase (i.e. first four months of the project from September 2008 till January 2009), there was still notable but decreasing number of statement deletions and property changes, which indicate corrections and an increasing familiarity of the domain experts with the system. Till June 2009 - the month of the official public announcement of CPL - the activities intensified, with regard to added statements, professors and changed properties. Due to feedback from other historians working in the field and the general public, the editing activity remained relatively high after the announcement, but decreased slowly. The number of editors (i.e. historian domain experts) ranged between 5 and 10.

For the development of the CPM we have chosen a very pragmatic approach. The development of a first version of the ontology was simplified due to the availability of

---

[19] http://www.w3.org/2003/01/geo/wgs84_pos#
[20] Genealogical Data Communication, The GEDCOM standard release 5.5, 1996.

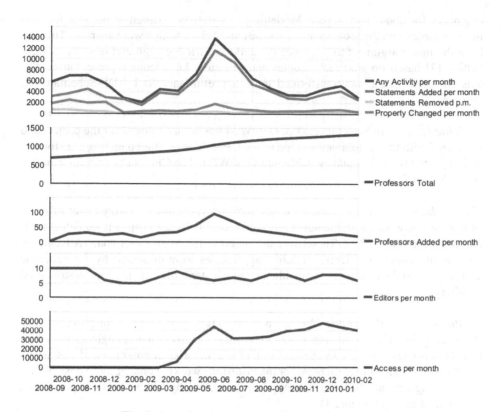

**Fig. 4.** Statistics about various CPL usage indicators

legacy data. The idea of using the FOAF and other existing vocabularies turned out inappropriate. The main reason for this being the absence of a precise understanding of the semantics to be represented within the catalog. Discussions between ontology engineers and historians about the overall ontology structure turned out to be very intricate due to the lack of understanding of the respective field of the other party. A solution to this dilemma was not to focus on the general ontology structure, but rather on small concrete representation issues and facilitate the evolution of the CPM and the CPL in an ontology/application co-design process. Overall, the engineering co-design methodology of the CPL can be characterized so far by 6 phases, which we describe in the sequel.

*Before CPL.* In 2006 the project was started using an single data table. This table was edited asynchronously by the project team. Since the database table become more complex and the number of needed columns reaches the column limit of 255 the project team requested technical support from database experts.

*(1) Information analysis.* Based on the existing database table, an analysis and remodeling of the database had been done. The resulting entity-relationship (ER) model was discussed and improved iteratively by the team of historians and knowledge

engineers for about half a year. Modelling this database structure became increasingly complex and discussions about entities and relationships were soaring. This has been the major argument to introduce the agile knowledge engineering method RapidOWL [2] based on wiki technologies and semantic knowledge representation. The existing ER schema was transformed into an preliminary RDF/RDF-Schema/OWL ontology.

*(2) Initialization.* In September 2008, OntoWiki has been deployed for the project and initialized with the preliminary version of the CPM. The existing data was transformed and imported from the database table into OntoWiki. The CPL starts with information about 700 professors.

*(3) Wiki-based knowledge acquisition and ontology refinement.* During the acquisition of new instance data with the now OntoWiki based CPL the project team detect limitations of the initial CPM. These were discussed within the project team. Advantages and disadvantages of different modeling approaches were presented by the Semantic Web experts to the historians and resulted in an substantially extended and restructured ontology.

*(4) Publication of the catalog.* As shown in the diagrams in Figure 4 the publication of the catalog data also results intensified knowledge acquisition and engineering activities. This is caused mainly by the feedback of the historian community and web users, but also result in a motivation boost for the historians working on CPL. The possibilities of accessing data has been permanently improved, e.g. by launching the experimental OntoWiki server (cf. Figure 1).

*(5) Interlinking other datasets.* The last phase of CPL so far was is the interlinking of the catalog with other databases. This will enhance working with the wiki and querying the data even more. Interlinking was performed with DBpedia [6], the German National Library using the unique PND [8] identifier and with the catalog of lecture directories[21] of the University of Leipzig.

*(6) Alignment to other ontologies.* As a result of the phases 4 and 5 the CPL gained quite much attention within the research community. As a consequence the interlinking with other knowledge bases became more important and there was a strong feedback with regard to aligning the CPM with other prosopographical knowledge bases.

## 6    Exploration and Access Interfaces

In order to facilitate the interaction of domain experts and interested people, CPL is accompanied with a number interfaces for accessing and exploring the information. In this section we briefly showcase the generic access interfaces Linked Data, Visual Query Builder and Relationship Finder, as well as the specifically developed public CPL website.

---

[21] http://histvv.uni-leipzig.de

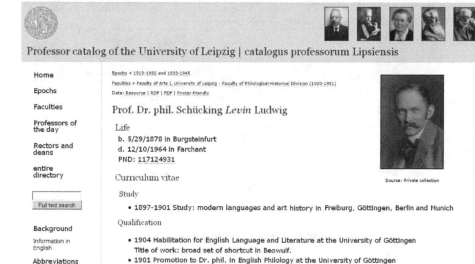

**Fig. 5.** Visualization of the resource representing "Prof. Dr. phil. Levin Ludwig Schücking'" on the public CPM website

**Public Website.** CPL is not just a tool for historians, but aims to showcase the results of historic research to the wider general public. For that purpose a special public website was created (as shown in Figure 5). The user interface of the public website is geared towards simplicity. The knowledge base can be explored by epochs, faculties, functions of professors (i.e. rector or dean) or alphabetically. Professors of the day are automatically selected based on important days in the life of a professor (i.e. birth or death). Furthermore, the public website comprises a full-text search, which searches within all literals stored in the CPL knowledge base.

**Linked Data.** The term Linked Data here refers to a set of best practices for publishing and connecting structured data on the Web[22]. These best practices have been adopted by an increasing number of data providers over the past three years, leading to the creation of a global data space that contains many billions of assertions Using OntoWiki's build-in endpoint functionality CPL is immediately available as Linked Data. Linked Data information is easy accessible e.g. using the Tabulator tool[23] [3]. Within the Linking Open Data effort, hundreds of data sets have already been connected to each other via `owl:sameAs` links. By interlinking CPL with other related datasets we aim at establishing CPL as a linked data crystallization point for academic prosopographical knowledge.

---

[22] http://www.w3.org/DesignIssues/LinkedData.html

[23] Resource "Schücking" in Tabulator is available at: http://dig.csail.mit.edu/2005/ajar/release/tabulator/0.8/tab.html?uri=http://catalogus-professorum.org/lipsiensis/Schuecking_144

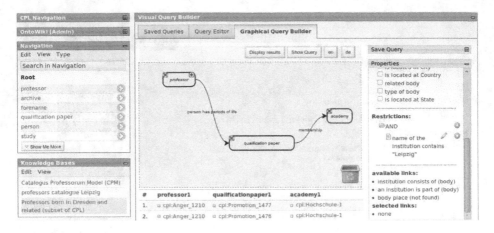

**Fig. 6.** *Visual Query Builder*

**Visual Query Builder.** OntoWiki also serves as a SPARQL endpoint, however, it quickly turned out that formulating SPARQL queries is too tedious for the historian domain experts. In order to simplify the creation of queries for the historians, we developed the *Visual Query Builder*[24] (VQB) as an OntoWiki extension, which is implemented in JavaScript and communicates with the triple store using the SPARQL language and protocol. VQB allows to visually create queries to the stored knowledge base and supports domain experts with an intuitive visual representation of query and data. Developed queries can be stored and added via drag-and-drop to the current query. This enables the reuse of existing queries as building blocks for more complex ones. VQB also supports the *set-based browsing* paradigm by visualizing different connectives, such as join, union, intersection, difference between queries. The incremental query building is facilitated by displaying results already during query creation. The VQB user interface is visualized in Figure 6. It consists of 5 panels, which visualize different aspects of the query creation:

- *Center panel:* the main workspace, where the query is visualized as a graph. Individual elements of the graph can be selected, deleted or moved on the canvas.
- *East panel:* displays information about the currently selected element. If the selected element is, for example, a class this panel contains the properties, which are used with instances of the class. The panel also contains controls for changing the query, such as via filter conditions.
- *West panel:* contains available classes in a tree display. Queries are grouped with the classes contained in the query.
- *South panel:* visualizes the result of the query.
- *North panel:* shows notifications such as usage hints, error or event descriptions and the generated SPARQL query.

The user interface can be adjusted by scaling or deactivating unused panels.

---

[24] http://aksw.org/Projects/OntoWiki/Extension/VQB

**Relationship Finder.** An important aspect of historical investigations is the search for relationships between different persons or entities of interest. An application supporting such investigations within RDF knowledge knowledge bases is *RelFinder*[25] [4]. With the help of *RelFinders* relationships between individual entities can be easily discovered and visualized. Figure 7, for example, visualizes the relationship between the entities "Schücking" and "München"[26]. In this example, three connections were found and visualized as paths through the knowledge base. *RelFinder* is a generic tool and can be used in conjunction with arbitrary SPARQL endpoints.

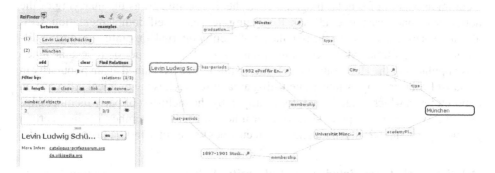

**Fig. 7.** Visualization of relationships in *RelFinder*

## 7  Prosopographical Use Cases

In this section we introduce some important prosopographical research use cases, which are facilitated by the ontological knowledge representation and the various exploration and access interfaces. These include in particular historical social network analysis, academic self-complementation analysis and the relationship between religion and university.

**Historical Social Network Analysis.** The analysis to what extend a certain professor influenced his students, colleagues and friends was previously only possible with a hardly justifiable manual effort. Although the CPL comprises primarily professors working in Leipzig, it reaches far beyond the limits of the Saxon state university, since all periods of life of a professor are included. Due to the semantic representation, it is easily possible to identify direct as well as indirect connections between individuals, such as, for example, an overlap in certain periods of life (such as a common school or university visit). Also, the detailed representation of qualification steps (such as doctoral and post-doctoral studies) facilitates the reconstruction of teacher-student relationships and thus the identification of certain schools of thought and on a more general level the establishment *academic genealogy*.

---

[25] Online at: http://relfinder.semanticweb.org

[26] More interesting relationships obtained from RelFinder are listed at: http://catalogus-professorum.org/tools/relfinder

**Academic Self-Complementation Analysis.** A crucial aspect of academic historic research is the analysis of the self-complementation functions of the different faculties. In particular cousin-hood was an important factor for chair appointments in German universities of the early modern period (i.e. from approx. AD 1500 to around AD 1800), which thus were heavily influenced by scholar dynasties. In the 19th century this practice changed dramatically, but still intellectual families aimed at preserving their social prestige by fostering the academic careers of their sons. CPL allows far reaching qualitative and quantitative research due to the fine-grained representation of family relationships. An interesting observation, for example, revealed by a query to CPL was that a common pattern of scholar biographies in the 19th century was the marrying of the daughter of ones academic advisor. Regarding academic self-complementation, CPL also allows to easily observe the popularity of faculty internal appointments during different periods of time: while very common in the early modern period it became much less popular after the Humboltanian reforms in the 1820s, but gained again popularity in socialist East Germany. Another interesting area of research is the political influence on appointments, which is facilitated, for example, by the inclusion of party memberships in CPL. The allows, in particular, to determine whether the membership in a certain political party statistically significantly affects ones career in a certain historic period.

**Relationship between Religion and University.** Religion was not only a founding factor of many European universities, but influenced academic live in one way or the other throughout the centuries. After the *Confessio Augustana* in the year 1543, Lutheranism was the obligatory confession at Universität Leipzig. Only in the 19th century it became possible to admit catholic members to the faculties and it is an interesting research question, facilitated by CPL, to investigate to what extend and in which fields this was actually the case. During the Third Reich period the destiny of Jewish professors is in the center of interest and later in, Eastern Germany, the non-confessionalism or reserves of religiosity in certain regime-distant fields is in the focus of interest.

# 8  Lessons Learned, Conclusions, Related and Future Work

**Lessons Learned.** A lesson we learned is, that such a project involving people with very different backgrounds and with very limited resources requires to establish a working *knowledge base / application co-design*, where both - knowledge bases and knowledge-based applications - are iteratively refined. Given the timely visibility of the knowledge base for a wider community, additional refinements are triggered by the interaction with the community. As we experienced with the historian domain expert team, the motivation boast due to the early public availability of the knowledge base project and the direct visibility of improvements and corrections can hardly be underestimated and by far outweighs initial maturity gains with longer development cycles. A growing added value for domain experts is the availability of background knowledge on the linked Data Web. CPL is one of the first prosopographical knowledge bases on the data web, but still the interlinking and fusing of information, for example, from DBpedia and Geonames is of great use for enhanced querying and exploration of the information. As more prosopographical knowledge bases become available this effect will be even more amplified, as we are currently already experiencing with applications of the CPL infrastructure for other prosopographical use cases.

**Conclusions.** With CPL we demonstrated a successful application of semantic knowledge representation techniques and an agile collaboration methodology in social sciences. For historians the followed knowledge base approach resulted in completely new research opportunities, compared to the book/lexicon based methodologies prevalent in prosopographical research. The use of knowledge bases and agile, web-based collaboration has the potential to trigger a paradigm shift in historic research: from individual centered research aiming to solve a specific research task towards collaborative research, which's results can be re-purposed in order to answer unanticipated research questions.

**Related Work.** *SemanticWeb for History*[27] (SWHi) is a project which was carried out from 2006 to 2008 at the University of Groningen in the Netherlands had the aim to combine available vocabularies to be able to model the early American history. A semantic browser[28] for this data has been developed. The vocabulary[29] developed during this project is not published so far. The *Historical Event Markup and Linking Project*[30] (HEML) primarily aims at representing early Greek history. The developed vocabulary [9] is based on events and periods. Central elements of the vocabulary[31] are the classes heml:Event, heml:Person, heml:Role and heml:Evidence.

Beyond history, wiki-based knowledge engineering has been successfully applied to other knowledge domains before (e.g. Brede Wiki for Neuroscience data [7] or the Modelling wiKi MoKi [10]).

**Future Work.** One hotspot of future work will be the expansion of the usage of the Catalogus Professorum. Currently, we are planning to extend the catalog to include the universities Munich and Berlin, which for a long time represented the academic center together with Universität Leipzig. In addition there are cooperations with the universities Utrecht and Zürich and a number of other prosopographical databases are currently investigating how to interlink with the CPL and the Linked Data Web in general. In addition we aim to adopt the developed techniques in order to support other historic knowledge engineering projects beyond prosopographical databases. Also, as for the Data Web in general we aim to increase the coherence by tighter interlinking of CPL with related and complementary knowledge on the Data Web.

## Acknowledgments

We would like to thank our colleague Christian Augustin and all of the young researchers, involved in archival work and interviews, for their helpful comments and

---

[27] http://americanimprints.ub.rug.nl/
[28] http://semweb.ub.rug.nl/
[29] http://semweb.ub.rug.nl/swhi#
[30] http://heml.mta.ca/
[31] http://www.heml.org/rdf/2003-09-17/heml/

inspiring discussions during the development of CPL. This work was supported by grants from the German Federal Ministry of Education and Research (BMBF), provided for the project LE4SW (project number 03WKP02A) and from the European Union's 7th Framework Programme provided for the project LOD2 (GA no. 257943).

# References

1. Auer, S., Dietzold, S., Riechert, T.: OntoWiki – A Tool for Social, Semantic Collaboration. In: Cruz, I., Decker, S., Allemang, D., Preist, C., Schwabe, D., Mika, P., Uschold, M., Aroyo, L.M. (eds.) ISWC 2006. LNCS, vol. 4273, pp. 736–749. Springer, Heidelberg (2006)
2. Auer, S.: The RapidOWL Methodology–Towards agile knowledge engineering. In: Proceedings of the 15th IEEE International Workshops on Enabling Technologies: Infrastructure for Collaborative Enterprises, pp. 352–357. IEEE Computer Society, Los Alamitos (2006)
3. Berners-lee, T., Chen, Y., Chilton, L., Connolly, D., Dhanaraj, R., Hollenbach, J., Lerer, A., Sheets, D.: Tabulator: Exploring and analyzing linked data on the semantic web. In: Procedings of the 3rd International Semantic Web User Interaction Workshop (SWUI 2006), p. 6 (2006)
4. Heim, P., Hellmann, S., Lehmann, J., Lohmann, S., Stegemann, T.: RelFinder: revealing relationships in RDF knowledge bases. In: Chua, T.-S., Kompatsiaris, Y., Mérialdo, B., Haas, W., Thallinger, G., Bailer, W. (eds.) SAMT 2009. LNCS, vol. 5887, pp. 182–187. Springer, Heidelberg (2009)
5. Heino, N., Dietzold, S., Martin, M., Auer, S.: Developing semantic web applications with the ontowiki framework. In: Networked Knowledge - Networked Media. SCI, vol. 221, pp. 61–77. Springer, Heidelberg (2009)
6. Lehmann, J., Bizer, C., Kobilarov, G., Auer, S., Becker, C., Cyganiak, R., Hellmann, S.: DBpedia - a crystallization point for the web of data. Journal of Web Semantics 7(3), 154–165 (2009)
7. Nielsen, F.Å.: Brede wiki: Neuroscience data structured in a wiki. In: 4th Semantic Wiki Workshop (SemWiki 2009) at ESWC 2009. Proceedings. CEUR WS, vol. 464 (2009)
8. Pfeifer, B., Senftleben, S.: Die Personennamendatei (PND). Leipziger Beiträge zur Informatik, vol. Band XXI, pp. 137–144. LIV (2010)
9. Robertson, B.: Exploring historical rdf with heml. Changing the Center of Gravity: Transforming Classical Studies Through Cyberinfrastructure 3(1) (Winter 2009)
10. Rospocher, M., Ghidini, C., Pammer, V., Serafini, L., Lindstaedt, S.N.: Moki: the modelling wiki. In: 4th Semantic Wiki Workshop (SemWiki 2009) at the 6th European Semantic Web Conference (ESWC 2009), Hersonissos, Greece. CEUR Workshop Proceedings, vol. 464 (June 1, 2009)

# Time-Oriented Question Answering from Clinical Narratives Using Semantic-Web Techniques

Cui Tao[1], Harold R. Solbrig[1], Deepak K. Sharma[1], Wei-Qi Wei[1], Guergana K. Savova[2], and Christopher G. Chute[1]

[1] Division of Biomedical Statistics and Informatics, Mayo Clinic, Rochester, MN
[2] Harvard Medical School, Boston, MA

**Abstract.** The ability to answer temporal-oriented questions based on clinical narratives is essential to clinical research. The temporal dimension in medical data analysis enables clinical researches on many areas, such as, disease progress, individualized treatment, and decision support. The Semantic Web provides a suitable environment to represent the temporal dimension of the clinical data and reason about them. In this paper, we introduce a Semantic-Web based framework, which provides an API for querying temporal information from clinical narratives. The framework is centered by an OWL ontology called CNTRO (Clinical Narrative Temporal Relation Ontology), and contains three major components: time normalizer, SWRL based reasoner, and OWL-DL based reasoner. We also discuss how we adopted these three components in the clinical domain, their limitations, as well as extensions that we found necessary or desirable to archive the purposes of querying time-oriented data from real-world clinical narratives.

## 1 Introduction

The rapid increase in the volume of electronic health records (EHR) available for research purposes provides new opportunities to create semantically interoperable healthcare applications and solutions for evidence-based medicine. An important aspect of EHR is the temporal ordering of clinical events. Time is essential in clinical research [20]. Exposing the temporal dimension in medical data analysis provides new research paths such as (1) uncovering temporal patterns at the disease and patient level to better understand the progression of a disease, (2) explaining past events such as the possible causes of a clinical situation, and (3) predicting future events such as possible complexities based on a patient's current status.

One important objective for enable meaningful use of EHR is to develop software applications "to realize the true potential of EHR to improve the safety, quality, and efficiency of care" [3]. In order to facilitate clinical researchers to expose the temporal dimension in medical data analysis, software platforms that allow users to ask free-form queries and retrieve temporal information automatically from clinical records are highly desired. First, the temporal information

P.F. Patel-Schneider et al.(Eds.): ISWC 2010, Part II, LNCS 6497, pp. 241–256, 2010.

interwoven in clinical narratives needs to be extracted and annotated to allow computer systems to be able to locate the information of interest. Second, temporal relations and assertions that are not explicitly expressed in the original documents need to be automatically inferred in order to enable the full capacity and true potential of secondary use of EHR for meaningful use. Third, temporal-oriented questions need to be captured in computer queries to query the annotated and inferred information.

The Semantic Web and the Web Ontology Language (OWL) [13] provide a suitable environment for modeling the temporal dimension of the clinical data, reasoning and inferring new knowledge, and querying for the information desired. The Semantic Web provides a standard mechanism with explicit and formal semantic knowledge representation, and automated reasoning capabilities. OWL is built on formalisms that adhere to Description Logic (DL) and therefore allows reasoning and inference. The Semantic Web Rule Language (SWRL) [23] can be used to add rules to OWL and enable Horn-like rules that can be used to infer new knowledge from an OWL based ontology and reason about OWL individuals. Once we have an ontology that can represent temporal assertions in the clinical domain precisely, we can annotate temporal expressions and relations with respect to the ontology and store the instances as RDF triples [17]. The information then become "machine-understandable". Tools and services such as reasoners, editors, querying systems, and storage mechanisms that have been developed by the Semantic Web community can be directly applied to the temporal data.

In this paper, we introduce a Semantic-Web based framework, which provides an API for querying temporal information from clinical narratives. The framework is centered by an OWL ontology called CNTRO (Clinical Narrative Temporal Relation Ontology), and contains three major components: time normalizer, SWRL based reasoner, and OWL-DL based reasoner. We also discuss how we adopted these three components in the clinical domain, their limitations, as well as extensions that we found necessary or desirable to archive the purposes of querying time-oriented data from real-world clinical narratives.

## 2    Related Work

Several approaches already exist for the modeling and query of temporal information. Most of these are research efforts that focus on temporal information stored in structured databases [32]. There are two existing temporal ontologies in OWL, the Time Ontology [29] and the SWRL Temporal ontology [25], the first of which is a general time ontology that defines basic time components and their relationships. And the second one is built for the SWRL Temporal Built-Ins library [24]. Both ontologies adopted Allen's Interval Based Temporal Logic [1], which provides a foundation of temporal logic for many temporal models. Tappolet and et al. [27] propose using time as an additional semantic dimension of data using RDF named graphs in combination with a temporal extension of the SPARQL query language called t-SPARQL. The SWRL Temporal Built-Ins library [24] defines a set of built-ins that can be used in SWRL rules to perform temporal operations and has been applied in clinical research such as the system described in [11]. These approaches, however, focus on the relationships between instances and intervals in time and it is not obvious how these relationships can be applied to actual events themselves.

There are also existing approaches that focus on the representations of free text narratives such as those encountered in clinical notes. Models such as Temporal Constraint Structure (TCS) [31] and the TimeML model [28] provide ways to represent temporal information in natural language. HL7 time specification [7] defines data types that can be used to specify the complex timing of events and actions such as those that occur in order management and scheduling system. While these models provide a good foundation, they are not currently compatible with OWL and other semantic-web based tools and do not support formal reasoning to infer new temporal knowledge.

## 3    Clinical Narrative Temporal Relation Ontology

We have developed an ontology in the Web Ontology Language (OWL) format for modeling temporal information in clinical narratives, and evaluated this ontology using real-world clinical notes [26]. In this section, we briefly introduce our OWL ontology for temporal relation reasoning in clinical narratives, which we call the Clinical Narrative Temporal Relation Ontology (CNTRO)[1]. This ontology can model the temporal information found both in structured databases and in natural-language based clinical reports. We investigated the existing conceptual models for temporal information cited in the previous section. CNTRO was developed based on these previous experiences combined with new ontological specifications that fit the needs of natural-language based clinical reports. We decided to first build a stand-alone model based on our requirements, which is what is described in this paper. Subsequent work will involve the integration of CNTRO and existing ontologies that cover time-related components such as the Time Ontology in OWL [29], and Basic Formal Ontology (BFO) [2].

Figure 1 shows the graphical view of the ontology. OWL classes are represented by a rectangles with rounded corners and data types are represented by an ovals. Subclass relationships are represented by hollow-headed arrows and object and data properties by solid-headed arrows.

The class, *Event* represents an occurrence, state, perception, procedure, symptom or situation that occurs on a time line in clinical narratives.

The *Time* class is the superclass of all the OWL temporal representation classes: *TimeInstant*, *TimeInterval*, *TimePhase*, and *TimePeriod*. An OWL *TimeInstant* is a specific point of time on the time line. In clinical reports, a time instant can be represented in different granularities such as year, month, and day. A time instant may also be represented in different formats. We implemented a normalizer that converts commonly used time notations to the XML dateTime

---

[1] http://www.cntro.org

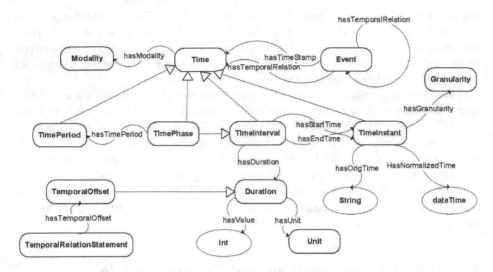

**Fig. 1.** A Graphical View of Clinical Narrative Temporal Relation Ontology

format. In the ontology, we defined two data properties *hasOrigTime* and *has-NormalizedTime* that keep track of the time instant in its original form and in the normalized form respectively. An OWL *TimeInterval* represents a duration of time. It could have two relations (OWL object properties), *hasStartTime* and *hasEndTime*. Each of them links to instances of *TimeInstant*. A *TimeInterval* could also have a *Duration*. An instance of the *Duration* class represents the time length of a *TimeInterval*. We use an OWL data type property *hasValue* and an OWL object property *hasUnit* to describe a *Duration*. Many clinical events recur periodically. Adopted and modified from the HL7 time specification [7], two OWL classes, *TimePhase* and *TimePeriod*, are defined in CNTRO to represent intervals of time that recur periodically. A *TimePhase* represents each occurrence of the repeating interval and a *TimePeriod* specifies a reciprocal measure of the frequency at which the *TimePhase* repeats. The class *TimePhase* is a subclass of *TimeInterval*, therefore, we can also specify a *StartTime*, an *EndTime*, and a *Duration*. In addition, a relation (OWL ObjectProperty), *hasTimePeriod*, is defined to specify the relation between a *TimePhase* and a *TimePeriod*. For example, "every 8 hours for 10 days starting from today" is a *TimePhase*. Its *StartTime* is "today". Its *Duration* is "10 days". And its *TimePeriod* is "every 8 hours". We also define the certainty of a *Time* instance. For example, a physician can describe a time notation with ambiguities such as "early next week" and "in approximately two weeks". In the CNTRO ontology, we defined a class called "Modality" which serves as a flag to indicate whether a time representation is approximated or not.

We can define the temporal relations between two events, or between an event and a time instance using the object property *hasTemporalRelation* and its subproperties. We also use Allen's temporal logic operators when defining our temporal relation properties: equal, before, after, during, meet, start, finish, and during. We have also defined their logical characteristics. For example, *before* is a transitive property, and its inverse property is *after*.

We can also use *TemporalRelationStatement* class to describe temporal relations between two events or between an event and a *Time* instance. The *TemporalRelationStatement* class is a sub-class of *rdf:Statement*, we can define temporal subject, object, and predicate of a *TemporalRelationStatement*. Using *TemporalRelationStatement* to describe a temporal relation enables defining properties of the relation by reification. For example, we can add an offset time frame to the relation by using an OWL object property called *hasTemporalOffset*. The domain of *hasTemporalOffset* is *TemporalRelationStatement* and the range of it is *Duration*. This offset defines the relative timing of a pair of events. In order to model the sentence "patient's bilirubin is elevated 2 weeks after the second cycle of chemotherapy", for example, we can use a *TemporalRelationStatement* to represent "patient's bilirubin is elevated" (object) *after* (predicate) "the second cycle of chemotherapy" (subject), and then add "2 week" as an instance of *TemporalOffset* to this *TemporalRelationStatement* instance.

We compared the expressiveness capabilities of the CNTRO ontology with the two existing temporal ontologies in OWL: the Time ontology [29] and the SWRL Temporal ontology [25]. Since these two ontologies are designed only for structured data in databases, they mainly focus on timing events with points anchored in absolute time. To cover the temporal assertions in natural-language based clinical narratives, we have added the following major expressiveness capabilities to the CNTRO ontology. (1) **Periodic Time Interval.** In clinical narratives, there are many events that recur periodically. It is important to be able to represent periodic time intervals. Two OWL classes, TimePeriod and TimePhase, have been defined to represent periodic time intervals in the CNTRO ontology. (2) **Relation between Two Events.** In many cases in clinical notes, physicians describe the relations between two events without indicating the time stamps of the events, i.e., (*patient's bilirubin is elevated after the second cycle of chemotherapy*). While the other two OWL ontologies defines that the temporal relations are only between Time entities themselves, CNTRO is able to capture this kinds of qualitative temporal relationships. (3) **Time Offset.** The CNTRO ontology defines a *TemporalOffset* class which enables representing the time offset of a relation using reification. (4)**Relative Time.** Relative time such as "today", "tomorrow", "two months ago", or "in 3 weeks" is very commonly used in clinical reports. The CNTRO ontology captures the relative time information in its original form and at the same is able to represent the calculated absolute time in the normalized form. (5) **Uncertainty.** Often temporal information is represent with uncertainty in clinical notes. CNTRO offers a property–*hasModality* to track of the uncertainty to make sure it can be taken into consideration in answering temporal questions.

# 4    Temporal Information Reasoning

CNTRO provides a conceptual model to represent temporal relations in clinical narratives. A lot of time qualitative and quantitative temporal relationships, however, are expressed implicit in the event occurrences. The answers of many time-oriented questions are not necessarily stated explicitly in clinical narratives, but rather need to be inferred. For example, here are three sentences from one patient's clinical notes: (1)*Patient's INR value is below normal* (Event_1) today. (note date: 01/26/07) (2)He has had the *chills and body aches* (Event_2) before *the abnormal test.* (Event_3)" (note date: 01/26/07) (3) On Jan. 30, 2007, patient started *Coumadin dosing plan of 1.0 mg* ((Event_4)).(note date: 02/09/07) To answer the question "did the patient experience body aches before the he started the Coumadin dosing plan?", we actually need a few different steps of inferences. We know that Event_1 has a time stamp "today" (time instant); Event_2 is *before* Event_3; and Event_4 has a time stamp which is a time interval that has start date "Jan. 30, 2007". We first infer that the date of "today" for Event_1 is the note date, which is "01/26/07". We then infer that Event_3 actually refers to Event_1. Therefore we know that Event_2 is before Event_1, which happened on "01/26/07". Hence, we know that Event_2 is before "01/26/07". Now we need to compare "01/26/07" and "Jan. 30, 2007" which is the Event_4's time stamp. In order to do that, we need to normalize the two dates, and infer that "01/26/07" is before "Jan. 30, 2007". Since that Event_2 is before "01/26/07", which is before "Jan. 30, 2007", which is the start time of Event_4, we then can finally infer that Event_2 is before Event_4.

This simple example illustrates how reasoners can help to infer temporal relations. In this section, we discuss three major components we need to do temporal relation inferences.

## 4.1    Temporal Representation Normalization

Temporal information in clinical text can be expressed in different ways [32]. In order to infer temporal relations in clinical narratives, our first step is to normalize the time expressions. Because the clinical records we are working with are from the US based Mayo Clinic, this research focuses on conversion of commonly used US temporal notations [4] to the xsd DateTime Data Type format [30]. We used the information extraction technology developed by the Brigham Young University (BYU) Data Extraction Group (DEG) [6] to recognize different time notations. The DEG group has developed a set of libraries that recognize when the same concept when represented in different formats, and we make use of their time recognition component to identify different representations of the same time. We then normalize the format and convert it into the xsd DateTime format associating both the original and normalized time with an instance of the *TimeInstance* class. The normalizer can also recognize the granularity of a time expression. The defined six different units of measures to represent different levels of granularity: year, month, day, hour, minute, and second. In this particular paper, the finest granularity we cover is day.

Temporal references often occur as relative terms within clinical text. Terms such as "today", "tomorrow", "last month", "two years ago", and "in two months" permeate the clinical document. The normalized form of a relative temporal reference can often be inferred from to its relationship to other absolute and relative temporal references. As an example, "today" is a relative expression since its value depends on the document context. As we always know date when a clinical narrative was written we can use it to convert "today" into an absolute equivalent date. Other relative temporal references can be converted to absolute equivalents with an accompanying granularity. As an example, if a clinical document was recorded on 2010-06-08, we can infer that "in 2 days" corresponds to 2010-08-10, with a granularity of *day*. The SWRL temporal built in library [24] provides functions to calculate to a time reference by adding or subtracting a duration from a given time point. Section 4.3 discusses how we adopt it in detail.

## 4.2   OWL DL Reasoning

**Logical Characteristics of Properties.** We leverage the logical definition properties to infer more temporal relations between events. For example, *before* is defined as being transitive, meaning that, if that event A is stated as occuring *before* event B, and event B *before* event C, we can infer that event A occurrs *before* event C. *Before* and *after* are defined as inverse properties. Therefore, given that event A is *before* event B, we can infer that event B is *after* event A, and vice versa. *equal* is defined as a symmetric property, meaning that, when event A is described as being *equal* with event B, we can infer that B is also *equal* with A. The temporal relations can be semantically defined using SWRL rules or computed using SWRL Built-Ins, which we will discuss in the next section.

CNTRO also provides the capability to define time offsets for temporal relations. Based on these time offsets, more temporal relations could be inferred. The RDF quads below provide an example

```
S1 e1 before e2
S1 hasTemporalOffset d1 (3 days)        [e1 occurred 3 days before e2]
S2 e2 before e3
S2 hasTemporalOffset d2 (2 days)        [e2 occurred 2 days before e3]
S3 e2 after e4
S3 hasTemporalOffset d3 (2 days)        [e2 occurred 2 days after e4]
```

Since *before* and *after* are transitive properties, we can use a reasoner such as Pellet [14] to infer that event *e1* is also *before* event *e3*. But Pellet does not provide the reasoning power to infer the temporal relation between events *e1* and *e4*. Based on the temporal offsets, however, we can calculate the time interval between these events using a pair of inverse operators $\alpha$ and $\beta$ to calculate time interval based on temporal offsets, where $\alpha$ is used when the temporal relation is *after* and $\beta$ is used when the temporal relation is *before*. To calculate the time interval between events *e1* and *e4*, we then have an operation, $\beta(3 \text{ days})\alpha(2 \text{ days})$. Since $\alpha$ and $\beta$ are inverse operators, the result of this operation is $\beta(1 \text{ day})$ meaning *e1* occurred 1 day before *e4*.

**Restriction Assertions.** With the temporal relations defined in CNTRO, we can use OWL restrictions to define known temporal relationships between different kinds of events. For example, we want define that a treatment of a condition must happen after it has been diagnosed. We define the temporal relations between the two SNOMED CT concepts: *CancerChemotherapy* and *CancerDiagnosisBasedOnClinicalEvidence* as

```
Class(sct:CancerChemotherapy partial
  restriction(CNTRO:after
    someValuesFrom (sct:CancerDiagnosisBasedOnClinicalEvidence)))
```

This definition allows as to restrict that a cancer chemotherapy must happen after a cancer diagnosis based on clinical evidence.

The above definition, however, is slightly different than what we need. A patient could have more than one diagnoses and treatments. We want to be able to specify that a treatment of a condition must happen after **the** diagnosis for this particular condition. We must consider the two relations *Treatment treats Condition WithDiagnosis* and *Treatment after Condition WithDiagnosis* together to ensure the correct semantic meaning. We need to be able to add the temporal relation property as a qualifier of the another relation. So that we can link a restriction to the class description to define a class of individuals x for which holds that if the pair (x,y) is an instance of P (the property concerned), then y should have certain temporal relation with x. So this is our preferred way to represent our example:

```
Class(sct:CancerChemotherapy partial
  restriction(treats
    CNTRO:after (sct:CancerDiagnosisBasedOnClinicalEvidence)))
```

This restriction describes the temporal qualification of a relation, if an instance of *CancerChemotherapy* a is for an instance of *CancerDiagnosisBasedOnClinicalEvidence* b, then a must happen after b. This definition can be described using SWRL rules, which we will discuss in the next section.

**Semantic Definition of Concepts.** With OWL DL, we can formally define clinical events or clinical-related temporal periods with temporal assertions, such as "infection after injection" and "before procedure". For example, SNOMED CT defines that "infection after injection" is a "infection as complication of medical care" that is after "injection". Using OWL DL, we define the *InfectionAfterInjection* class as fellow:

```
Class(InfectionAfterInjection partial
  intersectionOf
    (restriction(CNTRO:after someValuesFrom (Injection))
    InfectionAsComplicationOfMedicalCare))
```

With formal semantic definitions of clinical events, we can use the reasoners to automatically identify certain time-related events from patient records. This capability will potentially bring benefits to high throughput phenotyping, GWA (genome-wide association) studies, clinical trials, and epidemiology studies.

## 4.3   SWRL-Based Reasoning

**SWRL Temporal Built-In Library.** The SWRL Temporal Built-Ins Library is one of the SWRLTabBuiltInLibraries [24]. It defines a set of builtins that can be used in SWRL rules to perform temporal operations. It works with temporal information in the normalized form. Given two normalized time stamps, the Built-Ins provide basic functions such as calculating the durations, and comparing the two time stamps and checking if they satisfy certain temporal relations. It also can compare two durations and check if one is less than, equal to, or greater than the other. In addition, the Temporal Builtins provides an *add* function, which can calculate a new time stamp by adding (or subtracting) a duration from a given time stamp.

The temporal Builtins provides us the basic function blocks to build our temporal reasoner. After the temporal data has been normalized, many more information can be inferred or calculated using the function blocks. For example, with the *add* function, we can calculate the start/end time of a time interval given the end/start time and its duration. We can also calculate the time stamp of an event, given the time stamp of another event, and the temporal relation with time offset of the two events.

**SWRL RuleML.** SWRL is designed based on the combination of OWL DL and the unary/binary Datalog RuleML sub-language [23]. We can use SWRL to add semantic assertions and enable Horn-like rules that can be used to infer new knowledge from an OWL ontology and reason about OWL individuals. A rule composed by two or more shared variables is easily expressed in Datalog and corresponding decidable subsets of rule based languages. However, such role chains is hard to be expressed in OWL DL [8]. SWRL generalized OWL by allowing arbitrary patterns of variables and property conditionals expressions.

Using SWRL and the temporal relations defined in CNTRO, we can further define time events with complex temporal assertions and/or with more than two shared variables. For example, we can define that for a valid time interval, its start time must before its end time by the following rule:

```
TimeInterval(?t)^hasStartTime(?t, ?s)^hasNormalizedTime(?s, ?ns)
^hasEndTime(?t,?e)^hasNormalizedTime(?e, ?ne)^before(?ns,?ne)
--> ValidTimeInterval(?t)
```

We can also define temporal relation properties such as *meet*, *during*, *overlap*, *finish*, and *start*. For example, the temporal relation property *during* is defined as follow:

```
Event(?a1)^hasTimeStamp(?a1,?t1)^TimeInterval(?t1)^
hasStartTime(?t1,?s1)^hasNormalizedTime(?s1,?ns1)
hasEndTime(?t1,?e1)^hasNormalizedTime(?e1,?ne1)
Event(?a2)^hasTimeStamp(?a2,?t2)^TimeInterval(?t2)^
hasStartTime(?t2,?s2)^hasNormalizedTime(?s2,?ns2)
hasEndTime(?t2,?e2)^hasNormalizedTime(?e2,?ne2)
^before(?ns1,?ns2)^after(?ne1,ne2)
--> during(?a2,?a1)
```

```
Event(?a1)^hasTimeStamp(?a1,?t1)^TimeInterval(?t1)^
hasStartTime(?t1,?s1)^hasNormalizedTime(?s1,?ns1)
hasEndTime(?t1,?e1)^hasNormalizedTime(?e1,?ne1)
Event(?a2)^hasTimeStamp(?a2,?t2)^TimeInstant(?t2)^
hasNormalizedTime(?t2,?nt2)^
^before(?ns1,?nt2)^after(?ne1,nt2)
--> during(?a2,?a1)
```

We assume that if an event $A$ includes another event $B$, event $A$ must be associated with a time interval. Event $B$, however, could be associated with either a time instant or a time interval, each defined by one of the above rules. These temporal operators can also be expressed using SWRL Built-Ins that connect to Java methods.

We can combine the SWRL Built-Ins predicates and operators with SWRL rules to define clinical events and concepts. For example SNOMED CT has a concept "premature labor after 22 weeks but before 37 completed weeks of gestation without delivery", we can use SWRL rule expression to define the temporal part as:

```
Event(?p)^hasTimeStamp(?p, ?t) ^ hasDuration(?t, ?d)
^temporal:durationLessThan('154', ?d, temporal:Days)
^temporal:durationGreaterThan('259', ?d, temporal:Days)
```

In the above expression uses the two operators in *durationLessThan* and *durationGreaterThan* from SWRL temporal builtins to check if the duration of the event falls in the range specified in the concept. Since both the SWRL temporal builtins and CNTRO do not support the level of granularity on *week*, we have to convert *22 weeks* and *37 weeks* to 154 *days* and 259 *days*.

## 5    Implementation Status

We have designed and built a framework that embeds normalization, DL-based reasoning, and SWRL-based reasoning. The framework adopted the temporal computation components from the SWRL Temporal Built Ins library and uses Pellet [14] as the reasoning engine. It provides a query API for users to query data represented with respect to CNTRO. General search API parameters are:

- **findEvent(searchText)** returns a list of events that match the searching criteria. Currently we look for events based on text search. We are working on connecting our reasoning framework with Mayo Clinic's Text Analysis and Knowledge Extraction System (cTAKES) [12]. cTAKES can annotate clinical events with respect to standard ontologies such as SNOMED CT [21] (for clinical terms) or RxNorm [18] (for drug names). It annotates named entities expressed in different ways but have the same semantic meanings using the same concept code. We can then search by concept codes or labels instead.
- **GetEventFeature(event, featureflag)** returns a specific time feature for a given event. The parameter featureflag indicates which time feature the

user wants to retrieve: start time, end time, note taken time, or event time. All the time will be returned in the normalized format. If the specific time was not stated in the original file explicitly, it will call the reasoner and check if the time can be inferred. **Sample query**: When was the patient diagnosed with diabetes? When was the patient started his chemotherapy?

- **getDurationBetweenEvents(event1, event2)** returns the time interval between two events. The duration of the interval is either retrieved directly, calculated, or inferred from temporal relationships with offsets. **Sample query**: How long after the patient was diagnosed colon cancer did he start the chemotherapy?

- **getDuration(event)** returns the duration of a given event. The duration can be either retrieved directly or calculated. **Sample query**: How long did the symptoms of rectal bleeding last?

- **getTemporalRelationType(event1, event2)** returns the temporal relations between two events if it can be retrieved differently or inferred. **Sample query**: Was the PT scan after the colonoscopy?

- **getTemporalRelationType(event1, time)** returns the temporal relations between an event and a specific time if it can be inferred or retrieved. **Sample query**: Is there any behavior change within a week of the test?

- **getEventsTimeline(events)** returns the order (timeline) of a set of events. Optionally, when the order of the given list of events cannot be completely resolved, it returns a set lists with those events that cannot be sorted within the group. **Sample query**: What is the tumor status timeline as indicated in the patient's radiology note? What is the treatment timeline as recorded in oncology notes? When was the first colonoscopy done When was the most recent glucose test?

This temporal reasoning framework is an ongoing process. We are working on implementing and improving the features of the API, and evaluating the API with real world clinical data.

## 6 Discussions

**Instant vs. Interval.** Whether to view time as instants or intervals is a debate among a lot of researchers [32]. On one hand, a time instant can be viewed as a time interval with a very short duration. On the other hand, a time interval is a time instant on a coarse level of granularity. In medical text, both time instants and time intervals are used to describe clinical events. For example, a clinician may state that "patient's last cycle of chemotherapy was on Jan. 19", or "patient's last cycle of chemotherapy started from Jan. 10 and ended on Jan. 19". Currently we annotate the time stamp of an event simply based on the expressions themselves. When there is only one time expression stated, we consider it as a time instant. If duration, and/or start and end time were stated, we consider it as a time interval. Therefore, we consider that "Jan. 19" is a time instant with granularity *Day* whereas "started from Jan. 10 and ended on Jan. 19" is a time interval with both start time and end time indicated.

One might argue, however, that a cycle of chemotherapy should be a process with a duration instead of an occurrence that just happens on a specific point of time. We are currently investigating on how to further classify and specify events into different categories with different temporal characteristics. We then will be able to annotate the temporal information of an event based on the temporal characteristics of the event itself, instead of based on the temporal expressions used in the original documents. For example, for processes like chemotherapy or surgery, we use time intervals. But for occurrences like checking-in, we use time instants. Basic Formal Ontology and Medical Ontology [2] has defined different kinds of occurrences and process entities. We plan to adopt and expand the classes defined by BFO to our CNTRO ontology, so that the temporal information can be more properly annotated.

**Temporal Uncertainty and Temporal Imprecision.** There are different kinds of uncertainties we have encountered during both the annotation process and the reasoning process.

One kind of uncertainty is from the original source. CNTRO has defined a property called *hasModality* to capture uncertainties specified explicitly in the original documents. For example, "in approximately two weeks" or "about 3 hours", each is an approximated temporal expression with uncertainties. Temporal relations that are inferred based on this kind of temporal expressions will also be returned to users as approximated.

In clinical text, each time expression is stated on a certain level of granularity. But is that level of granularity sufficient enough for inferring temporal relations or calculate a duration? One example would be to get the duration between an event happened on Jan. and an event happened on June. Is that 5 months, 6 months, or 7 months? Another example is that an event $A$ has time stamp "Jan", and an event $B$ has time stamp "Jan 16". The reasoner could not infer a certain temporal relation between these two events. This kind of uncertainties was major caused by temporal imprecisions.

We also found that temporal information in clinical text can be expressed in a coarse notion that it is hard to use one of the pre-defined levels of granularity to describe it, i.e., "early next year", "middle of next week", "short after 11:30 PM", or "immediately after admission". This kind of imprecisions brings us problems for uncertainties on temporal relations and durations too. For example, given that an event $A$ has a time stamp "short after 11:30PM on Jan 16", and an event $B$ has a time stamp "Jan 17", how confident can we say that event $A$ is before event $B$?

In addition, sometimes one temporal expression can have different interpretations. For example, for the sentence "patient's last cycle of chemotherapy was on Jan. 19", there might be three different interpretations: (1) patient's last cycle of chemotherapy STARTED on Jan. 19; (2) patient's last cycle of chemotherapy ENDED on Jan. 19; or (3) patient's last cycle of chemotherapy STARTED and ENDED on Jan. 19. If we can specify the common duration of a cycle of chemotherapy, it might be helpful to disambiguate the confusions. For example, if we know the event usually lasts a few hours, but not a few days, we could interpret that patient's last cycle of chemotherapy STARTED and ENDED on Jan. 19.

How to describe the uncertainty in a systematic way while still support meaningful reasoning powers is a non-trivial problem. While OWL can provide means for including numeric uncertainty measures or level of uncertainties as data type properties, there is no standardized way of representing uncertainties. In order to adequately represent uncertainties in OWL, some language extension is necessary. For example, previous research has focused on extending OWL DL with fuzzy set theory [10,22], or using Bayesian networks as the underlying reasoning mechanism and probabilistic model [5,16]. We are currently investigating on adopting this previous work and using OWL to represent temporal uncertainties. In addition, we believe it will be useful to use ranges to represent imprecise temporal notions and currently working on extend the CNTRO ontology to reflect it.

**Negation.** SWRL and OWL's monotonicity assumption determines that negation as failure is not supported. But in practice, we need to have a *not* operator in both annotation and reasoning. In many cases, clinicians use negations of temporal relations in clinical narratives, such as "no later than", "not during", and "not before". Without a not operator, new temporal relation properties such as *not_before*, *not_after* have to been introduced and semantically defined, like what the SWRL Temporal Built-In Ontology does.

**Limitations with SWRL Built-Ins.** While SWRL Built-Ins provide a powerful extension mechanism that allows user-defined methods to be used in rules, and serve as important function blocks in our temporal relation reasoning framework, we found there are some limitations when using them. First, the Built-Ins do not use an input-output designation mechanism. Built-ins can assign (or bind) values to arguments. The implementation of the rule engine must detect the unbound arguments and assign values to them. The types or the positions of the unbound arguments cannot be defined through SWRL rules, therefore errors cannot be detected easily before run time. Therefore, we provided our own API for queries.

In addition, the SWRL Temporal Built-Ins implementation is not available as a stand-along program library yet. We have investigated two ways to leverage the Built-Ins library: (1) using the Protégé SWRL tab [15], and (2) using Pellet reasoner for SWRL Built-Ins. The first one can only be used in Protégé environment and the second has limited access to temporal operations. In our framework, we leveraged basic temporal Java classes implementation that comes with SWRL tab plug-in for Protégé, such as Instant, Period and Temporal to compute basic features and relations among events in patient's clinical note.

**Timing-Event-Dependent Change** It is important to monitor the changes between two time points or two timing events. For example, in "Most recent ultrasound in May 2007 showed no change comparing to Nov last year", we can annotate two timing events, "ultrasound in May 2007" and "ultrasound in

Nov last year". But with the current model, it is hard to annotate "no change" between these two events. BFO has explored two ways to representing changes: by comparing the discrepancies among the qualities at different time instants, or by capturing the continuous dynamic change over an interval of time. While measurement of change has been a topic widely covered by many researchers, currently there is no standard way for modelling it in OWL. OWL's monotonicity assumption precludes modelling the changes of property values over time without significant extra effort to circumvent the imposed constraints [9].

# 7    Conclusions and Future Work

In this paper, we introduce a Semantic-Web based framework for querying and inferring temporal information from clinical narratives. We have built an OWL ontology that models temporal information such as timing events, time instants, time intervals, durations, and temporal relations. Based on this ontology, temporal information in clinical narratives can be annotated and represented in RDF. This ontology also provides foundation pillars for us and users to define concepts and relations in the temporal aspects. Our framework embedded OWL DL-based reasoning, SWRL-based reasoning, and the SWRL Temporal Built-Ins library, combined these tools seamlessly to fit the needs of time-oriented question answering and inference from clinical narratives.

Several directions remain to be pursued. First, we would like to connect the reasoning framework to Mayo Clinic's Text Analysis and Knowledge Extraction System (cTAKES) [12]. We will extend and improve cTAKES and use it as an automatic annotator for temporal information [19] and annotate information with respect to the CNTRO ontology. We want to scale up the data collection and investigate more on reasoning temporal information in clinical narratives. We would also like to address the consistency issues and object identification problem over heterogeneous sources. Second, we would like to extend the CN-TRO ontology and embed more time-related semantic assertions as discussed in Section 4.2. We will also embed the SWRL rules discussed in Section 4.3 into the ontology itself. In addition, we will explore how to leverage the capabilities of Rule Interchange Format (RIF) and OWL2 for temporal information definition and reasoning. Third, we want to extend the CNTRO so that we can capture data with uncertainty and imprecision better as discussed in Section 6. Last, but not least, we want to implement a user-friendly user interface for health-care providers and clinical researchers to query the time-related information in clinical narratives.

*Acknowledgement.* This research is supported by the National Science Foundation under Grant #0937060 to the Computing Research Association for the CIFellows Project, the ONC Strategic Health IT Advanced Research (SHARP) award under Grant #90TR0002-01, and the Mayo Clinic eMERGE study under Grant #U01-HG04599.

# References

1. Allen, J.F.: Maintaining knowledge about temporal intervals. Communications of the ACM 26(11), 832–843 (1983)
2. Basic Formal Ontology (BFO), http://www.ifomis.org/bfo
3. Blumenthal, D., Tavenner, M.: The "meaningful use" regulation for electronic health records. The New England Journal of Medicine (NEJM) 363(6), 501–504 (2010)
4. Date and time notation by us, http://en.wikipedia.org/wiki/Date_and_time_notation_by_country#United_States
5. Ding, Z., Peng, Y.: A probabilistic extension to ontology language OWL. In: Proceedings of the 37th Hawaii International Conference on System Sciences, HICSS-37 (2004)
6. Embley, D.W., Campbell, D.M., Liddle, S.W., Smith, R.D.: Ontology-based extraction and structuring of information from data-rich unstructured documents. In: Proceedings of the 7th International Conference on Information and Knowledge Management (CIKM 1998), Washington D.C, pp. 52–59 (November 1998)
7. HL7 time specification, http://www.hl7.org/v3ballot/html/infrastructure-datatypes/datatypes.htm
8. Horrocks, I., Patel-Schneider, P.F.: A proposal for an owl rules language. In: Proceedings of the Thirteenth International World Wide Web Conference (WWW 2004), Manhattan, New York (2004)
9. Matheus, C.J., Baclawski, K., Kokar, M.M., Letkowski, J.J.: Using swrl and owl to capture domain knowledge for a situation awareness application applied to a supply logistics scenario. In: Adi, A., Stoutenburg, S., Tabet, S. (eds.) RuleML 2005. LNCS, vol. 3791, pp. 130–144. Springer, Heidelberg (2005)
10. Mazzieri, M., Dragoni, A.F.: A fuzzy semantics for the resource description framework. In: da Costa, P.C.G., d'Amato, C., Fanizzi, N., Laskey, K.B., Laskey, K.J., Lukasiewicz, T., Nickles, M., Pool, M. (eds.) URSW 2005 - 2007. LNCS (LNAI), vol. 5327, pp. 244–261. Springer, Heidelberg (2008)
11. O'Connor, M.J., Shankar, R.D., Parrish, D.B., Das, A.K.: Data integration for temporal reasoning in a clinical trial system. International Journal of Medical Informatics 78(1), S77–S85 (2009)
12. cTAKES on open health natural language processing (OHNLP) consortium, http://www.ohnlp.org
13. OWL Web Ontology Language Reference, http://www.w3.org/TR/owl-ref/
14. Pellet: Owl 2 reasoner for java, http://clarkparsia.com/pellet/
15. The Protégé Ontology Editor, http://protege.stanford.edu/
16. PR-OWL: A bayesian extension to the OWL ontology language, http://www.pr-owl.org/
17. Resource description framework (rdf), http://www.w3.org/RDF/
18. RxNorm, http://www.nlm.nih.gov/research/umls/rxnorm/
19. Savova, G., Bethard, S., Styler, W., Martin, J.H., Palmer, M., Masanz, J., Ward, W.: Towards temporal relation discovery from the clinical narrative. In: Proceedings in the American Medical Informatics Association (AMIA) Annual Symposium, San Francisco, California (November 2009)

20. Shahar, Y.: Timing is everything: Temporal reasoning and temporal data maintenance in medicine. In: Proceedings of Artificial Intelligence in Medicine. Joint European Conference on Artificial Intelligence in Medicine and Medical Decision Making (AIMDM 1999), Aalborg Denmark, pp. 30–46 (June 1999)
21. Systematized nomenclature of medicine–clinical terms (SNOMED CT), http://www.snomed.org
22. Stoilos, G., Stamou, G.: Extending fuzzy description logics for the semantic web. In: Proceedings of the 3rd International Workshop on Owl: Experiences and Directions (2007)
23. A Semantic Web Rule Language Combining OWL and RuleML, http://www.w3.org/Submission/SWRL/
24. SWRL temporal built-in library, http://protege.cim3.net/cgi-bin/wiki.pl?SWRLTemporalBuiltIns
25. The SWRLTab's valid-time temporal ontology, http://swrl.stanford.edu/ontologies/built-ins/3.3/temporal.owl
26. Tao, C., Wei, W.-Q., Savova, G., Chute, C.G.: A semantic web ontology for temporal relation inferencing in clinical narratives. In: Proceedings of the American Medical Informatics Association (AMIA) 2010 Annual Symposium, Washington DC (November 2010) (accepted)
27. Tappolet, J., Bernstein, A.: Applied temporal rdf: Efficient temporal querying of rdf data with sparql. In: Aroyo, L., Traverso, P., Ciravegna, F., Cimiano, P., Heath, T., Hyvönen, E., Mizoguchi, R., Oren, E., Sabou, M., Simperl, E. (eds.) ESWC 2009. LNCS, vol. 5554, pp. 308–322. Springer, Heidelberg (2009)
28. Markup language for temporal and event expressions, http://www.timeml.org/site/index.html
29. Time ontology in OWL, http://www.w3.org/TR/owl-time/
30. XML Schema Date/Time Datatypes, http://www.w3.org/TR/xmlschema-2/
31. Zhou, L., Melton, G., Parsons, S., Hripcsak, G.A.: A temporal constraint structure for extracting temporal information from clinical narrative. Biomedical Informatics 39(4), 424–439 (2006)
32. Zhou, L., Parsons, S., Hripcsak, G.: The evaluation of a temporal reasoning system in processing clinical discharge summaries. JAMIA 15(1), 99–106 (2008)

# Will Semantic Web Technologies Work for the Development of ICD-11?

Tania Tudorache, Sean Falconer, Csongor Nyulas, Natalya F. Noy, and Mark A. Musen

Stanford Center for Biomedical Informatics Research, Stanford University, US
{tudorache,sfalc,nyulas,noy,musen}@stanford.edu

**Abstract.** The World Health Organization is beginning to use Semantic Web technologies in the development of the 11th revision of the International Classification of Diseases (ICD-11). Health officials use ICD in all United Nations member countries to compile basic health statistics, to monitor health-related spending, and to inform policy makers. While previous revisions of ICD encoded minimal information about a disease, and were mainly published as books and tabulation lists, the creators of ICD-11 envision that it will become a multi-purpose and coherent classification ready for electronic health records. Most important, they plan to have ICD-11 applied for a much broader variety of uses than previous revisions. The new requirements entail significant changes in the way we represent disease information, as well as in the technologies and processes that we use to acquire the new content. In this paper, we describe the previous processes and technologies used for developing ICD. We then describe the requirements for the new development process and present the Semantic Web technologies that we use for ICD-11. We outline the experiences of the domain experts using the software system that we implemented using Semantic Web technologies. We then discuss the benefits and challenges in following this approach and conclude with lessons learned from this experience.

## 1  The International Classification of Diseases—A New Beginning

The International Classification of Diseases (ICD) is the standard diagnostic classification developed by the World Health Organization (WHO) to encode information relevant for epidemiology, health management, and clinical use. Health officials use ICD in all United Nations member countries to compile basic health statistics, to monitor health-related spending, and to inform policy makers. ICD is one of the most important classifications used for health care all over the world. ICD is created by a large collaborative effort among international medical experts. To keep up to date with scientific findings about diseases and to address new uses of the classification, the WHO publishes revisions of the classification approximately every decade. In 2007, the WHO started work on the 11th revision of ICD (ICD-11).

Our group is working closely with the WHO to support the collaborative development of ICD-11. The new requirements for ICD-11, which we describe in Section 3, call for a complete revamping of the classification representation in order to build a more solid and flexible formal foundation. ICD-11 will use OWL as the underlying representation language. The workflow for the new development process is also going

P.F. Patel-Schneider et al.(Eds.): ISWC 2010, Part II, LNCS 6497, pp. 257–272, 2010.

to change fundamentally. The process will become a Web-based open process that is powered by collaboration and social features.

This paper makes the following contributions:

- We analyzed the representational and functional requirements for supporting the new collaborative workflow for the development of ICD-11 (Section 3).
- We developed a customization of WebProtégé, a Web-based version of Protégé, to support distributed collaborative development of ICD-11 (Section 4).
- We performed a formative evaluation of the tool (Section 5).
- We analyzed lessons learned and the challenges and advantages of using Semantic Web technologies for the development of large medical terminologies (Section 6).

## 2    ICD History, Use, and Development

ICD traces its origins to the 19th century. The initial work on disease statistics actually began in the 16th century with the London Bills of Mortality that listed the number of burials as a warning against the onset of the bubonic plague. The London Bill of Mortality enumerated 81 causes of death and it is the predecessor of international mortality classifications.[1] Several governments and health organizations recognized the importance of this classification and became interested in it. In 1948, the World Health Organization (WHO) took over the responsibility for ICD and its creation and included for the first time the causes of morbidity, in addition to classifying causes of mortality.[2] Since then, ICD underwent revisions approximately every decade. The current revision of ICD, ICD-10, contains more than 20,000 terms and is used in over 100 countries around the world. ICD-10 is available in the six official languages of WHO (Arabic, Chinese, English, French, Russian, and Spanish) as well as in 36 other languages [11].

### 2.1    Uses of ICD

ICD is an essential resource for health care all over the world. Its strength comes from enabling researchers to undertake studies of temporal and spatial distributions of certain diseases and to make estimates of the effects of diseases on populations [3]. ICD also enables the study of numerous other epidemiological aspects of diseases in human populations. More recent uses include indexing and retrieving of medical records, or use in reimbursement, audit systems, and public policy. At its core, the most important contribution of ICD is the ability to exchange comparable data from different regions and allowing the comparison of different populations over long periods of time.

### 2.2    The Previous ICD Development Process

WHO publishes three types of ICD revisions and updates: Every decade, a *revision process* takes places and results in a new ICD revision, such as ICD-9 or ICD-10. To

---

[1] Mortality is the proportion of deaths to population, or the rate of death.

[2] Morbidity is defined as the incidence of disease, or the rate of disease.

keep up to date with new scientific findings that occurred between 2 subsequent ICD revisions, WHO makes *yearly updates* and *3-year major updates* of the classification [4]. The yearly update usually contains hundreds of changes, while the 3-year update involves more significant changes that impact the mortality and morbidity statistics.

The revision from ICD-9 to ICD-10 was done mainly via regular mail. Non-governmental organizations, statistical offices, and scientific societies proposed changes to ICD. WHO sent the proposed updated chapters to the involved parties for review and comments. Then, experts discussed controversial topics and agreed on the chapters in face to face meetings. In the last step of the process, WHO experts reviewed the changes for consistency across different chapters and for structural integrity of the overall classification.

Starting with ICD-10, the proposals for updates came from national stakeholders through the WHO collaborating centers [10]. WHO circulated the proposal for updates in a formal way via email. Experts then met in one or two teleconferences to seek agreement on edits or overall acceptance of the proposals, in addition to emailing of the proposals with lists of comments and originators. The centers usually went back to their national scientific societies to get analysis of the proposals from a scientific point of view. WHO made final decisions at the annual face to face meeting of the WHO Family of International Classifications Network, strongly believing that this was the best way to solve open issues with the proposals.

In an effort to streamline the ICD-10 update process, WHO developed a Web-based application, *ICD-10 Plus*,[3] to serve as the common platform for incremental updates to the ICD-10 revision. The goal of the platform was to make the development process transparent and to encourage the participation of external experts. ICD-10 Plus functions as a workflow engine that starts when an expert creates a change proposal and ends when the proposal is either removed from the system or implemented in ICD. Users also have different levels of authorization ranging from *standard users*, who can only submit proposals and participate in discussions, to *moderators* and *administrators*, who have more access permissions.

ICD-10 Plus is implemented as a database-backed system with a fixed scheme for storing the classification and the information attached to a disease. The types of structured proposals (e.g., a proposal to introduce a new category) are modeled as database tables with predefined fields. The workflow implementation is very specific to the ICD revision process and it is hard-coded in the tool.

## 3  Requirements for the New ICD Development

In 2007, WHO initiated the work on the 11th revision of ICD (ICD-11) with the mission "to produce an international disease classification that is ready for electronic health records that will serve as a standard for scientific comparability and communication."[4] ICD-11 will introduce major changes to ICD, which the WHO characterizes as (1) evolving from a focus on mortality and morbidity to a *multi-purpose and coherent classification* that can capture other uses, such as primary care and public health;

---

[3] http://extranet.who.int/icdrevision
[4] http://sites.google.com/site/icd11revision/home

(2) creating a *multilingual international reference standard* for scientific comparability and communication purposes; (3) ensuring that ICD-11 can function in electronic health records (EHRs) by *linking ICD to other terminologies and ontologies* used in EHRs, such as SNOMED CT; (4) introducing *logical structure and definitions* in the description of entities, and representing ICD-11 in OWL and SKOS. In addition to these changes in structure and content, the WHO is radically changing the revision process itself. Whereas the previous revisions were performed by relatively small groups of experts in face-to-face meetings and published only in English and in large tomes, development of ICD-11 requires a Web-based process with thousands of experts contributing to, evaluating, and reviewing the evolving content online.

Thus, the requirements for the new ICD revision fall in two categories (1) developing a richer and formal representation for ICD-11 that will support the new goals of the classification, and (2) designing and implementing an open social development environment to support the richer content acquisition.

### 3.1 Representation Requirements

ICD-10 is a statistical classification and it lacks a formal representation. A classification is a set of categories (buckets) into which one can place all the objects in the universe, for which the classification was designed. In the ICD case, the universe is represented by all diseases and health related problems. As such, ICD has to comply to the classification principles [1,4]:

- ICD must have a category for each (possible) disease.
- Categories cannot overlap, which means that a disease cannot be placed into two or more categories.
- Each category must have at least one disease; thus, a category cannot be empty.

To maintain its usefulness for statistical purposes, ICD must follow these principles. However, with its extension of goals, *to become a multi-purpose classification* for a much larger number of usages, the current ICD already faces a number of issues. The use of **different classification axes** by different branches of ICD allows the classification of a disease in two or more categories. For example, bacterial pneumonia is both an infectious and a respiratory disease. Such classification of a category in more than one branch violates one of the principles of classification.

Different uses will also require **different level of details** in the classification. For instance, in primary care it will be enough to have appendicitis as a category, but in a clinical-care setting we will need a much higher level of detail, and even more details in a research setting. The ICD-11 representation will have to encompass all its uses with their possibly different properties. From this representation, we will have to be able to extract valid classifications for the different uses at the appropriate level of detail, while maintaining the coherence among them.

If previous ICD revisions contained only minimal information about a disease, usually just a code, WHO will significantly extend the ICD-11 **representation of a disease** to cover different aspects of diseases, such as clinical description, causal mechanisms, risk factors, treatment, functional impact, and so on. These aspects can serve as different classification axes. We will need to devise representation patterns for the new attributes of a disease and find ways of **linking them to predefined value sets**.

ICD is in use in many countries around the world. Some of the countries, including the USA, Canada, Germany, and Australia, found ICD to be insufficient for the level of detail that they needed for clinical and administrative uses, and created extensions of the classification, known as *Clinical Modifications* [4]. For example, the ICD-10-CM in use by USA has more than 60,000 categories. As a consequence, there are now multiple extensions of ICD with no formal linkages among them, restricting the compilation of international statistics only to certain cases. ICD-11 will try to **integrate the clinical modifications** into one consistent representation. In the initial step, ICD-11 will merge the clinical modifications into one representation, which medical and classification experts will curate in a second step. From the all-encompassing representation, we should be able to generate the country-specific classifications that will represent a subset of the original. To support this requirement, we need to **maintain the metadata about the provenance** of the country specific categories. Our representation will also need to be able to **model the relevant usages for a category** (for example, a disease is relevant for morbidity use, but not for mortality).

The content of ICD-11 will also be evidence-based: for each piece of information stored in ICD (e.g., the risk factors of a disease), the experts will have to provide scientific evidence in form of links to publications or official documents. The ICD representation will have to **store the evidence** in form of metadata attached to each assertion.

Furthermore, ICD-11 has to be **language-agnostic** and provide translations of the labels used in the classification in several languages. WHO also intends to maintain a mapping between the ICD-10 code and ICD-11 code of a disease to support the **migration of existing medical software** to the new classification.

One important WHO desiderata is to make ICD useful in electronic health records by linking it to other standard terminologies and biomedical ontologies, such as SNOMED CT. We must develop a representation and methodology for creating **references to terms in external resources**. For example, the description of a disease will include a body part. Rather than creating its own anatomy taxonomy, ICD will reference a term in the *Anatomy* branch of SNOMED CT. The reference will have to store metadata, such as the source of the term, terminology version, identifier, and link to the term.

### 3.2 Development Process Requirements

While the previous ICD development happened mostly in face-to-face meetings behind closed doors, WHO envisions to use an open social process for ICD-11 that will involve a large international community of experts. The process will be similar to Wikipedia, where a large number of people contribute to the content. In the ICD case, WHO hopes that a large number of medical experts will contribute to the content of ICD-11.

The development of ICD-11 will happen in several phases. The *alpha phase* is open only to WHO experts. The goal of this phase is to develop the new representation, to test it internally, and to fill the content of the *alpha draft*, the initial draft of ICD-11. The alpha phase will end in May 2011. We performed the work presented in this paper as part of the alpha phase. In the *beta phase*, WHO will open ICD-11 to a large community of experts for feedback, and will also use it for field trials. In the last phases, ICD will be open to the entire public for viewing (2014). Following the approval by the World Health Assembly (WHA), which is planned for 2014, the implementation of ICD-11 in health care systems across the world will start.

The workflows that each of the development phases needs differ significantly. In the alpha phase, the main focus is on finding an agreement on the ICD formal representation, and having WHO experts pre-fill a large part of the content. In this phase, it is critical to enable many experts to fill in effectively as much content as possible. In the beta phase, the process will change completely. The number of users will increase from around a hundred to thousands. In such a situation, enforcing access policies becomes a priority. The focus changes from having an effective editing platform to having a platform where users can make change proposals and discuss issues in the classification. Collaboration and workflow support becomes one of the most important features. WHO envisions a reviewing process of the ICD-11 content by external domain experts similar to the scientific peer review process that will ensure a high quality of the classification. WHO will also define a quality assurance process for ICD that should become an integral part of the development cycle.

The main challenge in our experience so far was the lack of a well defined collaboration workflow. The different groups of WHO experts could not agree on a concrete workflow for the alpha phase, the roles of users, their access policies and the sequence of steps and responsibilities. However, proceeding to the next phases of ICD will be impossible without a well defined workflow.

## 4    The Semantic Web Approach

As the representation requirements on ICD-11 have become more complex (Section 3.1), the WHO decided to use OWL as the underlying formal representation language. WHO created a committee—the Health Informatics Modeling Topic Advisory Group (HIM-TAG), to design an appropriate OWL representation for ICD-11, the *content model*. One of the paper authors (Prof. Mark A. Musen) serves as a chair of the HIM-TAG. The Revision Steering Committee (RSG) serves as the planning and steering authority in the update and revision process. The RSG in collaboration with the HIM-TAG are in charge of defining workflows for the different phases of the ICD development.

Our group has been involved in both committees from the beginning of the ICD-11 revision process, and has contributed to the development of the content model. We have also designed and implemented a Semantic Web tool that the WHO domain experts use to edit the ICD-11 content in the alpha phase. The tool is based on WebProtégé—a lightweight ontology editor for the Web [9], which extends the popular Protégé platform. Besides browsing and editing support, WebProtégé supports collaboration processes and has a highly customizable user interface for knowledge acquisition.

We describe the design of the ICD-11 OWL ontology in Section 4.1, and WebProtégé in Section 4.2.

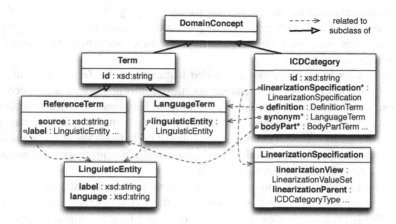

**Fig. 1.** A snippet of the ICD Ontology. The *ICDCategory* is the top-level class in the ICD disease hierarchy. The usage of a class is modeled using the *linearizationSpecification* property. The property values of a disease class are instances of the class *Term*.

## 4.1 The ICD-11 OWL Ontology

The ICD ontology[5] should serve as the underlying representation for all information related to diseases, including the definition of disease characteristics, linkages to external terminologies, as well as linguistic information for translation in multiple languages. One of the major challenges in designing the ontology was supporting the different usages of ICD that have to conform to valid classification principles (see Section 3.1). The HIM-TAG proposed a layered model of the ICD Ontology. The *Foundation Layer* will contain an all-encompassing model of all the usages of ICD and it will allow multiple parents of a category. The *Linearization Layer* will provide a view on the Foundation Layer, called a *Linearization*, for each specific usage. The main characteristic of a Linearization is that it is *linear*—each category will have exactly one parent, which satisfies the most important principle of a classification. There will be linearization corresponding to mortality, morbidity, primary care, and so on. Tu and colleagues [7] provide a full description of the ICD Ontology and the content model. The Foundation Layer ensures that all linearizations provide a consistent representation of diseases, and that they are coherent with one another.

Figure 1 shows the main components of the ICD Ontology that correspond to the Foundation Layer. A class in the ontology represents an ICD category. The *ICDCategory* is the root of the ICD disease hierarchy. We used a metaclass level to describe the different classification axes corresponding to the properties in the ontology. For example, a disease has one or more associated body parts. The representation in the ontology is as follows: A disease class has a property *bodyPart* that is prescribed by the *ClinicalDescription* metaclass[6]. In other words, we associate a property *bodyPart*

---

[5] Available online at: http://tinyurl.com/icd-ontology

[6] The disease classes have a number of other metaclasses corresponding to different classification axes.

to each disease class, rather than to its instances. Other properties include textual definition, synonyms, clinical descriptions (body part, body system), manifestation properties (signs and symptoms, investigations), causal properties, temporal and functional properties, treatment, and so on.

The values for these properties are reified instances of the *Term* class and its subclasses. Reification enables us to encode additional information for a property value, such as scientific evidence, translations in multiple languages, or metadata about the linkages to external terminologies. For example, a synonym for a disease name is an instance of *LinguisticTerm* that has a unique identifier, but also provides labels for different languages. Similarly, the value for the *causalMechanism* property is an instance that must contain links to scientific evidence.

The values for most of the properties should come from predefined value sets. These property values are represented as instances of the *ReferenceTerm* class and its subclasses. The ICD ontology uses two types of reference terms: terms in external terminologies and terms in value sets defined locally in the ICD ontology. A reference term instance has metadata associated with it, such as a unique identifier, a preferred label, the source terminology, a direct URL link to the term (if available). For example, the *Myocardial Infarction* class has as a value for the *bodyPart* property a reference to the SNOMED CT term for *Entire Heart*. The value of *bodyPart* is therefore an instance that has as property values: *id*=302509004 (identifier in SNOMED CT), *source*=SNOMED CT, *preferredLabel*=Entire heart, a direct URL[7] to the term in BioPortal, and other properties to store additional metadata.

The ICD ontology is built in a modular way. The upper ontology (i.e, the content model) provides the basic structures and is imported directly or indirectly by all other modules. The right hand side of Figure 2 shows the import hierarchy of the ontology modules that make up the ICD-11 ontology.

## 4.2   WebProtégé—The Knowledge Authoring Platform Used for ICD-11

Our group developed WebProtégé—a highly customizable Web interface for browsing and editing ontologies, which provides support for collaboration. We have created a specific customization of WebProtégé for the content acquisition in ICD-11.[8] In the remainder of the section, we describe the architecture and the features of the tool.

### Architecture

Figure 2 shows an overview of the WebProtégé architecture. The core of the system is the **Collaboration Framework** [8], which provides all the collaboration and ontology access services that the client applications need through the *Ontology Access API*, *Notes and Discussions API* and the *Change Tracking API*. The *Workflow API* provides the workflow support, including access policies and user management. The clients of the Collaboration Framework are the Protégé desktop application, WebProtégé, and any other application that access its services either directly through the Java APIs or through the remote RMI interfaces.

---

[7] Direct URL in BioPortal to *Entire Heart*: http://tinyurl.com/bp-heart

[8] The ICD demo platform is available at: http://icatdemo.stanford.edu

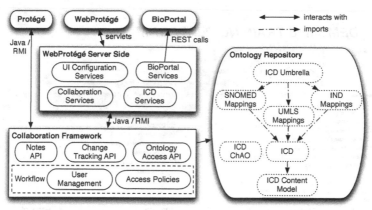

**Fig. 2.** An architecture diagram of WebProtégé used in the ICD context. WebProtégé UI uses the services on the server side to display information. The WebProtégé server side and the desktop client Protégé connect to the *Collaboration Framework* to access the ontology and the collaboration services. The *Ontology Repository* stores the ontologies available to the clients. The modules and the import hierarchy of the ICD ontologies are also shown.

The WebProtégé server connects to the Collaboration Framework to access the ontology content and the collaboration services. The WebProtégé server also provides services for the user-interface configuration and for creating links to external terminologies stored in BioPortal—a web-based repository for biomedical ontologies and terminologies [6]. In the WebProtégé deployment for the ICD-11 authoring, we have used the Collaboration Framework and implemented additional services for the ICD-specific functionality.

**Features**

The main feature of WebProtégé is the support for Web-based **browsing and editing** of ontologies. Medical experts from around the world are using the system to edit the ICD-11 content simultaneously. Every change that one editor makes is immediately committed to a shared copy of the ontology. The change is then propagated to all other clients in the collaboration framework and other editors see the changes in real time.

Another key feature of WebProtégé is the support for **collaboration**. The support for notes and discussion ensures that domain experts can raise questions and discuss different issues that arise during the authoring. Several note types are available in the Web interface. Users can attach notes to any entity in the ICD ontology (class, property, individual), or even to a specific property-value assertion (for instance, to one of the synonyms of a disease class). The notes and discussions are threaded and can contain any HTML formatted text. **Change tracking** is one of the most important features that WHO uses in the quality assurance. Indeed, for each action that the user performs, there may be several granular changes that happen in the ontology (e.g., creating a reference to another term involves creating a new instance, adding property values to it, and setting a property value in the container class to refer to that instance). We store granular changes for analysis and debugging purposes, but present the user only with the user-friendly high-level descriptions of changes that correspond to the user actions.

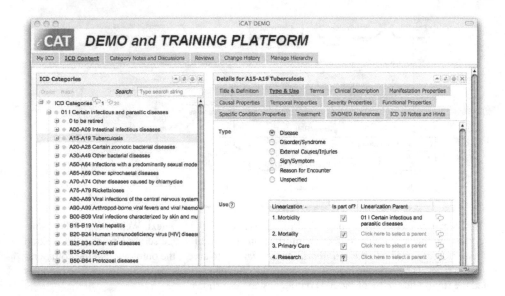

**Fig. 3.** The ICD authoring tool using WebProtégé. Each tab contains one or more panels, called *portlets* that can be arranged by drag-n-drop. The left hand-side portlet shows the disease class hierarchy of the ICD ontology. The right panel shows the uses (linearizations) of the selected disease in the tree, in this case *Tuberculosis*.

Protégé stores both notes and individual changes that the authors make as instances in the *Changes and Annotation Ontology (ChAO)* [5]. The changes are hierarchical, meaning that a change may be composed of other sub-changes, similar to nested database transactions. We have implemented a tool that uses the structured changes log to analyze the activity in the ICD ontology and to generate statistics.[9]

The new development process for ICD also envisions that the content of ICD-11 will be reviewed in a manner similar to scientific paper reviews (see Section 3.2). We have implemented a rudimentary **reviewing feature** that allows a user with the appropriate privileges to request the review of a disease description. WebProtégé sends a review request to the reviewer, who can later log into the system and enter her review. As any other note types, reviews can be attached to a class representing a disease, or to individual property values.

Figure 3 shows the user interface of WebProtégé as deployed for the ICD-11 authoring. We have implemented a **declarative user interface** that allows us to define the user interface components and layout in an XML configuration file. The configuration file declares the binding of the user interface elements to the underlying ontology entities. For example, a text field can be used to edit the values of a string property, or a radio button for functional properties. By simply changing the XML configuration file, we can quickly change the user interface without the need to compile or re-deploy the application. This feature provides great flexibility for projects such as the ICD

---

[9] The Change Analysis Plugin is available at: `http://tinyurl.com/ch-analysis`

development, in which the underlying ontology structure (even the upper ontology) is still under discussion and active development. We implemented the user interface as a portal, in which users can arrange *portlets*—components providing independent pieces of functionality, simply by drag-n-drop. WebProtégé is **extensible** and has a **plugin infrastructure**. We implement some ICD-specific portlets for the ICD infrastructure. The high customizability of the user interface also allows us to define different layouts for different users based on their interest and domain of expertise. One of our main reasons for developing a highly configurable user interface was to be able to hide from domain experts ontology details that are not relevant to them. We have spent a lot of effort in ensuring that the user interface does not look like an ontology editing environment, but is **customized for the domain experts**.

One of the important requirements in the ICD-11 development is to support the **linkages to terms in external terminologies**. We implemented a *Reference Portlet*, which allows users to import terms from terminologies stored in BioPortal with a single click. For instance, if the user wants to import a reference to a term *Heart* from the SNOMED CT Anatomy branch, we invoke a RESTful service call to BioPortal and fetch the results. Then, the user can simply click on an import link next to one of the search results. On the backend, we create an instance of a *ReferenceTerm*, and present it in a table format to the user.

## 5  Usage of the Semantic Web Platform

During the alpha phase of the ICD-11 revision project, editors have been updating and making changes to the ontology from November 2009 to present day. Figure 4a displays the changes that the editors have made during this time period. Different colors in the stacked area chart correspond to changes made by different editors. The x-axis is the months of the project and the y-axis is the number of changes made by a given author.

After six months of use, there has been a total of 15,025 changes made by 16 different editors out of 48 users who have logged into the system. The platform does not require a sign in for read access, and therefore we did not keep track of other users of the system. Contributions have ranged greatly from as little as a single change to as many as 7,709. The average number of changes per editor is 684, while the median is 85. The editors have created a total of 483 definitions for terms, added 2,464 completely new terms to the hierarchy, removed 149 ICD-10 terms and moved 1,415 ICD-10 terms to a new location in the ontology hierarchy. While previous versions of ICD did not support multiple inheritance, there is already 464 terms with multiple parents within ICD-11.

Editors have also been actively participating in discussions. Figure 4b displays a stacked area chart representing editor note contributions per month in the alpha phase. The editors created a total of 5,035 notes. Similar to change contributions, the amount of note contribution has varied greatly amongst the editors. Contributions range from as few as 1 to as many as 2,422. The average is 315, while the median is 105.

These statistics demonstrate that the tool is being used actively by the editors, but it does not provide details about their feelings on the usability of the tool. To gather this information, we carried out a web-based survey, which we describe in the next section.

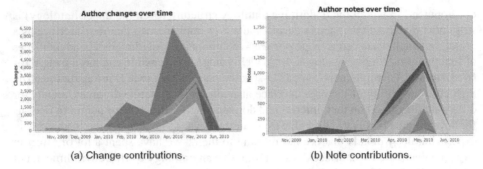

(a) Change contributions.                    (b) Note contributions.

**Fig. 4.** Visual representation of editor changes (a) and notes (b) contributions per month during the alpha phase of the ICD-11 revision process

**Table 1.** Survey questions. iCAT stands for "The ICD Collaborative Authoring Tool" and is the customized WebProtégé platform used for editing ICD.

| # | *Question* |
|---|---|
| 1 | **Rate your experience with iCAT?** |
| | I thought iCAT was easy to use. |
| | I found iCAT to be unnecessarily complex. |
| | I think I would need technical support to be able to use iCAT˙ |
| | I found the various features of iCAT to be well integrated. |
| | I thought there was too much inconsistency in iCAT˙ |
| | I think most people would learn iCAT quickly. |
| | I found iCAT cumbersome to use. |
| | I felt very confident using iCAT˙ |
| | I needed to learn a lot about iCAT before I could effectively use it. |
| 2 | **What did you like about iCAT?** |
| 3 | **What did you dislike about iCAT?** |
| 4 | **Did you find the discussion and commenting features helpful?** |
| 5 | **Do you have any additional comments not already covered?** |

## 5.1   Survey Feedback

To solicit feedback from our users, we conducted a web-based survey consisting of five questions (see Table 1). The first question consisted of completing nine sub-questions, each evaluated on a 5-point Likert scale ranging from "Strongly disagree" to "Strongly agree". These sub-questions were adapted from the System Usability Scale (SUS) [2]. The other five questions were all open-ended.

**Survey results**

Thirteen experts responded to the survey[10]. The respondents were content experts (medical doctors) and classification experts. None of the respondents had ontology expertise. Some of the experts took part in a training session for WebProtégé in September 2009, but others had to start using the tool with no prior training. The results of the first

---
[10] The results of the survey are available at: http://tinyurl.com/icat-survey

question shows a range of answers. Five of the editors found the tool easy to use, while 4 editors had difficulties with it, and 4 others were neutral. Half of the respondents found the tool too complex and some felt that they needed training or support to use it. On the positive side, the respondents found the tool to be well integrated and consistent.

The second and third questions provide more detail about these ratings. The second question asked about the features that the editors liked. The answers indicate that it is "logical," has "easy structure and clear layout", "easy navigation" and the good "integration of features." Other features mentioned were the easiness in performing hierarchical changes and the integration with BioPortal.

On the negative side, 7 of the 10 responses highlighted the complexity and time needed to enter information. Some of the respondents found that the tool was too related to the informatics side, forcing users to enter too much detail. One respondent also indicated that they wanted to be able to display more information, in report form, on a single screen and be able to export this information for discussion.

The fourth question asked about the usefulness of the notes feature. Five of the 9 respondents indicated that they found the feature useful. The other respondents either had not used the feature, or indicated they had trouble finding new comments.

The final question asked whether the respondents had any comments not already covered by the survey. Only four respondents provided further feedback. Two indicated great concern about having an open editing process. They felt that too many users contributing without classification expertise could lead to inconsistencies in the hierarchy and continual changes to certain areas. The other two respondents re-iterated the time needed to enter information and felt that there needed to be support for submitting a less structured form of the content model.

## 6   Discussions and Future Work

The results of the survey provide important feedback both to the developers of the collaboration tools and to the team developing the content model itself (the HIM-TAG). Indeed, WebProtégé only exposes the underlying ontology in a form-based interface. Most users, however, do not distinguish the tool and the model that the tool exposes. The editors in our survey felt the representation was very complex and that they did not understand many of the fields. Indeed, ICD-11 is enormously more descriptive than previous ICD revisions. If in previous revisions, a category contained mainly a code and a handful of properties, in ICD-11 editors can fill in over 40 fields attached to a disease. The description of the fields is also not well defined in the content model yet and there are fierce discussions in the HIM-TAG related to representation of disease characteristics. For example, including a severity in a disease description can be problematic because the severity may vary in different stages of disease or based on other factors. Given this newly added complexity of the model and the constant changes that it is still undergoing, we believe that some of the issues that the editors raised in our survey, were actually related to the content model itself and not to the tool.

Of course, there were also issues related to the tool itself. We implemented the initial version of the tool in a very short period of time with no formal requirements. The experts and HIM-TAG were able to formulate the requirements only after they saw the first prototype of the tool in September 2009. Ever since then, the tool, the ICD ontology, and requirements evolve together. Supporting the ICD-11 process as it is today was possible only by having a flexible system that can be easily adapted to changes in the underlying ontology. Any application that would hard-code the binding between the user interface and the ontology would not have worked in this setting.

We are aware of several user interface issues in the tool that make the work of the editors more difficult. In several discussions with the editors and also from the survey, we know that the lack of synchronization between the class trees in the different tabs causes a change of context and editors can make mistakes much easier. We are also aware of some performance problems due to the concurrency support in the ontology APIs on the backend side that use a coarse-grain lock at the level of an ontology. Meanwhile, we have developed an implementation of the ontology APIs with a finer-grained locking mechanism that addresses this issue.

Some problems are related to the backend side and the fact that the Changes and Annotation ontology grows at a much more rapid rate than the domain ontology. Therefore, we must address the question of whether to store the change tracking information as instances in an ontology makes sense. The crucial question is whether the structure and type information of the changes provide any advantage. We believe that it does. The Change Analysis plugin generates different kind of statistics and makes use of the type structure information for some of the statistics (for example, how many classes have been moved from one branch of the ontology to the other). We need to find better technical solutions to address this scalability problem.

Our experience with ICD-11 made us realize that developing and providing content to formal models is just cognitively difficult. This process is even harder for domain experts who have no prior knowledge of ontologies or knowledge representation. In developing the web platform, we tried to "hide" as much of the underlying formal representation as possible. We implemented simplified editing widgets, such as instance tables, to present and support the acquisition of reified relationships. Most of the relations in the ontology are reified, but domain experts are not aware of their underlying representation. We also tried to model the entire user interface as a form-based interface using common editing widgets (texfields, radio buttons, check boxes, etc.) that were already familiar to the users, and did the "heavy-lifting" on the server side. However, we think that there may be a limit to how much the user interface will be able to simplify a task that is similar to ours.

Initially, we developed the ICD ontology using a frame-based formalism and soon after we migrated the representation to OWL. Although in the current model we use very few DL-specific constructs, there are several benefits of using OWL. First, it is a W3C recommendation and has a well defined formal semantics, which will enable the reuse of ICD-11 in other ontologies and by other tools. Second, we make use of two of the most common modeling patterns in OWL—defining value sets and reified relationships—which form the basis for the entire representation. Third, we make use of inverse properties, cardinalities, domain and ranges, which are important in the model and are enforced in the user interface. Fourth, we plan to use OWL annotations for storing the metadata and linguistic information, which play a central role in the model. Fifth, in order to manage manage poly-hierarchies and multiple inheritance, we plan to convert the current representation into a DL form using OWL defintions and restrictions. We will then be able to check the consistency of the manually created polyhierarchies using a DL reasoner.

One concern that the editors raised is whether the development of the ICD-11 can be open to the community at large, as planned by WHO. It is not yet clear whether a development à la Wikipedia would work for ICD. The main issue is whether a domain expert with no understanding and training in the new content model will be able to contribute in a significant way to the development of ICD. We do not have answers to these questions yet. For the beta phase, we plan to implement a much stricter access policy mechanism (access rights at the level of branches and depending on many work-flow variables). The collaboration workflow will also be significantly different. It will switch its focus from editing to a proposal-based process. External experts will be able to submit structured change proposals. However, the exact workflow and quality assurance for the beta phase is still undefined. As in many other cases, the technology and implementation is only secondary, and the main challenge is social: the various teams of experts that are developing the workflow are yet to agree on one.

The general evaluation of WebProtégé as a platform for the collaborative authoring of ICD is positive. We are currently working with WHO to build two other WebProtégé deployments to be used for the collaborative development of two other WHO classifications: the International Classification of Patient Safety (ICPS) and the International Classification of Traditional Medicine (ICTM). The flexibility and versatility of WebProtégé allows very quick customization of the tool for different ontologies. In fact, we have built a first prototype[11] for the Traditional Medicine classification in less than 2 weeks. We spent most of this time adapting the upper level ontology to the ICTM content model, and then we were able to configure the new user interface showing Chinese characters in a very short time. For the near future, we will add collaboration features needed for the beta phase, and will also work on providing internationalization support. We also plan to perform further evaluations of the tool, and we are particularly excited to witness the use of the platform in a much larger setting.

## 7  Conclusions

Developing ontologies is a cognitively hard process and we do not yet have a good grasp of simple interfaces for this type of development. Even though defining a new class in ICD-11 is as "simple" as filling out a number of pre-defined terms, with value sets for many of the fields also pre-defined, users still found the process difficult and cumbersome. And while some of this difficulty was in fact the difficulty in understanding the meaning of the fields themselves, some of it was from the amount of information that we must present on the screen. We need to consider ways to custom-tailor interfaces dynamically, based on the role that a particular user is playing in the workflow, based on the parts of the class tree that he is interested in, and parts of the content model that he is either qualified or interested in filling out.

Our experience working with WHO on the ICD-11 alpha draft also helped both teams understand better what the advantages of the semantic technologies were. The

---

[11] The ICTM prototype is available online at: http://icatdemo.stanford.edu/ictm/

need to reference other ontologies, the need for multiple inheritance along with single-inheritance linearizations, the ability to integrate labels in multiple languages are exactly the strong points of semantic-web technologies such as RDF and OWL.

**Acknowledgments.** We are very grateful to Jennifer Vendetti, Timothy Redmond, Jack Elliott, and Martin O'Connor for their help with the design and implementation of WebProtégé, and to Bedirhan Üstün, Robert Jakob, Can Çelik, and Sara Cottler from WHO for their work in developing the requirements for the project and for providing documentation on the previous ICD development process. We would like to thank Alan Rector, Samson Tu and all the members of the HIM-TAG and RSG for their invaluable work in designing the ICD content model. The work presented in this paper is supported by the NIGMS Grant 1R01GM086587-01. Protégé is a national resource supported by grant LM007885 from NLM.

# References

1. Bailey, K.D.: Typologies and taxonomies: An introduction to classification techniques. Sage Publications, Inc., Thousand Oaks (1994)
2. Brooke, J.: Usability evaluation in industry. In: Jordan, P.W., et al. (eds.) SUS: a 'quick and dirty' usability scale, pp. 184–194. Taylor & Francis, Abington (1996)
3. Israel, R.A.: The International Classification of Disease. Two Hundred Years of Development. Public Health Rep. 93(2), 150–152 (1978)
4. Jakob, R.: Disease Classification. International Encyclopedia of Public Health 2, 215–221 (2008)
5. Noy, N.F., Chugh, A., Liu, W., Musen, M.A.: A framework for ontology evolution in collaborative environments. In: Cruz, I., Decker, S., Allemang, D., Preist, C., Schwabe, D., Mika, P., Uschold, M., Aroyo, L.M. (eds.) ISWC 2006. LNCS, vol. 4273, pp. 544–558. Springer, Heidelberg (2006)
6. Noy, N.F., Shah, N.H., Whetzel, P.L., Dai, B., Dorf, M., Griffith, N., Jonquet, C., Rubin, D.L., Storey, M.-A., Chute, C.G., Musen, M.A.: BioPortal: ontologies and integrated data resources at the click of a mouse. Nucleic Acids Research, 10.1093/nar/gkp440 (2009)
7. Tu, S.W., et al.: A content model for the ICD-11 revision. Technical Report BMIR-2010-1405, Stanford Center for Biomedical Informatics Research (2010)
8. Tudorache, T., Noy, N.F., Tu, S., Musen, M.A.: Supporting Collaborative Ontology Development in Protégé. In: Sheth, A.P., Staab, S., Dean, M., Paolucci, M., Maynard, D., Finin, T., Thirunarayan, K. (eds.) ISWC 2008. LNCS, vol. 5318, pp. 17–32. Springer, Heidelberg (2008)
9. Tudorache, T., Vendetti, J., Noy, N.: WebProtege: A lightweight OWL ontology editor for the Web. In: OWL: Experiences and Directions, 5th Intl. Workshop, OWLED 2008, Karlsruhe, Germany (2008)
10. World Health Organization: Collaborating Centres for the WHO Family of International Classifications (WHO-FIC),
    http://www.who.int/classifications/network/collaborating/
    (last accessed: June 2010)
11. World Health Organization: International Classification of Diseases (ICD),
    http://www.who.int/classifications/icd/ (last accessed: June 2010)

# Using SPARQL to Test for Lattices: Application to Quality Assurance in Biomedical Ontologies

Guo-Qiang Zhang[1] and Olivier Bodenreider[2]

Case Western Reserve University, Cleveland, OH 44106, USA
National Library of Medicine, Bethesda, MD 20892, USA

**Abstract.** We present a scalable, SPARQL-based computational pipeline for testing the lattice-theoretic properties of partial orders represented as RDF triples. The use case for this work is quality assurance in biomedical ontologies, one desirable property of which is conformance to lattice structures. At the core of our pipeline is the algorithm called *NuMi*, for detecting the *Nu*mber of *Mi*nimal upper bounds of any pair of elements in a given finite partial order. Our technical contribution is the coding of *NuMi* completely in SPARQL. To show its scalability, we applied *NuMi* to the entirety of SNOMED CT, the largest clinical ontology (over 300,000 conepts). Our experimental results have been groundbreaking: for the first time, all non-lattice pairs in SNOMED CT have been identified exhaustively from 34 million candidate pairs using over 2.5 billion queries issued to Virtuoso. The percentage of non-lattice pairs ranges from 0 to 1.66 among the 19 SNOMED CT hierarchies. These non-lattice pairs represent target areas for focused curation by domain experts. RDF, SPARQL and related tooling provide an efficient platform for implementing lattice algorithms on large data structures.

## 1 Introduction

Lattices arise naturally from many disciplines. We speak of lattices because of their familiarity and their elegant structural properties. In the Semantic Web, lattices are intimately related to conceptual structures, since good ontologies often have a lattice structure [23]. The deeper philosophical and mathematical reason for lattice to be a desirable structural property for the taxonomy relation (e.g. IS-A) in ontologies can be elucidated using a theory called Formal Concept Analysis (FCA [3,11,15]). Starting from two very basic types, objects and attributes, with the assumption that intension and extension are fundamental adjoining facets of the notion of *concept*, one arrives at the mathematical structure of *(complete) lattices* automatically using FCA. The upshot of this is that if we encounter a non-lattice fragment in a taxonomic hierarchy, then somewhere upstream the notion of intension and extension has not been rigorously enforced, revealing gaps in conceptual modeling which can be subtle to detect otherwise.

Though desirable, the lattice property of ontologies is not always found in most biomedical ontologies. For example, SNOMED CT, the largest clinical ontology, is only a lattice "for the most part" (assuming superficial top and bottom elements). As can be seen from Fig. 1, the double-circled concepts "Tissue specimen from breast" and "Tissue specimen from heart" legitimately share the two features of being a kind

P.F. Patel-Schneider et al.(Eds.): ISWC 2010, Part II, LNCS 6497, pp. 273–288, 2010.

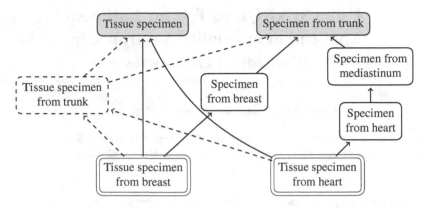

**Fig. 1.** Non-lattice fragment from the Specimen hierarchy in SNOMED CT. Dashed lines represent a possible remedy by adding the concept "Tissue specimen from trunk".

of tissue specimen and a kind of specimen from trunk. The current representation in SNOMED CT involves two minimal upper bounds shared by the two double-circled concepts, "Tissue specimen" and "Specimen from trunk", which is the reason why this fragment is not part of a lattice. The corresponding lattice-conforming representation would require the creation of the concept "Tissue specimen from trunk" (the dashed component in Fig. 1), which would be the single minimal (least) upper bound of the two double-circled concepts. Incidentally, the concept pair "Tissue specimen from breast" and "Tissue specimen from heart" is one of the 28,464 non-lattice pairs found by this work in the Specimen hierarchy (see Table 1).

Since the majority of biomedical ontologies are manually curated, automatic testing for lattices plays an important role in auditing and quality assurance of these ontological systems. The absence of a lattice structure can be indicative of issues including missing concepts, misaligned concepts and inconsistent use of pre-coordination [7,19]. However, it is also possible that the clinical utility of some concepts was deemed insufficient to warrant their creation.

We present a scalable, SPARQL-based computational pipeline for testing the lattice-theoretic properties of partial orders represented as RDF triples. The basic idea is to simply follow the definition of a lattice and check that each pair of elements has a least upper bound. If this is true, then each pair of elements also has a greatest lower bound, by pure mathematical reasoning (Proposition 1). Thus to check for lattices, we only need to check either the existence of least upper bounds, or the existence of greatest lower bounds, for the whole partial order. For convenience, we will focus on least upper bounds in this paper, but all results translate directly to the situation of (testing for existence of) greatest lower bounds. In practice, for finite partial orders, including the taxonomic backbones of ontological systems, the existence of a least upper bound for a pair is equivalent to the uniqueness of minimal upper bounds of the given pair.

With brute force, checking for the number of minimal upper bounds can be computationally daunting for large ontological systems. For example, the July 31, 2009 version of SNOMED CT comprises 307,754 (N) active concepts, with maximal depth of 30. If for each of the 47,356,108,381 [(N*(N-1))/2] pairs we (1) find all their upper bounds and (2) detect the minimal ones among the upper bounds, the computation quickly become intractable. Assuming each pair takes 10 ms (a reasonable estimate), processing the whole SNOMED CT would take about 15 years.

The main contribution of this paper is a demonstration of the suitability of Semantic Web technologies, namely RDF and SPARQL, for quality assurance in large biomedical ontologies by testing their lattice-theoretic properties. We have discussed the clinical significance of our work in [19]. Here, we focus on general technical aspects. We have developed an algorithm called *NuMi*, for detecting the *Nu*mber of *Mi*nimal upper bounds of any pair of elements in a given partial order, which we have implemented completely in SPARQL. We also propose further optimization by applying a reverse version of the algorithm. The experimental results reported here are groundbreaking: for the first time, we have been able to exhaustively check the entirety of SNOMED CT for its lattice-theoretic properties within a time frame of 2 months sequential computation. The percentage of non-lattice pairs ranges from 0 to 1.66 (Table 2) among the 19 SNOMED CT hierarchies among over 34 million candidate pairs. These pairs represent potential target areas for focused curation by domain experts. The reason that testing the lattice-property allows us to efficiently and systematically identify targets for curation stems from the rationale broadly captured in FCA [3,11,15] and indirectly through the work of Jiang and Chute [7] as well.

## 2    Background

### 2.1    SNOMED CT

SNOMED CT is a comprehensive concept system for healthcare, distributed and maintained by the International Health Terminology Standard Development Organization (IHTSDO) [6]. SNOMED CT provides broad coverage of clinical medicine, including findings, diseases, and procedures, and is used in electronic medical records [1].

The development of SNOMED CT is supported by an infrastructure based on description logics. From a structural perspective, SNOMED CT can be seen as a series of large directed acyclic graphs, one for each of its 19 "hierarchies": Procedure, Physical force, Event, Staging and scales, Substance, Environment or geographical location, Situation with explicit context, Body structure, Observable entity, Pharmaceutical / biologic product, Physical object, Qualifier value, Special concept, Specimen, Social context, Clinical finding, Organism, Linkage concept, and Record artifact. No concept is shared across hierarchies except for the root. Each concept comes with a SNOMED CT ID, which is an integer such as those given in the first column of Table 1. SNOMED CT concepts are linked by hierarchical relations, within each hierarchy (e.g., "Tissue specimen from heart" IS-A "Tissue specimen"). Associative relations (across hierarchies) form the basis of the logical definitions (e.g., "Nephrectomy" procedure site "Kidney"), but are not used in this work. The version of SNOMED CT used in this study is dated July 31, 2009 and comprises 307,754 active concepts.

The motivation for this work comes in part from the application of Formal Concept Analysis (FCA) to a limited subset of SNOMED CT by Jiang and Chute [7]. These authors used FCA as an auditing tool by constructing local contexts from normal form presentations in SNOMED CT (i.e., logical definitions in description logic) and compared with the resulting lattices for anonymous (unlabeled) nodes. They showed that given a small SNOMED CT fragment, this method can automatically identify a candidate pool of missing concepts for further examination by domain experts. However, constructing lattices from contexts is so computationally expensive that it is hardly scalable [10,15]. FCA-based analysis is therefore not applicable to the entirety of large ontologies such as SNOMED CT. Moreover, for ontological systems without rich logical definitions, the need for background contexts would render the FCA-based approach inapplicable.

## 2.2 Lattices, Complete Lattices and Quasi-primes

We first review basic definitions in lattice theory. Our main references are [5,20].

A partially ordered set (poset) is a set $L$ with a reflexive, transitive relation $\leq \subseteq L \times L$. If $(L, \leq)$ is a poset, then its dual is the poset $(L, \geq)$. We denote posets by their carrier set as long as the partial order is clear from the context. An element $u$ is called a *upper bound* of a subset $X \subseteq L$, if for each $x \in X$ we have $x \leq u$. For convenience, we write $\mathsf{ub}(X)$ for the set of upper bounds of $X$. An element $m$ is called an *minimal upper bound* of a subset $X \subseteq L$, if $m$ is an upper bound of $X$, and for any $n \leq m$ such that $x \leq n$ for each $x \in X$, we have $m = n$. We write $\mathsf{mub}(X)$ for the set of minimal upper bounds of $X$. When $\mathsf{mub}(X)$ is a singleton, the unique minimal upper bound is called the least upper bound, or join, of $X$. The notion of lower bound, maximal lower bound, greatest lower bound (meet), is defined dually. Specifically, $\mathsf{mlb}(X)$ represents the set of maximal lower bounds of $X$.

A poset $L$ is a *lattice* if every two elements of $L$ have a join and a meet. These meets and joins of binary sets will be written in infix notation: $\bigvee \{x, y\} = x \vee y$ and $\bigwedge \{x, y\} = x \wedge y$. A poset $L$ is a *complete lattice* if every subset $S \subseteq L$ has a least upper bound $\bigvee S$ (join) and a greatest lower bound (meet) $\bigwedge S$.

Fig. 2 contains the diagram of a small poset which is not a lattice. Note that in this poset, we have $\mathsf{ub}\{a, b\} = \{x, y, \top\}$, and $\mathsf{mub}\{a, b\} = \{x, y\}$, i.e. $x$ and $y$ are minimal upper bounds of $a$ and $b$. Hence the pair $\{a, b\}$ does not have a least upper bound and this diagram does not represent a lattice. When the size of $\mathsf{mub}\{a, b\}$ is greater than 1, we call $a, b$ a non-lattice pair (e.g., $a, b$ in Fig. 2).

In connection with ontology, one can think of concepts as elements of a poset, and the ordering relation as the subsumption relation [8]. If $x, y$ are concepts, we write $x \leq y$ to mean $x$ IS-A $y$, or $y$ subsumes $x$. The join $x \vee y$ of two concepts $x, y$ is the lowest common ancestor of $x$ and $y$, and the meet $x \wedge y$ is their greatest common descendant.

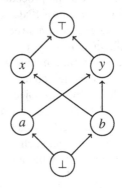

**Fig. 2.** A poset which is not a lattice

Every finite lattice is a complete lattice. Therefore, every finite lattice has a top (largest) element, denoted as $\top$, and a bottom (least) element, denoted as $\bot$. One can think of a tree as a lattice by adjoining a superficial bottom. In this sense, lattices are more general than trees, but posets are more general than lattices. Multiple inheritance is not permitted in trees: each node in a tree can have at most one parent (the node immediately above it). Lattices permit multiple inheritance but insist on the existence of (unique) join and meet for any pair of nodes. Since adding top and bottom elements can be globally achieved either conceptually or materialized, we assume that posets come with top $\top$ and bottom $\bot$ in the remainder of the paper. By convention, we have $\bigvee \emptyset = \bot$ and $\bigwedge \emptyset = \top$.

If every pair of elements in a finite poset has a least upper bound $\vee$, then the greatest lower bound of a pair $a, b$ can be obtained as $a \wedge b := \bigvee \{x \mid x \le a \text{ and } x \le b\}$.

This leads to the well-known property of "half-implies-whole" in the next proposition, which forms the basis of our computational strategy later.

**Proposition 1.** *Let $(L, \le)$ be a finite poset. If any pair of elements in $L$ has a least upper bound, then $L$ is a lattice.*

In a poset $(L, \le)$, an element $x$ is *covered* by $y$ if $x \le y$, $x \ne y$, and for any $z$ such that $x \le z \le y$, we have either $z = x$ or $z = y$. To reduce storage, it is often the case that only the coverage relation is stored explicitly for ontological systems such as SNOMED CT.

The notion of quasi-prime [21] will be useful later on for selecting candidate pairs for our SPARQL query. An element $q$ in a poset $(L, \le)$ is called a *quasi-prime* if for any subset $X \subseteq L$, $q \in \mathsf{mlb}(X)$ implies $q \in X$ (when $\bigwedge$ exists for $X$, we have $\mathsf{mlb}(X) = \{\bigwedge X\}$). Since we will be concerned with finite posets with top and bottom elements, the combinatorial interpretation of quasi-primes for finite posets will be useful. Note that the top element of a poset cannot be a quasi-prime, since $\top = \bigwedge \emptyset$ but $\top \notin \emptyset$. The notion of coquasi-prime can be defined similarly.

**Proposition 2.** *An element $q$ in a poset $(L, \le)$ is a quasi-prime if and only if there exists $q^* > q$, such that for any $x > q$, we have $x \ge q^*$.*

Proposition 2 implies that for each quasi-prime $q$, there is a unique element $(q^*)$ covering $q$, or $q$ has a single parent node. Moreover, for any $a \in L$ such that neither $a \le q$ nor $a \ge q$ holds, if $x$ is an upper bound of $\{a, q\}$, then $x$ must already be an upper bound of $\{a, q^*\}$. As a consequence, we have $\mathsf{mub}\{a, q\} = \mathsf{mub}\{a, q^*\}$.

*Proof.* We provide a general proof without assuming $L$ to be finite.

(Only If). For a quasi-prime $q \in L$, consider the set $U_q := \{x \in L \mid x > q\}$. Since the top element is not a quasi-prime, $U_q$ is not empty. If no lower bound of $U_q$ is strictly above $q$, then $q \in \mathsf{mlb}(U_q)$ but $q \notin U_q$, which is impossible for a quasi-prime $q$. Therefore, there exists a lower bound $q'$ of $U_q$ such that $q < q'$. Since $q' \in U_q$ by definition of $U_q$, $q'$ must be the least element in $U_q$. We take the required $q^*$ to be $q'$.

(If). Suppose there exists $q^* > q$, such that for any $x > q$, we have $x \ge q^*$. Suppose $q \in \mathsf{mlb}(X)$ for some $X \subseteq L$. $X$ cannot be empty since otherwise $q = \top$, and there cannot be an element $q^*$ strictly above top. If $q \in X$, then there is nothing to prove. Otherwise, for any $x \in X$ we have $q < x$, and so $q^* \le x$ by assumption. This means $q^*$ is a lower bound of $X$ strictly dominating $q$, which contradicts the assumption that $q$ is a maximal lower bound ($q \in \mathsf{mlb}(X)$). Therefore, we must have $q \in X$.

As a consequence of Proposition 2, quasi-primes are not needed in detecting non-lattice properties, since if there is any violation of the uniqueness of minimal upper bounds involving a quasi-prime $q$, there must already be such a violation involving $q^*$. In Fig. 1, the concepts "Specimen from breast", "Specimen from heart" and "Specimen from mediastinum" are quasi-primes with visually identifiable $q^*$'s in the same diagram. This leads to the following definition.

**Definition 1.** *A pair $a, b$ in a poset $(L, \leq)$ is called a probe pair (probe, for short) if*

1. *neither $a$ nor $b$ is a quasi-prime;*
2. *$a$ and $b$ are not comparable, i.e., neither $a \leq b$ nor $b \leq a$ holds.*

In Section 4, our quality assurance use case will significantly reduce the candidate pairs to be tested by running *NuMi* on probes only.

## 2.3  RDF and SPARQL

The Resource Description Framework (RDF) is a directed, labeled graph data format for representing information in the Web [12]. Based on (subject, predicate, object) triples, RDF is well suited for the representation of graphs in general, including posets and lattices. Because of its origins in the Semantic Web, RDF uses Unified Resource Identifiers (URIs) as names for the nodes and the links in the graph.

SPARQL is a query language for RDF graphs [13]. SPARQL queries are expressed as constraints on graphs, and return RDF graphs or sets as results. For example, SPARQL can be used for retrieving the set of common ancestors of two nodes in a graph, i.e., to compute the upper bounds for a pair of nodes from the graph. In practice, as shown in Fig. 3, the list of direct ancestors common to two nodes $a$ and $b$ can be easily obtained by querying the nodes of which both classes are a subclass. The same variable (?upper) is used in the constraints imposed to the graph for the two nodes $a$ and $b$. This pattern forms the basis for testing the lattice-theoretic properties in the queries we developed for this work.

Although developed in a description logic environment, SNOMED CT is distributed as a set of relational tables and there is no version of SNOMED CT available in RDF. We transformed SNOMED CT into RDF using a simple script. A base URI was added to SNOMED CT

```
SELECT ?upper
WHERE {
        :a rdfs:subClassOf ?upper.
        :b rdfs:subClassOf ?upper.}
```

**Fig. 3.** SPARQL query for common ancestors of $a$ and $b$

identifiers in order to create URIs for concepts and predicates. The fully-specified name was used as the label for the concepts. All relations among concepts were transformed into triples, using `rdfs:subClassOf` for representing the native IS-A relationship.

Additionally, we precomputed the transitive closure of `rdfs:subClassOf`, because a transitively-closed graph was assumed in some aspects of our algorithm to both improve speed and avoid computing transitive closure on the fly, which is not supported in most existing SPARQL environments.

# 3   Methods

In this section we first present a simple algorithm in the conventional style for computing the mub set (see Section 2.2) mub$\{a, b\}$ for a given pair $a, b$ within a finite poset $(L, \leq)$. We then introduce a SPARQL implementation of the algorithm with the input poset viewed as a graph, represented as RDF triples. Finally, we revisit the original algorithm and propose an optimization.

## 3.1   Algorithm for Identifying Minimal Upper Bounds

The following algorithm finds the number of minimal upper bounds of $a, b$ by keeping track of "counts" of the number of times an element in $L$ occurs as an upper bound of $a, b$. After *count* is initialized (lines 1-3), every upper bound of $a, b$ gets *count* incremented by 1 (lines 4-8). So by line 8, the counts for each upper bound of $a, b$ is precisely 1, and we have ub$\{a, b\} = \{x \mid count(x) = 1\}$, at this point. The third iteration (lines 9-13) increases counts for those members in ub$\{a, b\}$ that are *not* minimal within ub$\{a, b\}$ (i.e., those elements that are strictly above some other element in ub$\{a, b\}$). At the end of this iteration (line 13), we have mub$\{a, b\} = \{x \mid count(x) = 1\}$.

---

**Data**: Elements $a, b$ in a finite poset $(L, \leq)$
**Output**: The size of mub$\{a, b\}$

1  **for** *each $x \in L$* **do**
2  $\quad \mid \quad$ $count(x) := 0$
3  **end**
4  **for** *each $u \in L$* **do**
5  $\quad \mid \quad$ **if** *$u$ is an upper bound of $\{a, b\}$* **then**
6  $\quad \mid \quad \mid \quad$ $count(u) := count(u) + 1$
7  $\quad \mid \quad$ **end**
8  **end**
9  **for** *each $v \in$ ub$\{a, b\}$* **do**
10 $\quad \mid \quad$ **if** *$v > u$ for some upper bound $u$ of $\{a, b\}$* **then**
11 $\quad \mid \quad \mid \quad$ $count(v) := count(v) + 1$
12 $\quad \mid \quad$ **end**
13 **end**
14 Count the number of $x \in L$ with $count(x) = 1$ and output this number

---

**Algorithm 1.** *NuMi* in the conventional procedural style, for finding the number of minimal upper bounds.

The correctness of *NuMi* is self-evident. To check for lattices, one runs *NuMi* on every pair $a, b$ from an input poset $L$ for potential violations of the lattice property. If all pairs have exactly one minimal upper bound, then $L$ is a lattice; otherwise, pairs with more than one minimal upper bound will be identified as non-lattice pairs.

## 3.2  Implementing *NuMi* in SPARQL

We describe a method to implement *NuMi* completely in SPARQL. This allows us to take advantage of the highly optimized storage and access environment of RDF stores in existing systems such as Virtuoso. To implement Algorithm 1, we construct a two-part SPARQL query, corresponding to two iterations in Algorithm 1: (1) to compute ub$\{a, b\}$ (lines 4-8), and (2) to compute mub$\{a, b\}$ (lines 9-13).

Although aggregation operators such as count are not part of the SPARQL 1.0 specification, they are already available in many SPARQL environments, including Virtuoso. In contrast, ancestor tracing and computing transitive closure are not supported in most of existing SPARQL environments. Both for this reason and for saving computational time, we decided to *precompute the transitive closure* of rdfs:subClassOf in RDF store for the input graph for *L*. In other words, the RDF store, for which the SPARQL queries are made, is assumed to be transitively closed.

The first part of the query finds all upper bounds ?u, and tracks the result by having each upper bound to receive 1 as count. This is straightforward using the query fragment indicated in the middle of Fig. 4.

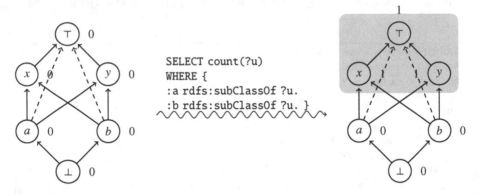

**Fig. 4.** SPARQL query for finding ub$\{a, b\}$ indicated in the shaded area. Dashed edges represent those due to the effect of transitive closure.

The second part of the query finds elements ?u that are strictly above some elements in ub$\{a, b\}$. This can be achieved by the query fragment indicated in the middle of Fig. 5.

Joining the two query fragments with the union operator, we obtain a complete SPARQL query for finding the minimal upper bounds mub$\{a, b\}$. Fig. 6 displays a complete working SPARQL query.

## 3.3  Optimization: Reverse SPARQL Query

In earlier sections, we described optimization strategy by skipping pairs of concepts that (1) do not belong to the same SNOMED CT sub-hierarchy; (2) are in hierarchical relationship to one another (i.e., comparable); (3) involve a quasi-prime. This subsection describes a strategy that checks for the existence of greatest lower bounds, rather than least upper bounds, achieved by running SPARQL queries "in reverse."

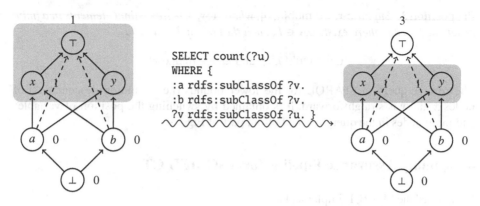

```
SELECT count(?u)
WHERE {
 :a rdfs:subClassOf ?v.
 :b rdfs:subClassOf ?v.
 ?v rdfs:subClassOf ?u. }
```

**Fig. 5.** SPARQL query for finding mub$\{a, b\}$ indicated in the shaded area on the right side

```
SELECT   ?sb   count(?sb) as ?sb_links
FROM <http://newton.case.edu/TEST>
WHERE {
    {
        <http://mor.nlm.nih.gov/SNOMEDCT#256889002> rdfs:subClassOf ?sb.
        <http://mor.nlm.nih.gov/SNOMEDCT#258462005> rdfs:subClassOf ?sb.
    }
    union
    {
        <http://mor.nlm.nih.gov/SNOMEDCT#256889002> rdfs:subClassOf ?sa.
        <http://mor.nlm.nih.gov/SNOMEDCT#258462005> rdfs:subClassOf ?sa.
        ?sa rdfs:subClassOf ?sb.
    }                    }
ORDER BY ASC (?sb_links)
```

**Fig. 6.** Example SPARQL query for SNOMED CT concepts 256889002 and 258462005

On average, concepts in ontological hierarchies tend to have fewer upper level concepts representing more general and abstract entities, and more lower level concepts.

Since lattices can be viewed either top-down or bottom-up, only one direction (i.e., among least upper bounds or greatest lower bounds) needs to be tested (see Proposition 1). For distinct elements $u, v, x, y$ in a finite poset, if $x, y \in$ mub$(\{u, v\})$ then $u, v \in$ mlb$(\{x, y\})$ in most cases, as illustrated in the top part of Fig. 7. Since there are generally more concepts lower in a hierarchy (as in SNOMED CT's taxonomic backbone), this motivates us to test for maximal bounds (mlb) in the original order $(L, \leq)$, or equivalently, minimal upper bounds (mub) in the reverse order $(L, \geq)$, to reduce the number of non-lattice pairs. The following proposition ensues that even though the lattice and non-lattice status of a given pair is usually different when the order is reversed, a non-lattice pair in the original order is guaranteed to appear in a "fragment" generated by some non-lattice pair in the reverse.

**Proposition 3.** *Suppose $x, y \in \mathsf{mub}\{a, b\}$, where $x, y, a, b$ are distinct elements in a finite poset $(L, \leq)$. Then there exists $u, v \in L$, such that $a \leq u, b \leq v$, and*

$$x, y \in \mathsf{mub}\{u, v\} \text{ and } u, v \in \mathsf{mlb}\{x, y\}.$$

The reverse query in SPARQL for the lower bounds (i.e., common descendants) of nodes $x$ and $y$ is straightforward. It is obtained by switching the position of variables and input nodes in the query (at the bottom part of Fig. 7).

# 4    Quality Assurance Pipeline for SNOMED CT

We applied the SPARQL implementation of *NuMi* to auditing SNOMED CT by systematically identifying all non-lattice fragments. The following phases were involved in this quality assurance study: (1) acquiring SNOMED CT data; (2) selecting probes; (3) testing probes; (4) summarizing and analyzing results.

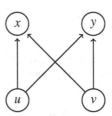

```
SELECT ?lower
WHERE {
        ?lower rdfs:subClassOf :x.
        ?lower rdfs:subClassOf :y.}
```

**Fig. 7.** Template for reverse SPAQRL query to obtain lower bounds

## 4.1    Acquiring SNOMED CT Data

From the 307,754 active concepts in SNOMED CT, we created RDF triples for representing the IS-A relations. We created URIs for all SNOMED CT concepts and used the `rdfs:subClassOf` predicate to represent the IS-A relationship. Then we computed the transitive closure of the IS-A relation and created a distinct set of triples for it. The graph of hierarchical relations contains a total of 439,733 `rdfs:subClassOf` triples, while the transitively-closed graph contains 1,191,796 triples. The two sets of triples were loaded into two separate graphs using the open source Virtuoso triple store [16].

## 4.2    Selecting Eligible Probes

Not every pair of SNOMED CT concepts needs to be tested for its lattice properties. For efficiency reasons, we start by selecting eligible probes for further testing. Since the 19 hierarchies in SNOMED CT do not share any concepts (except the root), only pairs within the same hierarchy require testing. In addition, only probe pairs (see Definition 1) need to be tested for the lattice property, since pairs in which each concept does not have at least two parent concepts will always have their evidence as part of a non-lattice structure exhibited by pairs without involving a quasi-prime (see Subsection 2.2). Moreover, any pair in which one concept is an ancestor of the other does not need to be tested because the ancestor concept in such pairs is the unique common ancestor of the two, with reflexivity assumed. In practice, we constructed SPARQL queries to implement the eligibility criteria described above and ran them in order to establish the list of probes to be tested for lattice properties. Because reverse queries operate on the graph of descendants instead of that of ancestors, eligibility criteria for probes are slightly different. More specifically, probe concepts are required to have at least two child concepts (instead of at least two parent concepts).

### 4.3  Testing Probes Using SPARQL

The algorithms presented earlier for selecting the probes to be tested, and testing them, were implemented without any *ad hoc* programming. Generic queries were created for each algorithm and subpart thereof and loaded as stored procedures in Virtuoso. Stored procedures are used to compensate for the fact that SPARQL queries could not be "prepared" (i.e., the query plan cannot be cached by the ODBC driver in our environment), which is a serious limitation, as about half of the query execution time is generally devoted to building the query plan.

We used a simple script to compute the cartesian product of all pairs of concepts within a given hierarchy of SNOMED CT. Each pair was evaluated as a potential probe by querying a stored procedure instantiated with the pair. If qualified as a probe, the pair was then tested using *NuMi* by querying a second stored procedure. The results were stored in text files for further processing. The open source Virtuoso RDF store version 06.00.3123 was used for this experiment, running on a Dell 2950 server (Dual Xeon processor) with 32GB of memory. A total of 500,000 9kB buffers were allocated to Virtuoso. For benchmarking purposes, we tested both direct and reverse queries on all eligible probes.

## 5  Results

Although all 19 hierarchies of SNOMED CT were processed, due to space constraints, we only report results on the 7 largest and most clinically relevant hierarchies, covering 84% of all concepts. Table 1 provides summary statistics of our analysis of the lattice-theoretic property of SNOMED CT by the "direct" approach (Section 3.1). Table 2 contains the corresponding results for the "reverse" approach (Section 3.3). The first column contains names of the hierarchies and their SNOMED CT ID numbers. The second column (Size) displays the number of concept nodes for each hierarchy. The PP column represents the total number of probe pairs (see Definition 1). The NL column is the total number of non-lattice pairs found. PP% is the percentage of probe pairs among all pairs, NL% is the percentage of non-lattice pairs among all probe pairs, i.e. NL/PP. AT(ms) is the average query time for all probe pairs in milliseconds and TT(h) is the total query time for all probe pairs in hours.

The SPARQL implementation of *NuMi* was run on each probe, whose total number for each hierarchy is displayed in the PP column. Over 1.6 billion queries were issued to Virtuoso for testing the probes. Note that the proportion of probe pairs (PP%) and non-lattice pairs (NL%) is significantly lower in the reverse approach compared to the direct approach. Overall, about 1.9% of all pairs are non-lattice pairs with the direct approach, but only 0.009% are non-lattice pairs with the reverse approach.

**Table 1.** Summary of query results (direct approach)

| Hierarchy | Size | PP | NL | PP% | NL% | AT(ms) | TT(h) |
|---|---|---|---|---|---|---|---|
| Body Structure (123037004) | 31,309 | 84,809,733 | 57,134,031 | 17.3 | 67.4 | 13.463 | 317.16 |
| Clinical Finding (404684003) | 101,027 | 871,780,229 | 53,988,609 | 17.1 | 6.2 | 6.758 | 1,636.53 |
| Organism (410607006) | 30,149 | 682,168 | 74,460 | 0.2 | 10.9 | 2.657 | 0.50 |
| Pharmaceutical (373873005) | 16,718 | 24,204,248 | 1,034,394 | 17.3 | 4.3 | 2.569 | 17.27 |
| Procedure (71388002) | 55,328 | 307,032,527 | 35,802,137 | 20.1 | 11.7 | 9.950 | 848.60 |
| Specimen (123038009) | 1,209 | 227,941 | 28,464 | 32.1 | 12.5 | 2.657 | 0.17 |
| Substance (105590001) | 23,514 | 9,695,574 | 3,291,344 | 3.5 | 33.9 | 3.371 | 9.08 |
| Total | 259,254 | 1,298,432,420 | 151,353,439 | 16.2 | 11.7 | 7.84 | 2,829.31 |

**Table 2.** Summary of query results (reverse approach)

| Hierarchy | Size | PP | NL | PP% | NL% | AT(ms) | TT(h) |
|---|---|---|---|---|---|---|---|
| Body Structure (123037004) | 31,309 | 29,934,856 | 91,787 | 6.1 | 0.31 | 11.710 | 97.37 |
| Clinical Finding (404684003) | 101,027 | 243,428,748 | 251,662 | 4.8 | 0.1 | 5.069 | 342.76 |
| Organism (410607006) | 30,149 | 6,739,818 | 1,040 | 1.5 | 0.02 | 2.425 | 4.54 |
| Pharmaceutical (373873005) | 16,718 | 5,190,080 | 6,446 | 3.7 | 0.1 | 2.020 | 2.91 |
| Procedure (71388002) | 55,328 | 65,800,235 | 174,574 | 4.3 | 0.26 | 8.239 | 150.59 |
| Specimen (123038009) | 1,209 | 53,397 | 889 | 7.3 | 1.66 | 2.244 | 0.033 |
| Substance (105590001) | 23,514 | 4,666,656 | 17,340 | 1.7 | 0.4 | 2.368 | 3.07 |
| Total | 259,254 | 355,813,790 | 543,738 | 4.5 | 0.2 | 6.08 | 601.27 |

What is more significant is the total number of non-lattice pairs. With the direct approach, a total of 151,353,439 pairs are non-lattice pairs, while "only" 543,738 pairs are non-lattice pairs with the reverse approach (i.e., 0.36% of the pairs in the direct approach). It is also observed that in the direct approach, the size of the set $\mathsf{mub}\{a, b\}$ for probe $(a, b)$ is moderate, ranging from two to a few hundreds. In contrast, the size of the set $\mathsf{mlb}\{x, y\}$ (reverse approach) can reach several thousands.

In summary, the average query time with the direct approach is about the same as with the reverse approach. However, the number of probe pairs to be tested differs significantly between the two approaches (16.2% vs. 4.5% of all possible within each hierarchy). This difference is attributable to the structure of the ontology, not the algorithm. Nonetheless, the reverse approach represents a significant saving by focusing on pairs that do not involve quasi-primes and are not in a hierarchical relationship. The total computational time of about 600 hours is in stark contrast to over 2,800 hours. In practice, domain experts curating the ontology will review either the (non-single) upper bounds resulting from the analysis by the direct approach, or the non-lattice pairs (i.e., pairs with no greatest lower bound) resulting from the analysis by the reverse approach. Therefore, while the reverse approach offers superior computational performance over the direct approach.

# 6  Discussion

## 6.1  Technical Significance

To our knowledge this work is the first systematic analysis of the lattice-theoretic properties of a large ontology. Unlike other similar techniques such as Formal Concept

Analysis (FCA), our approach is scalable and can be applied to the entirety of the ontology. In contrast, in their analysis of the same ontology (SNOMED CT) with FCA, Jiang and Chute had to rely on stratified sampling of a limited subset of hierarchies in order to estimate the proportion of anonymous nodes [7].

The availability of efficient tools for Semantic Web technologies including RDF stores and SPARQL query engines primarily enables the integration of large datasets in the framework of the Semantic Web, as illustrated by the increasing amount of linked data available [9]. It also provides alternative implementation options for problems that can be construed as constraints on graphs, including the analysis of lattice-theoretic properties of ontologies. This is particularly important when algorithms for such problems need to be applied to large ontologies, created for real-life applications. While it was somewhat challenging (and not completely intuitive at first) to express the *NuMi* algorithm entirely in a declarative query language such as SPARQL, the alternative, i.e., implementing the algorithm procedurally with a traditional programming language, would have required far more ad hoc coding and have posed different challenges due to the sheer size of SNOMED CT. The only coding required for this work was the scripting needed for choreographing the various elements of our quality assurance pipeline (probe selection, probe testing and limited post-processing of the results for statistical analysis purposes). The bulk of the processing was supported by Virtuoso, the triple store and SPARQL query engine, to which some 2.5 billion queries were sent (counting probe selection and testing).

We also take advantage of the mathematical properties of lattices for optimization purposes. We show that the most intuitive algorithm designed for testing the least upper bond (Section 3.1) is equivalent to the "reverse" algorithm testing the greatest lower bound (Section 3.3). While the complexity of the two algorithms is the same, we show that the reverse algorithm is significantly more efficient due to the structure of the data, reducing total execution time by 80% and reducing the total number of non-lattice pairs by more than 2 orders of magnitude.

## 6.2 Ontological Significance

Quality assurance in ontologies is an active field for research (see, for example, [22] for quality assurance in biomedical ontologies). The quality of SNOMED CT has been examined from several perspectives.

On the one hand, SNOMED CT is developed using an environment based on description logic. Therefore, all the relations made available through the relational database in which SNOMED CT is distributed are guaranteed to be logically consistent. However, the limited expressiveness of the description logic dialect used by SNOMED CT, $\mathcal{EL}$, severely limits the types of inconsistency discoverable by the DL classifier [14].

Structural approaches have been developed for analyzing SNOMED CT, including the Abstraction Network methodology [17], and have been contrasted with description logics [18]. Examples of errors identified by the Abstraction Network methodology and invisible to the DL classifier include missing IS-A relations and duplicate concepts. The analysis of SNOMED CT using this methodology was restricted to the Specimen hierarchy.

The application of Formal Concept Analysis (FCA) to SNOMED CT was discussed earlier [7]. Its strongest limitation is the lack of scalability. We also showed that our results (based on the lattice-theoretic properties) were generally consistent with that of [7] based on FCA. A benefit of FCA is that, unlike our approach, it provides some explanatory information for missing concepts (anonymous nodes).

Overall, we believe that these various methods provide complementary prospectives on the quality of SNOMED CT and have different strengths and limitations. None of these approaches provides an automated solution to auditing ontologies. In fact, each method identifies deviation from properties assumed to be desirable (e.g., logical consistency in DL, structural properties of lattices). Further analysis of the consequences of such deviation generally requires manual review by domain experts. These approaches, however, are important elements of quality assurance as they point out potential problems and help focus the review of the ontology. In addition to scalability, one advantage of our approach is applicable to underspecified ontologies consisting mostly of a taxonomic backbone, while the other three approaches require a richer set of source data (e.g., associative relations across hierarchies) for maximal efficiency.

## 6.3  Biomedical Significance

Using our approach, we have been able to reproduce the findings of Jiang and Chute [7], which we have discussed in [19]. As shown through the example of the procedure concept "Hypophysectomy" used in [7], the presence of anonymous nodes revealed by FCA generally corresponds to the presence of non-lattice fragments in the ancestors of this concept. The trade-off between the two approaches is in part more explanatory power (FCA) vs. scalability to the entire ontology.

One unresolved question is the extent to which the concepts identified as "missing" (anonymous nodes in FCA, missing least upper bound in the lattice analysis) have clinical utility, i.e., whether their absence from SNOMED CT is detrimental to some of its uses, including clinical documentation and clinical decision support. With over 300,000 concepts, SNOMED CT is the largest clinical ontology currently available and both its developers and the user community are reluctant to increase its size unless this is really needed to satisfy some use case.

SNOMED CT can be extended through post-coordination, i.e., by refining existing concepts and by combining existing concepts in a controlled way. For example, while anatomical structures are lateralized in SNOMED CT (e.g., "Left kidney"), procedure concepts are generally not lateralized (e.g., "Nephrectomy", but no "Left nephrectomy"). However, lateralized procedure concepts can be created by refining the non-lateralized procedure concept with a lateralized anatomical structure (e.g., creating "Left nephrectomy" by making "Left kidney" the procedure site of "Nephrectomy"). Similar concepts could be created through post-coordination for variants of nephrectomy (e.g., "Partial nephrectomy" and "Cadaver nephrectomy").

While these concepts could also be created in SNOMED CT as pre-coordinated concepts, creating concepts for all possible combinations of variants would likely result in combinatorial explosion and management issues for both developers and users. Therefore, it is unclear whether the concepts identified as missing from a structural perspective have enough clinical utility (defined by the editorial guides for SNOMED CT) to warrant their creation as pre-coordinated concepts.

It is beyond the scope of this paper to assess clinical utility. Our study simply identifies areas from which concepts are potentially missing. The availability of models (not just guidelines) for desirable levels of pre-coordination (vs. excessive pre-coordination) would enable us to filter out cases were the absence of a least upper bound corresponds to a feature in SNOMED CT, rather than a bug.

## 6.4    Limitations, Generalization and Future Work

As mentioned earlier, our algorithm can help guide the manual curation of SNOMED CT by domain experts, but is incomplete by itself for quality assurance purposes. In particular, our algorithm is known to identify false positives, i.e., concepts missing from a the structural perspective of lattices, but lacking clinical utility for them to be created as pre-coordinated concepts in SNOMED CT. In future work, we will work on the formalization of criteria for excessive pre-coordination and filter out these cases from the output of our algorithm. It would be desirable to develop computable metrics to assess the quality of ontological systems.

The strengths of our approach and FCA could be combined. Our approach could be used for efficient identification of non-lattice pairs, and followed by a limited analysis of the local fragments around the non-lattice pairs by FCA. The combined methods are expected to provide a more efficient and powerful pipeline for quality assurance of SNOMED CT. This combination is also a possible direction for future work.

Unlike FCA and other structural methodologies, our approach relies solely on the taxonomic backbone of ontologies and could therefore be applied to other ontologies, including ontologies lacking a rich network of associative relations (e.g., the Gene Ontology [4]).

Our results summarized in Table 1 and Table 2 were carried out by running SPARQL queries coding *NuMi* and instantiated for all probe pairs. The queries can be parallelized and run independently by partitioning the probe pairs into smaller groups. If we spread the queries to $n$ processors, each with its own local, independent SNOMED CT RDF triple store, then an $n$-fold reduction of the computational time can be achieved.

Finally, we also plan to extend our analysis to other order-theoretic properties such as "part of", which is a key relation in anatomical ontologies such as the Foundational Model of Anatomy [2]. This may suggest the formulation of other order-theoretic properties in SPARQL than the lattice-theoretic one presented in this study.

## Acknowledgements

This research was supported in part by the Intramural Research Program of the National Library of Medicine (NLM), and in part by Clinical and Translational Science Award (UL1 RR024989) from the National Center for Research Resources (NCRR) and Comprehensive Cancer Center Support Grant (5P30 CA043703-16) from National Cancer Institute (NCI), all of the National Institutes of Health (NIH).

# References

1. Donnelly, K.: SNOMED-CT: The advanced terminology and coding system for eHealth. Stud. Health Technol. Inform. 121, 279–290 (2006)
2. FMA, http://sig.biostr.washington.edu/projects/fm/
3. Ganter, B., Wille, R.: Formal Concept Analysis. Springer, Heidelberg (1999)
4. Gene Ontology, http://www.geneontology.org/
5. Gierz, G., Hofmann, K.H., Keimel, K., Lawson, D.J., Mislove, M., Scott, D.S.: Continuous Lattices and Domains. Encyclopedia of Mathematics and its Applications, vol. 93. Cambridge University Press, Cambridge (2003)
6. International Health Terminology Standard Development Organization (IHTSDO), http://www.ihtsdo.org/
7. Jiang, G., Chute, C.G.: Auditing the semantic completeness of SNOMED CT using formal concept analysis. J. Am Med. Inform. Assoc. 16(1), 89–102 (2009)
8. Joslyn, C.: Poset Ontologies and Concept Lattices as Semantic Hierarchies. In: Wolff, K.E., Pfeiffer, H.D., Delugach, H.S. (eds.) ICCS 2004. LNCS (LNAI), vol. 3127, pp. 287–302. Springer, Heidelberg (2004)
9. Linked data, http://linkeddata.org/
10. Kuznetsov, S.O., Obiedkov, S.A.: Comparing performance of algorithms for generating concept lattices. J. Exp. Theor. Artif. Intell. 14(2-3), 189–216 (2002)
11. Priss, U.: Formal Concept Analysis as a Tool for Linguistic Data Exploration. In: Hitzler, P., Scharfe, H. (eds.) Conceptual Structures in Practice. Chapman & Hall/CRC Studies in Informatics Series, pp. 177–198 (2009)
12. RDF, http://www.w3.org/RDF/
13. SPARQL, http://www.w3.org/TR/rdf-sparql-query/
14. Suntisrivaraporn, B., Baader, F., Schulz, S., Spackman, K.: Replacing SEP-Triplets in SNOMED CT using tractable description logic operators. In: Bellazzi, R., Abu-Hanna, A., Hunter, J. (eds.) AIME 2007. LNCS (LNAI), vol. 4594, pp. 287–291. Springer, Heidelberg (2007)
15. Troy, A., Zhang, G.Q., Tian, Y.: Faster concept analysis. In: Priss, U., Polovina, S., Hill, R. (eds.) ICCS 2007. LNCS (LNAI), vol. 4604, pp. 206–219. Springer, Heidelberg (2007)
16. Virtuoso, http://virtuoso.openlinksw.com/
17. Wang, Y., Halper, M., Min, H., Perl, Y., Chen, Y., Spackman, K.A.: Structural methodologies for auditing SNOMED. J. Biomed. Inform. 40(5), 561–581 (2007)
18. Wei, D., Bodenreider, O.: Using the Abstraction Network in complement to Description Logics for quality assurance in biomedical terminologies - A case study in SNOMED CT Medinfo 2010 (in press, 2010)
19. Zhang G.Q and Bodenreider O. Large-scale, exhaustive lattice-based structural auditing of SNOMED CT. In: American Medical Informatics Association (AMIA) Fall 2010 Symposium (in press, 2010)
20. Zhang, G.Q.: Logic of Domains. Birkhäuser, Basel (1991)
21. Zhang, G.Q.: Quasi-prime algebraic domains. Theoretical Computer Science 155, 221–264 (1996)
22. Zhu, X., Wei, J.W., Baorto, D., Weng, C., Cimino, J.: A review of auditing methods applied to the content of controlled biomedical terminologies. J. Biomedical Informatics 42, 412–425 (2009)
23. Zweigenbaum, P., Bachimont, B., Bouaud, J., Charlet, J., Boisvieux, J.F.: Issues in the structuring and acquisition of an ontology for medical language understanding. Methods Inf. Med. 34(1-2), 15–24 (1995)

# Exploiting Relation Extraction for Ontology Alignment

Elena Beisswanger

Jena University Language and Information Engineering (JULIE) Lab,
Friedrich-Schiller-Universität Jena,
Fürstengraben 30, 07743 Jena, Germany
elena.beisswanger@uni-jena.de

**Abstract.** When multiple ontologies are used within one application system, aligning the ontologies is a prerequisite for interoperability and unhampered semantic navigation and search. Various methods have been proposed to compute mappings between elements from different ontologies, the majority of which being based on various kinds of similarity measures. As a major shortcoming of these methods it is difficult to decode the semantics of the results achieved. In addition, in many cases they miss important mappings due to poorly developed ontology structures or dissimilar ontology designs. I propose a complementary approach making massive use of relation extraction techniques applied to broad-coverage text corpora. This approach is able to detect different types of semantic relations, dependent on the extraction techniques used. Furthermore, exploiting external background knowledge, it can detect relations even without clear evidence in the input ontologies themselves.

**Keywords:** Ontology Alignment, Relation Extraction, Wikipedia.

## 1 Background and Problem Statement

Ontologies specify the major terms and concepts (also called classes) of a domain and their relations in a formal manner. An increasing number of information systems in different application domains rely on ontologies to organize data. While in case of the Semantic Web they are used to define the semantics of (Web) documents, in biomedicine they serve as vocabulary to semantically annotate huge literature collections and factual data stores. In biomedical natural language processing (bio-NLP), in turn, ontologies support (amongst others) information extraction and semantic search applications.

However, especially in the field of biomedicine conceptual knowledge is scattered over various different, often disconnected ontologies. While some of them topically overlap (such as two different anatomy ontologies), others complement each other rather by design (such as ontologies for anatomical structures, cells, proteins, biological processes, drugs and diseases) [19]. Both, extraction patterns and search queries easily transcend the conceptual coverage of a single ontology. As a consequence, missing links between ontologies hamper effective information extraction and search, besides generally limiting data interoperability. The

P.F. Patel-Schneider et al.(Eds.): ISWC 2010, Part II, LNCS 6497, pp. 289–296, 2010.

process of linking related ontology elements (*viz.*, classes, relations, and class instances) is called ontology alignment (OA) (cf. [8]).

OA has become an active field of research. Various methods have been proposed, most of them grounded in the intuition that elements with similar features (string-based, structural, extensional or semantics-based ones, cf. [8]) tend to be semantically related. Typically, given a certain similarity measure, similarity values are computed for pairs of elements and a threshold is chosen to decide which value is needed for a pair to be accepted as being "semantically related".

However, besides being sensitive to differing naming conventions and poorly developed or dissimilar ontology structures (cf., e.g., [1]), a major drawback of many similarity-based approaches is that the interpretation of their results is rather difficult. This applies both, to the type of relation existing between elements found to be similar (commonly, an equivalence relation is assumed[1]), and the similarity scores themselves. As a consequence, the lack of clear semantics hampers the incorporation of such alignments in reasoning applications and cross-ontology search. Generally, for the alignment of (complementary) biomedical ontologies other relation types than equivalence are critical, for example, *subClassOf*, *partOf*, and less common ones, such as *locatedIn*, *treats*, or *regulates*.

A completely different approach to OA with the potential to detect many different types of semantic relations is looking for relation evidences not within the given ontologies themselves, but in large-size, broad-coverage text corpora. This requires both, a suitable text corpus and an appropriate relation extraction (RE) machinery. Regarding the latter, in the field of NLP, there is a large body of work that could be exploited, targeting the extraction of various different relation types from text (cf., e.g., [11,9,5]). Concerning the text corpus, Wikipedia excels as a good candidate, for several reasons. First, it is a huge conglomerate of collaboratively assembled encyclopedic knowledge that currently seems to be unmatched in its size, broad coverage and up-to-dateness. Second, besides the free-text parts packed with definition phrases, it comes with a wide range of additional, more structured relation sources that could be used to support and complement results from free-text-based RE. These include the Wikipedia infoboxes, holding a multitude of conceptual relations in terms of implicit subject-predicate-object triples, the category system and articles linked to it, forming a huge concept graph with untyped semantic relations as edges, and the cross-links between articles, representing association-type like relations.

Along these lines the following research questions were derived:

1. How can established RE approaches contribute to the alignment of ontologies? How can they be adapted to the alignment use case? In particular, how can ontology class mentions be detected in text?[2]
2. Given Wikipedia as data source, how can relations extracted from free-text parts be integrated with relations extracted from structured parts of articles?
3. How can corpus-based ontology alignment methods be evaluated, in particular if they target relations other than equivalence, such as *subClassOf*?

---

[1] Very few systems also detect *subClassOf* relations (cf., e.g., [7]).

[2] Note that in my work I focus on the alignment of ontology classes only.

In the following I will discuss related work, outline expected contributions, present my working plan and conclude with an overview on the current state of my work and the next steps to be taken.

## 2   Related Work

**Automatic ontology alignment** is hampered by the fact that in many existing ontologies the meaning of classes and relations is insufficiently specified. To compensate for this shortcoming, alignment approaches have been developed incorporating various kinds of external background knowledge (e.g., [1,16,22]). Structured resources, such as ontologies (e.g., [1]) or WordNet (e.g., [16]) are preferably used, due to their easily accessible semantics. However, their coverage is generally limited and for many domains such resources lack completely. The opposite is the case for unstructured text. It is available in large quantities across many domains. However, relations are hidden in natural language phrases and an appropriate NLP system is required to access them. I am aware of only few alignment approaches exploiting relation extraction from text. One example is the work by van Hage et al., experimenting with basic linguistic methods (Hearst pattern matching on the Web and parsing definition phrases from an online dictionary) to discover *subClassOf* relations in the domain of food [22].

In **Ontology learning** (OL), a neighboring field of OA, relation extraction from text is much more common. OL is concerned with the automatic construction (or extension) of ontologies from given data sets, such as text corpora or databases. A typical text-based OL system extracts relevant terms and term variants, groups them to concepts and subsequently identifies *subClassOf* relations forming the backbone of the ontology (cf., [6]). The population of ontologies with instances is also widespread. For example, the SOFIE Framework extracts facts from free-text parts of Wikipedia articles to extend ontologies with instance data [21]. Only few systems go further and extract other relations than *subClassOf* and *instanceOf* (cf., [23]). In the case of ontology extension, as in the case of OA, a major challenge is to recognize the linguistic appearance of known concepts in text.

**Concept recognition** comprises two (not necessarily separate) steps: candidate detection and candidate disambiguation. While the first step influences recall of the RE procedure, the second one, tackling lexical ambiguity arising from homonymy and polysemy of words, has an impact on precision. Several concept recognition tools have been released, most of them relying on matching concept labels against text. The techniques used range from simple string matching procedures to advanced forms incorporating detailed linguistic analysis and synonym enrichment, as in the case of MetaMap (a system frequently used in the field of bio-NLP) [2]. Some terms in text qualify as mapping target for more than one concept. While simpler systems typically enumerate all candidates, more sophisticated solutions employ word sense disambiguation (WSD) techniques to identify the correct mapping (for a comprehensive survey, cf., [14]). Recently some new WSD approaches have been proposed exploiting Wikipedia specific information, such as page links and disambiguation pages (cf., e.g., [12]).

**Automatic extraction of semantic relations from text** is a broad research field in NLP. A plethora of statistical, rule-based, and machine learning-based approaches has been proposed targeting different types of relations, ranging from hypernymy (cf., e.g., [11,20,17]) and meronymy (cf., e.g., [9]), denoting the *subClassOf* and *partOf* relation on the linguistic level, to domain-specific relations (cf., e.g., [5]). The first version of the alignment system I am developing will focus on the detection of *subClassOf* relations between classes. Thus I am particularly interested in work on automatic hypernym extraction. Most common approaches either rely on lexico-syntactic patterns (cf., e.g., [11,20]), or exploit the distributional similarity or co-occurrence of terms (cf., e.g., [17]). As a unique feature, pattern-based approaches detect hypernymy relations explicitly mentioned in text. While Hearst utilizes a small set of hand crafted patterns (such as "$term_1$ *is a* $term_2$") [11], Snow et al. achieve a major improvement in recall by automatically deriving a much larger set of patterns from text and using them as features in a machine learning approach [20].

In recent years, **Wikipedia** has become a popular resource for RE and other NLP tasks and applications (for a survey, cf., [12]). So far relation extraction efforts mainly concentrate on structured facets of Wikipedia, such as infoboxes, page links, and the category system. Amongst others, Ponzetto and Strube created a taxonomy based on the Wikipedia category system by refining the previously untyped semantic relations [15]. Bizer et al. built DBpedia, consisting of over 4.5 million RDF triples mainly derived from Wikipedia infobox templates [4]. Recently WikiNet was published, a collection of 3 million concepts and over 36 million relations mainly extracted from Wikipedia infoboxes and the category system [13]. Both, results and extraction machinery of some of these projects have been made publicly available. Fewer efforts target the full-text body of Wikipedia articles, an example is [21].

## 3    Expected Contributions

1. The main contribution of my work will be an **ontology alignment system** exploiting conceptual relations entangled in unstructured and structured parts of a huge text corpus, *viz.* Wikipedia. While the current version is restricted to the extraction of *subClassOf* relations from free-text, two extensions of the system are scheduled: the detection of other types of semantic relations critical for the alignment of biomedical ontologies and the incorporation of relations extracted from structured parts of Wikipedia.
2. The UIMA-based[3] **NLP tools** I am developing for the analysis of Wikipedia articles are designed to work independently from the alignment system. Thus, they can be deployed in other application scenarios, too.
3. Amongst the few existing RE-based alignment systems, my system will be distinguished by a proper **concept recognition** step. To assess the state-of-the-art in concept recognition, a thorough investigation of existing approaches will be carried out. Concept recognition also is a key issue in other

---

[3] Unstructured Information Management Architecture (http://uima.apache.org/)

tasks involving both, ontologies and textual data (such as semantic search, semantic annotation of text, or text-based OL). Thus, the intended study is of potential interest even outside the OA community.

4. Finally, my work will cover **evaluation strategies** for alignments holding relations of other types than equivalence (a first step in this direction was taken in terms of the "Oriented matching" task of the 2009 Campaign of the Ontology Alignment Evaluation Initiative [7]), as well as an **alignment algorithm** that can even cope with large-sized ontologies, avoiding an exhaustive analysis of class and label pairs.

# 4   Alignment System Design and Development

The alignment system will consist of the following components:

1. **Alignment algorithm.** It decides on which class comparisons to be made, queries the index, filters the query result, invokes the RE module and the relation repository, and integrates results.
2. **Lucene index.** The index contains Wikipedia articles, sentence-wise.
3. **Relation extraction module.** The RE module extracts relations between class pairs from free-text parts of Wikipedia.
4. **Relation repository.** The relation repository contains relations extracted from structured parts of Wikipedia.
5. **Result store.** The result store saves the output of the RE module across different runs of the system, to avoid duplicate work.

**Procedure.** The alignment of two ontologies will proceed as follows. First, the system imports the ontologies. Next, the alignment algorithm starts selecting pairs of classes to be compared. For each selected class pair that is not yet in the result store label pairs are formed. For each label pair that is not yet in the result store the index is queried for sentences containing normalized forms of both labels. A filtering step eliminates those sentences in which at least one label refers to a wrong word sense. The remaining sentences and the label pair are handed over to RE module, which, in turn, searches the sentences for relation evidences. Sentences with overlapping labels are dealt with separately. Next, the relation repository is queried for additional relation evidences. Based on all evidences found, the alignment algorithm decides whether the class pair is related or not. Results for newly analyzed class and label pairs are saved in the result store. If no more class pairs need to be analyzed, results are integrated and cleaned up and the final alignment is exported.

**Relation extraction.** The RE module of the alignment system will incorporate a dependency feature-based relation classifier, similar to the one proposed in [20]. In the current version, it predicts *subClassOf* relations only (as in [20]). However, in principle the classifier could be enabled to detect also other relation types, given that sufficient train and test data is available (see Section 5). As baseline for the *subClassOf* extraction, a second RE module is used, relying on the original Hearst patterns [11] (in this respect, it is similar to [22]).

In the alignment system, relation extraction is preceded by a two-stage **concept recognition** step. First, class labels are detected in text by means of an extended string matching procedure. It involves lower-casing, stop word removal, removal of special characters, stemming, and a filtering step evaluating part-of-speech tags and syntactic information of class labels and text. In the second stage (which is not implemented yet), ambiguities will be resolved considering both, the context of candidate classes in the respective ontology (e.g., the labels of adjacent classes and relations, as in [21]) and Wikipedia specific information. For the latter, existing Wikipedia-based WSD approaches will be evaluated.

**Alignment algorithm.** The overall alignment process is governed by the alignment algorithm. There are two major tasks to perform: the selection of class pairs to be analyzed, and the decision about mappings between classes. For the first task, a brute force approach is used in the current version of the system. All classes in one ontology are compared to all classes in the other one, considering all possible label pairs. Since for large ontologies this implies high computational costs, the adoption of a new, optimized procedure is scheduled. Inspired by existing work (such as the Anchor-Flood algorithm, looking for mappings between previously defined blocks of similar classes only [18]), for the selection of class pairs it will consider both, the structure of the input ontologies and already computed mappings. The second task is solved based on two input streams: relation evidences from free-text parts of Wikipedia articles delivered by the RE module, and relation evidences originating from structured parts as they will be available from the relation repository. While a first version of the RE module is already in place, the relation repository is still pending.

## 5    Status and Next Steps

So far I have developed an UIMA-based NLP application for relation extraction from Wikipedia. It comprises two text processing pipelines (one for creating the Wikipedia index required by the OA system, the other for the RE task itself), and a scheduling system that allows to run several pipeline instances simultaneously (a prerequisite to efficiently process a large data collection such as Wikipedia). Besides existing text processing components from JCoRe [10] (sentence splitter, tokenizer, POS tagger, chunker, etc.), the two pipelines include the following newly developed components: a UIMA Collection Reader for Wikipedia [3] (it makes Wikipedia articles accessible for subsequent UIMA analytics by parsing the MediaWiki mark-up and filtering relevant contents), a new indexing component, a UIMA-based Hearst pattern matcher, and a second, more advanced RE module, incorporating a dependency feature-based relation classifier. To build the classifier, I basically parsed sentences extracted from Wikipedia abstracts (the first paragraph, before the table of contents), extracted noun phrases as anchor pairs, labeled them as being hyponym/hypernym pairs in WordNet[4], extracted the dependency paths between all anchor pairs, took "frequent" paths

---

[4] http://wordnet.princeton.edu/

as features, generated feature vectors for the labeled anchor pairs (taking the frequency of occurrence of a path between an anchor pair as feature value), and trained the classifier with all anchor pairs of which the feature vector contained a minimum number of non-zero values. Currently, a manual *subClassOf* annotation project is running that will deliver gold standard data required to evaluate the RE modules.

There are two immediate next steps when the gold standard has been completed. First, the performance of the RE module will be evaluated. Second, the concept recognition step will be refined (which precedes the actual relation extraction) by implementing the scheduled disambiguation stage. The evaluation will be rerun to assess which impact it has on RE results. Thereafter, the next major steps will be to enhance the alignment algorithm and to prepare the RE module for the detection of new relation types. For each relation type, train and test data must be provided to retrain and evaluate the included classifier. Finally, the relation repository will be populated with relations extracted from structured parts of Wikipedia (e.g., incorporating harmonized results of [15,4,13]).

In conclusion, this doctoral project lies at the junction of two different avenues of research, *viz.* NLP-based relation extraction and ontology alignment. The main challenge is to properly integrate these currently almost unrelated approaches, in order to open up the rich reservoir of conceptual relations entangled in natural language texts for OA. Furthermore, it requires to respond to the methodological requirements of aligning concrete, large-sized (bio-)ontologies. Up until now, I have implemented a first simple version of an alignment system working along these lines. Although it still lacks many of the envisaged sophisticated features, it can already discover *subClassOf* relations between ontology classes applying a well established RE approach to the English Wikipedia.

## Acknowledgments

I would like to thank my supervisors Udo Hahn and Heiner Stuckenschmidt for their assistance, and the anonymous reviewers for their helpful comments.

## References

1. Aleksovski, Z., Klein, M.C.A., ten Kate, W., van Harmelen, F.: Matching unstructured vocabularies using a background ontology. In: Staab, S., Svátek, V. (eds.) EKAW 2006. LNCS (LNAI), vol. 4248, pp. 182–197. Springer, Heidelberg (2006)
2. Aronson, A.R.: Effective mapping of biomedical text to the UMLS Metathesaurus: The METAMAP program. In: Proceedings of the AMIA 2001, pp. 17–21 (2001)
3. Beisswanger, E., Hahn, U.: JULIE Lab's UIMA Collection Reader for Wikipedia. In: Proceedings of the LREC 2010 Workshop on New Challenges for NLP Frameworks, pp. 15–19 (2010)
4. Bizer, C., Lehmann, J., Kobilarov, G., Auer, S., Becker, C., Cyganiak, R., Hellmann, S.: DBpedia - a crystallization point for the web of data. Web Semantics 7(3), 154–165 (2009)

5. Ciaramita, M., Gangemi, A., Ratsch, E., Saric, J., Rojas, I.: Unsupervised learning of semantic relations between concepts of a molecular biology ontology. In: Proceedings of the IJCAI 2005, pp. 659–664 (2005)
6. Cimiano, P.: Ontology Learning and Population from Text: Algorithms, Evaluation and Applications. Springer, Heidelberg (2006)
7. Euzenat, J., Ferrara, A., Hollink, L., Isaac, A., Joslyn, C., Malaisé, V., Meilicke, C., Nikolov, A., Pane, J., Sabou, M., Scharffe, F., Shvaiko, P., Spiliopoulos, V., Stuckenschmidt, H., Sváb-Zamazal, O., Svátek, V., dos Santos, C.T., Vouros, G.A., Wang, S.: Results of the Ontology Alignment Evaluation Initiative 2009. In: Proceedings of the ISWC 2009 Workshop on Ontology Matching (2009)
8. Euzenat, J., Shvaiko, P.: Ontology matching. Springer, Heidelberg (2007)
9. Girju, R., Badulescu, A., Moldovan, D.: Automatic discovery of part-whole relations. Computational Linguistics 32(1), 83–135 (2006)
10. Hahn, U., Buyko, E., Landefeld, R., Mühlhausen, M., Poprat, M., Tomanek, K., Wermter, J.: An overview of JCoRe, the JULIE Lab UIMA component repository. In: Proceedings of the LREC 2008 Workshop on UIMA for NLP, pp. 1–7 (2008)
11. Hearst, M.A.: Automatic acquisition of hyponyms from large text corpora. In: Proceedings of the ACL 1992 Conference, pp. 539–545 (1992)
12. Medelyan, O., Milne, D., Legg, C., Witten, I.H.: Mining meaning from Wikipedia. International Journal of Human-Computer Studies 67(9), 716–754 (2009)
13. Nastase, V., Strube, M., Boerschinger, B., Anas, E.: WikiNet: A very large scale multi-lingual concept network. In: Proceedings of the LREC 2010 (2010)
14. Navigli, R.: Word sense disambiguation: A survey. ACM Computing Surveys 41(2), 1–69 (2009)
15. Ponzetto, S.P., Strube, M.: Deriving a large scale taxonomy from Wikipedia. In: Proceedings of the AAAI 2007 Conference, pp. 1440–1445 (2007)
16. Reynaud, C., Safar, B.: Exploiting WordNet as background knowledge. In: Proceedings of the ISWC 2007 Workshop on Ontology Matching (2007)
17. Sanderson, M., Croft, W.B.: Deriving concept hierarchies from text. In: Proceedings of the SIGIR 1999 Conference, pp. 206–212 (1999)
18. Seddiqui, M.H., Aono, M.: An efficient and scalable algorithm for segmented alignment of ontologies of arbitrary size. Web Semantics 7(4), 344–356 (2009)
19. Smith, B., Ashburner, M., Rosse, C., Bard, J., Bug, W., Ceusters, W., Goldberg, L., Eilbeck, K., Ireland, A., Mungall, C.J., Leontis, N., Rocca-Serra, P., Ruttenberg, A., Sansone, S.A., Scheuermann, R.H., Shah, N., Whetzel, P.L., Lewis, S.E.: The OBO Foundry: coordinated evolution of ontologies to support biomedical data integration. Nature Biotechnology 25(11), 1251–1255 (2007)
20. Snow, R., Jurafsky, D., Ng, A.Y.: Learning syntactic patterns for automatic hypernym discovery. In: Advances in Neural Information Processing Systems 17, pp. 1297–1304. MIT Press, Cambridge (2005)
21. Suchanek, F.M., Sozio, M., Weikum, G.: SOFIE: A Self-Organizing Framework for Information Extraction. In: Proceedings of the WWW 2009 Conference (2009)
22. Van Hage, W.R., Katrenko, S., Schreiber, G.: A method to combine linguistic ontology-mapping techniques. In: Gil, Y., Motta, E., Benjamins, V.R., Musen, M.A. (eds.) ISWC 2005. LNCS, vol. 3729, pp. 732–744. Springer, Heidelberg (2005)
23. Völker, J., Haase, P., Hitzler, P.: Learning expressive ontologies. In: Proceedings of the 2008 Conference on Ontology Learning and Population, pp. 45–69 (2008)

# Towards Semantic Annotation Supported by Dependency Linguistics and ILP

Jan Dědek

Department of Software Engineering, Charles University,
Prague, Czech Republic
dedek@ksi.mff.cuni.cz

**Abstract.** In this paper we present a method for semantic annotation of texts, which is based on a deep linguistic analysis (DLA) and Inductive Logic Programming (ILP). The combination of DLA and ILP have following benefits: Manual selection of learning features is not needed. The learning procedure has full available linguistic information at its disposal and it is capable to select relevant parts itself. Learned extraction rules can be easily visualized, understood and adapted by human. A description, implementation and initial evaluation of the method are the main contributions of the paper.

**Keywords:** Semantic Annotation, Dependency Linguistics, Inductive Logic Programming, Information Extraction, Machine Learning.

## 1 Introduction

Automated semantic annotation (SA) is considered to be one of the most important elements in the evolution of the Semantic Web. Besides that, SA can provide great help in the process of data and information integration and it could also be a basis for intelligent search and navigation.

In this paper we present main results and reflections of our ongoing PhD project, a method for classical and semantic information extraction and annotation of texts, which is based on a deep linguistic analysis and Inductive Logic Programming (ILP). This approach is quite novel because it directly combines deep linguistic parsing with machine learning (ML). This combination and the use of ILP as a ML engine have following benefits: Manual selection of learning features is not needed. The learning procedure has full available linguistic information at its disposal and it is capable to select relevant parts itself. Extraction rules learned by ILP can be easily visualized, understood and adapted by human.

A description, implementation and initial evaluation of the method are the main contributions of the paper.

## 2 Related Work

There are many users of ILP in the linguistic and information extraction area. Authors of [12] summarized some basic principles of using ILP for learning from

P.F. Patel-Schneider et al.(Eds.): ISWC 2010, Part II, LNCS 6497, pp. 297–304, 2010.

text without any linguistic preprocessing. One of the most related approaches to ours can be found in [1]. The authors use ILP for extraction of information about chemical compounds and other concepts related to global warming and they try to express the extracted information in terms of ontology. They use only the part of speech analysis and named entity recognition in the preprocessing step. But their inductive procedure uses also additional domain knowledge for the extraction. In [17] ILP was used to construct good features for propositional learners like SVM to do information extraction. It was discovered that this approach is a little bit more successful than a direct use of ILP but it is also more complicated. The later two approaches could be also employed in our solution.

There are other approaches that use deep parsing, but they often use the syntactic structure only for relation extraction and either do not use machine learning at all (extraction rules have to be handcrafted) [19], [9], [4] or do some kind of similarity search based on the syntactic structure [8], [18] or the syntactic structure plays only very specific role in the process of feature selection for propositional learners [3].

There is also a long row of information extraction approaches that use classical propositional learners like SVM on a set of features manually selected from input text. We do not cite them here. We just refer to [13] – using machine learning facilities in GATE. This is the software component (Machine Learning PR) to that we have compared our solution. Our solution is also based on GATE (See next sections.)

Last category of related works goes in the direction of semantics and ontologies. Because we do not develop this topic in this paper, we just refer to the ontology features in GATE [2], which can be easily used to populate an ontology with the extracted data. We discus this topic later in Section 4.4.

## 3    Exploited Methods – Linguistics and ILP

In our solution we have exploited several tools and formalisms. These can be divided into two groups: linguistics and (inductive) logic programming. First we describe the linguistic tools and formalisms, the rest will follow.

### 3.1    GATE

GATE[1] [5] is probably the most widely used tool for text processing. In our solution the capabilities of document and annotation management, utility resources for annotation processing, JAPE grammar rules [6], machine learning facilities and performance evaluation tools are the most helpful features of GATE that we have used.

### 3.2    PDT and TectoMT

As we have started with our native language – Czech (a language with rich morphology and free word order), we had to make tools for processing Czech

---

[1] http://gate.ac.uk/

available in GATE. We have implemented a wrapper for the TectoMT system[2] [20] to GATE. TectoMT is a Czech project that contains many linguistic analyzers for different languages including Czech and English. We have used a majority of applicable tools from TectoMT: a tokeniser, a sentence splitter, morphological analyzers (including POS tagger), a syntactic parser and the deep syntactic (tectogrammatical) parser. All the tools are based on the dependency based linguistic theory and formalism of the Prague Dependency Treebank project [10]. So far our solution does not include any coreference and discourse analysis.

### 3.3 Inductive Logic Programming

Inductive Logic Programming (ILP) [16] is a machine learning technique based on logic programming. Given an encoding of the known background knowledge (in our case linguistic structure of all sentences) and a set of examples represented as a logical database of facts (in our case tokens annotated with the target annotation type are positive examples and the remaining tokens negative ones), an ILP system will derive a hypothesized logic program (in our case extraction rules) which entails all the positive and none of the negative examples.

As an ILP tool we have used "A Learning Engine for Proposing Hypotheses" (Aleph v5)[3], which we consider very practical. It uses quite effective method of inverse entailment [15] and keeps all handy features of a Prolog system (we have used YAP Prolog[4]) in its background.

From our experiments (Section 5) can be seen that ILP is capable to find complex and meaningful rules that cover the intended information.

## 4   Implementation

Here we just briefly describe implementation of our system. The system consists of several modules, all integrated in GATE as processing resources.

### 4.1   TectoMT Wrapper (Linguistic Analysis)

First is the TectoMT wrapper, which takes the text of a GATE document, sends it to TectoMT linguistic analyzers, parses the results and converts the results to the form of GATE annotations.

### 4.2   ILP Wrapper (Machine Learning)

After a human annotator have annotated several documents with desired target annotations, machine learning takes place. This consists of two steps:

1. learning of extraction rules from the target annotations and
2. application of the extraction rules on new documents.

---

[2] http://ufal.mff.cuni.cz/tectomt/
[3] http://www.comlab.ox.ac.uk/activities/machinelearning/Aleph/
[4] http://www.dcc.fc.up.pt/~vsc/Yap/

In both steps the linguistic analysis has to be done before and in both steps background knowledge (a logical database of facts) is constructed from linguistic structures of documents that are being processed. We call the process of background knowledge construction as *ILP serialization*. Although this topic is quite interesting we do not present details here because of space limitations.

After the ILP serialization is done, in the learning case, positive and negative examples are constructed from target annotations and the machine learning ILP inductive procedure is executed to obtain extraction rules.

In the application case a Prolog system is used to check if the extraction rules entail any of target annotation candidates.

The learning examples and annotation candidates are usually constructed from all document tokens (and we did so in the present solution), but it can be optionally changed to any other textual unit, for example only numerals or tectogrammatical nodes (words with lexical meaning) can be selected. This can be done easily with the help of *Machine Learning PR* (LM PR) from GATE[5].

ML PR provides an interface for exchange of features (including target class) between annotated texts and propositional learners in both directions – during learning as well as during application. We have used ML PR and developed our *ILP Wrapper* for it. The implementation was a little complicated because complex linguistic structures cannot be easily passed as propositional features, so in our solution we use the ML PR interface only for exchange of the class attribute and annotation id and we access the linguistic structures directly in a document.

### 4.3   Root/Subtree Preprocessing/Postprocessing

Sometimes annotations span over more than one token. This situation complicates the process of machine learning and this situation is often called as "chunk learning". Either we have to split a single annotation to multiple learning instances and after application we have to merge them back together, or we can change the learning task from learning annotated tokens to learning borders of annotations (start tokens and end tokens). The later approach is implemented in GATE in *Batch Learning PR* in the 'SURROUND' mode.

We have used another approach to solve this issue. Our approach is based on syntactic structure of a sentence and we call it "root/subtree preprocessing/postprocessing". The idea is based on the observation that tokens of a multi-token annotation usually have a common parent node in a syntactic tree. So we can

1. extract the parent nodes (in dependency linguistics this node is also a token and it is usually one of the tokens inside the annotation),
2. learn extraction rules for parent nodes only and

---

[5] *Machine Learning PR* is an old GATE interface for ML and it is almost obsolete but in contrast to the new *Batch Learning PR* the LM PR is easy to extend for a new ML engine.

3. span annotations over the whole subtrees of root tokens found during the application of extraction rules.

We call the first point as *root preprocessing* and the last point as *subtree post-processing*. We have successfully used this technique for the 'damage' task of our evaluation corpus (See Section 5 for details.)

### 4.4   Semantic Interpretation

Information extraction can solve the task "how to get documents annotated", but as we aim on the semantic annotation, there is a second step of "semantic interpretation" that has to be done. In this step we have to interpret the annotations in terms of a standard ontology. On a very coarse level this can be done easily. Thanks to GATE ontology tools [2] we can convert all the annotations to ontology instances with a quite simple JAPE [6] rule, which takes the content of an annotation and saves it as a label of a new instance or as a value of some property of a shared instance. For example in our case of traffic and fire accidents, there will be a new instance of an accident class for each document and the annotations would be attached to this instance as values of its properties. Thus from all annotations of the same type, instances of the same ontology class or values of the same property would be constructed. This is very inaccurate form of semantic interpretation but still it can be useful. It is similar to the GoodRelation [11] design principle of *incremental enrichment*[6]: "...you can still publish the data, even if not yet perfect. The Web will do the rest – new tools and people."

But of course we are not satisfied with this fashion of semantic interpretation and we plan to further develop the semantic interpretation step as a sophisticated "annotation → ontology" transformation process that we have proposed in one of our previous works [7].

### 4.5   How to Download

So far we do not provide our solution as a ready-made installable tool. But a middle experienced Java programmer can build it from source codes in our SVN repository[7].

## 5   Evaluation

We have evaluated our state of the art solution on a small dataset that we use for development. It is a collection of 50 Czech texts that are reporting on some accidents (car accidents and other actions of fire rescue services). These reports come from the web of Fire rescue service of Czech Republic[8]. The labeled corpus

---

[6] http://www.ebusiness-unibw.org/wiki/Modeling_Product_Models#Recipe:
_.22Incremental_Enrichment.22
[7] Follow the instructions at http://czsem.berlios.de/
[8] http://www.hzscr.cz/hasicien/

**Table 1.** Evaluation results

| task/method | matching | missing | excessive | overlap | prec.% | recall% | **F1.0%** |
|---|---|---|---|---|---|---|---|
| damage/ILP | 14 | 0 | 7 | 6 | 51.85 | 70.00 | 59.57 |
| damage/ILP – lenient measures | | | | | 74.07 | 100.00 | 85.11 |
| dam./ILP-roots | 16 | 4 | 2 | 0 | 88.89 | 80.00 | 84.21 |
| damage/Paum | 20 | 0 | 6 | 0 | 76.92 | 100.00 | 86.96 |
| injuries/ILP | 15 | 18 | 11 | 0 | 57.69 | 45.45 | 50.85 |
| injuries/Paum | 25 | 8 | 54 | 0 | 31.65 | 75.76 | 44.64 |
| inj./Paum-afun | 24 | 9 | 38 | 0 | 38.71 | 72.73 | 50.53 |

is publically available on the web of our project[9]. The corpus is structured such that each document represents one event (accident) and several attributes of the accident are marked in text. For the evaluation we selected two attributes of different kind. The first one is 'damage' – an amount (in CZK - Czech Crowns) of summarized damage arisen during a reported accident. The second one is 'injuries', it marks mentions of people injured during an accident. These two attributes differ. Injuries annotations always cover only a single token, while damage annotations usually consist of two or three tokens – one or two numerals express the amount and one extra token is for currency.

To compare our solution with other alternatives we took the Paum propositional learner from GATE [14]. The quality of propositional learning from texts is strongly dependent on the selection of right features. We obtained quite good results with features of a window of two preceding and two following token lemmas and morphological tags. The precision was further improved by adding the feature of *analytical function* from the syntactic parser (see the last row of Table 1).

Results of a 10-fold cross validation are summarized in Table 1. We used standard information retrieval performance measures: precision, recall and $F_1$ measure and also theirs lenient variants (overlapping annotations are added to the correctly matching ones, the measures are the same if no overlapping annotations are present).

In the first task ('damage') the methods obtained much higher scores then in the second ('injuries') because the second task is more difficult. In the first task also the root/subtree preprocessing/postprocessing improved results of ILP such that afterwards, annotation borders were all placed precisely. The ILP method had better precision and worse recall than the Paum learner but the $F_1$ score was very similar in both cases.

In Figure 1 we present some examples of the rules learned from the whole dataset. The rules demonstrate a connection of a target token with other parts of a sentence through linguistic syntax structures. For example the first rule connects a root numeral ($n.quant.def$) of 'damage' with a mention of 'investigator' that stated the mount. In the last rule only a positive occurrence of the verb 'injure' is allowed.

---

[9] http://czsem.berlios.de/

```
[Rule 1] [Pos cover = 14 Neg cover = 0]
damage_root(A) :- lex_rf(B,A), has_sempos(B,'n.quant.def'), tDependency(C,B),
    tDependency(C,D), has_t_lemma(D,'investigator').
[Rule 2] [Pos cover = 13 Neg cover = 0]
damage_root(A) :- lex_rf(B,A), has_functor(B,'TOWH'), tDependency(C,B),
    tDependency(C,D), has_t_lemma(D,'damage').

[Rule 1] [Pos cover = 7 Neg cover = 0]
injuries(A) :- lex_rf(B,A), has_functor(B,'PAT'), has_gender(B,anim),
    tDependency(B,C), has_t_lemma(C,'injured').
[Rule 8] [Pos cover = 6 Neg cover = 0]
injuries(A) :- lex_rf(B,A), has_gender(B,anim), tDependency(C,B),
    has_t_lemma(C,'injure'), has_negation(C,neg0).
```

**Fig. 1.** Examples of learned rules, Czech words are translated

# 6  Conclusion and Future Work

From our experiments can be seen that ILP is capable to find complex and meaningful rules that cover the intended information. But in terms of the performance measures the results are not better than those from a propositional learner. This is quite surprising observation because Czech is a language with free word order and we would expect much better results of the dependency approach than those of the position based approach, which was used by the propositional learner.

Our method is still missing an intelligent semantic interpretation procedure and it should be evaluated on bigger datasets (e.g. MUC, ACE, TAC, CoNLL) and other languages. So far we also do not provide a method for classical relation extraction (like e.g. in [3]). In the present solution we deal with relations implicitly. The method has to be adapted for explicit learning of relations in the form of "subject predicate object".

Our method can also provide a comparison of linguistic formalisms and tools because on the same data we could run our method using different linguistic analyzers and compare the results.

## Acknowledgments

This work was partially supported by Czech projects: GACR P202/10/0761, GACR-201/09/H057, GAUK 31009 and MSM-0021620838. The author would like to thank his supervisor Peter Vojtáš for the guidance of the PhD thesis.

## References

1. Aitken, S.: Learning information extraction rules: An inductive logic programming approach. In: van Harmelen, F. (ed.) Proceedings of the 15th European Conference on Artificial Intelligence. IOS Press, Amsterdam (2002)
2. Bontcheva, K., Tablan, V., Maynard, D., Cunningham, H.: Evolving GATE to Meet New Challenges in Language Engineering. Natural Language Engineering 10(3/4), 349–373 (2004)
3. Bunescu, R., Mooney, R.: Extracting relations from text: From word sequences to dependency paths. In: Kao, A., Poteet, S.R. (eds.) Natural Language Processing and Text Mining, ch. 3, pp. 29–44. Springer, London (2007)

4. Buyko, E., Faessler, E., Wermter, J., Hahn, U.: Event extraction from trimmed dependency graphs. In: BioNLP 2009: Proceedings of the Workshop on BioNLP, pp. 19–27. ACL, Morristown (2009)
5. Cunningham, H., Maynard, D., Bontcheva, K., Tablan, V.: GATE: A framework and graphical development environment for robust NLP tools and applications. In: Proceedings of the 40th Anniversary Meeting of the ACL (2002)
6. Cunningham, H., Maynard, D., Tablan, V.: JAPE: a Java Annotation Patterns Engine. Tech. rep., Department of Computer Science, The University of Sheffield (2000), http://www.dcs.shef.ac.uk/intranet/research/resmes/CS0010.pdf
7. Dědek, J., Vojtáš, P.: Computing aggregations from linguistic web resources: a case study in czech republic sector/traffic accidents. In: Dini, C. (ed.) Second International Conference on Advanced Engineering Computing and Applications in Sciences, pp. 7–12. IEEE Computer Society, Los Alamitos (2008), http://www2.computer.org/portal/web/csdl/doi/10.1109/ADVCOMP.2008.17
8. Etzioni, O., Banko, M., Soderland, S., Weld, D.S.: Open information extraction from the web. ACM Commun. 51(12), 68–74 (2008)
9. Fundel, K., Küffner, R., Zimmer, R.: Relex—relation extraction using dependency parse trees. Bioinformatics 23(3), 365–371 (2007)
10. Hajič, J., Hajičová, E., Hlaváčová, J., Klimeš, V., Mírovský, J., Pajas, P., Štěpánek, J., Vidová-Hladká, B., Žabokrtský, Z.: Prague dependency treebank 2.0 CD-ROM. In: Linguistic Data Consortium LDC2006T01, Philadelphia (2006)
11. Hepp, M.: Goodrelations: An ontology for describing products and services offers on the web. In: Gangemi, A., Euzenat, J. (eds.) EKAW 2008. LNCS (LNAI), vol. 5268, pp. 329–346. Springer, Heidelberg (2008)
12. Junker, M., Sintek, M., Sintek, M., Rinck, M.: Learning for text categorization and information extraction with ILP. In: Cussens, J., Džeroski, S. (eds.) LLL 1999. LNCS (LNAI), vol. 1925, pp. 84–93. Springer, Heidelberg (2000)
13. Li, Y., Bontcheva, K., Cunningham, H.: Adapting SVM for Data Sparseness and Imbalance: A Case Study on Information Extraction. Natural Language Engineering 15(02), 241–271 (2009), http://journals.cambridge.org/repo_A45LfkBD
14. Li, Y., Zaragoza, H., Herbrich, R., Shawe-Taylor, J., Kandola, J.S.: The perceptron algorithm with uneven margins. In: ICML 2002: Proceedings of the Nineteenth International Conference on Machine Learning, pp. 379–386. Morgan Kaufmann Publishers Inc., San Francisco (2002)
15. Muggleton, S.: Inverse entailment and progol. New Generation Computing, Special issue on Inductive Logic Programming 13(3-4), 245–286 (1995)
16. Muggleton, S.: Inductive logic programming. New Generation Computing 8(4), 295–318 (1991), http://dx.doi.org/10.1007/BF03037089
17. Ramakrishnan, G., Joshi, S., Balakrishnan, S., Srinivasan, A.: Using ilp to construct features for information extraction from semi-structured text. In: Blockeel, H., Ramon, J., Shavlik, J., Tadepalli, P. (eds.) ILP 2007. LNCS (LNAI), vol. 4894, pp. 211–224. Springer, Heidelberg (2008)
18. Wang, R., Neumann, G.: Recognizing textual entailment using sentence similarity based on dependency tree skeletons. In: RTE 2007: Proceedings of the ACL-PASCAL Workshop on Textual Entailment and Paraphrasing, pp. 36–41. ACL, Morristown (2007)
19. Yakushiji, A., Tateisi, Y., Miyao, Y., Tsujii, J.: Event extraction from biomedical papers using a full parser. In: Pac. Symp. Biocomput., pp. 408–419 (2001)
20. Žabokrtský, Z., Ptáček, J., Pajas, P.: TectoMT: Highly modular MT system with tectogrammatics used as transfer layer. In: Proceedings of the 3rd Workshop on Statistical Machine Translation, pp. 167–170. ACL, Columbus (2008)

# Towards Technology Structure Mining from Scientific Literature

Behrang QasemiZadeh

Unit for Natural Language Processing, DERI
National University of Ireland, Galway
behrang.qasemizadeh@deri.org

**Abstract.** This paper introduces the task of Technology-Structure Mining to support Management of Technology. We propose a linguistic based approach for identification of Technology Interdependence through extraction of technology concepts and relations between them. In addition, we introduce Technology Structure Graph for the task formalization. While the major challenge in technology structure mining is the lack of a benchmark dataset for evaluation and development purposes, we describes steps that we have taken towards providing such a benchmark. The proposed approach is initially evaluated and applied in the domain of Human Language Technology and primarily results are demonstrated. We further explain plans and research challenges for evaluation of the proposed task.

## 1 Introduction

We are drowning in the sea of data and effective intelligent-contextual information retrieval systems have turned out to be strategic tools in different disciplines, among them interdisciplinary field of Management of Technology [1](MoT). The role technology plays in shaping our lives, and its critical role in an increasingly competitive knowledge based economy is a matter of fact. Technology is developed and propagates globally with a surprising velocity, and managing the accelerated rate of technology development becomes a universal challenge. MoT tries to bring efficiency in technology organization mainly through the process of Technology Watch. Technology Watch in general is the process of extracting tactical information about technology. However, the manual process of extracting such information is tedious and time consuming considering the gigantic amount of information. [2]

A long discussed topic in MoT is Technology-structure relationships [3]. One empirical research aspect of technology-structure relationship deals with *interdependence of technologies* i.e. how technologies are related to each other. We propose a linguistic based approach to facilitate the process of extracting information about technologies by proposing a methodology for extracting information about interdependencies of technologies -e.g. how technologies are built on top of each other. We have named the proposed task "Technology Structure Mining".

The proposed research involves several established research challenges in Information Extraction and Natural Language Processing such as Named Entity

P.F. Patel-Schneider et al.(Eds.): ISWC 2010, Part II, LNCS 6497, pp. 305–312, 2010.

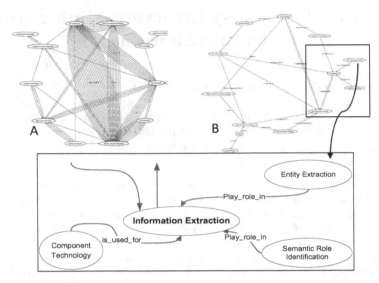

**Fig. 1.** In the above figure, ellipses show technologies and each labeled edge shows a relationship between pairs of technologies. The represented figure above has been generated from a part of publications in the ACL anthology reference corpus. Graph A illustrates state of the art in most text mining/ontology learning systems where co-occurrences of terms are usually considered as a measure for relating concepts. Graph B illustrates the goal of our proposed research where concepts are related to each other by help of natural language processing techniques for relation extraction. Graph A is generated automatically by help of our proposed method, while graph B is extracted and annotated by a careful study of graph A.

Recognition [4], Semantic Role Identification [5], and Relation Extraction [6],[7]. Considering technology as applied science, then scientific publications can be considered as a primary source of information about technologies and emerging technological trends. Figure 1 illustrates an example of the result of the proposed task after analysis of publications in the domain of Human Language Technology from ACL Anthology Reference Corpus (ACL ARC)[8] and offers a graphical representation of the outcome of analysis.

Evaluation and Understanding of the outcome of any task like the one proposed here remains a research challenge. In addition, while any task like the one we will introduce here tackles the problem of knowledge acquisition and tries to engineer the bottleneck of knowledge acquisition through automated methodologies and algorithms, the development and evaluation of such methods relies closely on the provided dataset for testing and training e.g. [9],[10]. In other words such research is more task-driven rather than fact-driven. We address and target these issues in our proposed research.

The rest of the paper is organized as follows. The next section briefly introduces related work. In section 3, we propose a formal definition for the proposed task and explain our goals through some examples. The applied methodology for approaching the task is briefly explained in section 4. In section 5, we report statistical information of our analysis on our reference corpus. Finally we conclude and give the direction of our future work in section 6.

## 2  Related Work

There has been number of research directions for supporting MoT and the task of Technology Watch. Most of the reported research is focusing on the task of patent mining e.g. [11], assisting Intellectual Property Management [12], and technology road-mapping [13]. However, as to the knowledge of the author there is no research reported on mining information specifically from scientific publications for the task of technology interdependency mining.

We classify the task of Technology Structure Mining as an activity situated between two emerging research areas: Ontology Learning (OL)[14] and Open (Domain) Information Extraction (OIE)[15]. OL tries to extract *related* concepts and relations from a given corpus automatically. In [14], Cimiano et al give a survey of current methods in ontology construction and discuss the relation between ontologies and lexica as well as ontology and natural language. However, OIE is an extraction paradigm that extracts a large set of relational tuples from a given corpus without requiring any human input e.g. TextRunner System [16]. As defined, OIE gets a corpus as an input and it generates a list of relational tuples as output. Although it is claimed that the sole input to an OIE system is a corpus, these systems still use self-supervised learners that rely on a classifier that needs to be trained prior to full scalable applications. Evaluation of both OL and OIE remains to be a research challenge and unclear.

Finally, we consider much of the work in BioNLP as the closest to the proposed task here. Bio texts are usually written for describing a specific phenomenon e.g. gene expression, protein pathways etc. in a very specific context. Extracting such information, e.g. extracting instances of specific relations or interactions between genes and proteins, from Bio-literature is similar to the task of technology structure mining. However, despite the proposed application here, Bio-Text Mining is well supported by ontologies, and language resources; the context and concepts are usually clearly defined and tools which are tuned for the domain are available. The availability of knowledge resources such as well defined ontologies in this domain enables Bio-Text miners to build new semantic layers on top of already existing semantic resources (ontologies).

## 3  Task Definition

We identify the task of technology structure extraction to comprise of four major processes: identification of technology terms at the lexical level, mapping the lexical representation of technologies into a termino-conceptual level, extracting relations between pairs of termino-conceptual technologies at the lexical level (i.e. at sentence surface structure), and finally mapping/grouping relations at the lexical level into canonical relation classes at the conceptual level.

At the lexical layer the representation of an identical technology may comprise of lexical variants e.g. Human Language Technology may be signaled by HLT, Human Language Technology, Natural Language Processing, and NLP. However, at the conceptual level all these lexical variations refer to the same concept i.e. HLT. In a similar way, a semantic relation between pairs of technologies can be conveyed by different lexical representation e.g. lexical relations such as *used in*, *applied in*, and *employed by* are expressing the same conceptual relation *DEPEND ON*.

We name the result of the above processes the *Technology Structure Graph* (TSG). Therefore, we define the task of technology structure extraction as the process of mapping a scientific corpus into a *TSG* graph with the following definition:

**Definition 1.** *A* Technology Structure Graph (TGS) *is a tuple* $G = \langle V, P, S, \Sigma, \alpha, \beta, \omega \rangle$ *where:*

1. *V is a set of pairs $\langle W, T \rangle$ where $\langle W, T \rangle$ is a uniquely identifiable terminology from a set of identifiers N and T is the terminology semantic type, e.g., $\langle NLP, \mathsf{TECHNOLOGY} \rangle$ or $\langle Lexicon, \mathsf{RESOURCE} \rangle$ or $\langle Quality, \mathsf{PROPERTY} \rangle$. To support different level of granularity of information abstraction we also consider V can contain pairs $\langle G_i, \mathsf{GRAPH} \rangle$ where $G_i$ has the same definition as G above.*
2. *P is a set of technology terms at lexical level, uniquely identifiable from a set of identifiers R, e.g.,* Natural Language Processing, NLP, Human Language Technology.
3. *S is a set of lexical relations, uniquely identifiable from a set of identifiers Q, e.g.,* used by, applied for, is example of.
4. *$\Sigma$ is a set of relations, i.e., the canonical relations vocabulary, e.g.,* $\{\mathsf{DEPEND\_ON}, \mathsf{KIND\_OF}, \mathsf{HAS\_A}\}$.
5. *$\alpha$ is a partial function that maps $\langle W, T \rangle$ to a label of $\Sigma$ annotated by a symbol from a fixed set M, i.e., $\alpha : V \times V \to \Sigma \times M$. M can be, e.g., the symbols $\{\Box, \Diamond\}$ from modal logic.*
6. *$\beta$ is a function that maps P to a tuple in V i.e., $\beta : P \to V$.*
7. *$\omega$ is a function that maps S to a term in $\Sigma$ i.e., $\omega : S \to \Sigma$.*

Considering the following input sentence:

"There have been a few attempts to integrate a speech recognition device with a natural language understanding system." [17]

with $M$ defined as *possible* and *certain* modalities, i.e., $\{\Box, \Diamond\}$, then the expected output of analysis will be as follows:

$V = \{\langle \mathsf{NLU}, \mathsf{TECHNOLOGY} \rangle, \langle \mathsf{SR}, \mathsf{TECHNOLOGY} \rangle\}$
$P = \{\text{natural language understanding}, \text{speech recognition}\}$
$\Sigma = \{\mathsf{MERGE}\}$
$S = \{integrate\ with\}$
$\beta = \text{natural language understanding} \mapsto \langle \mathsf{NLU}, \mathsf{TECHNOLOGY} \rangle$

speech recognition $\mapsto \langle$SR, TECHNOLOGY$\rangle$

$\omega =$ integrate with $\mapsto$ MERGE

$\alpha = \langle\langle$SR, Technology$\rangle, \langle$NLU, Technology$\rangle\rangle \mapsto \langle$MERGE, $\Diamond\rangle$

In our proposed definition, we have considered the computational cost and complexity of the processes that are involved in the automatic generation of structured representation from natural language text. Therefore, in the proposed definition above the expressiveness of the model is not the only concern but also the practical computational aspect of converting natural language text into a structured model like the one we have proposed here.

As a step towards the proposed research goals in this paper, we have used the provided baseline in *Definition 1* for annotating a development dataset comprising of 486 sentences from the domain of Human Language Technology. Further information about the annotated corpus can be found in [18].

## 4   Proposed Methodology

Figure 2 presents a schematic view of the proposed methodology. The proposed method comprises of 5 major steps. (1) Text extraction deals with identification and extraction of text from scientific publications, (2) Indexing and storage

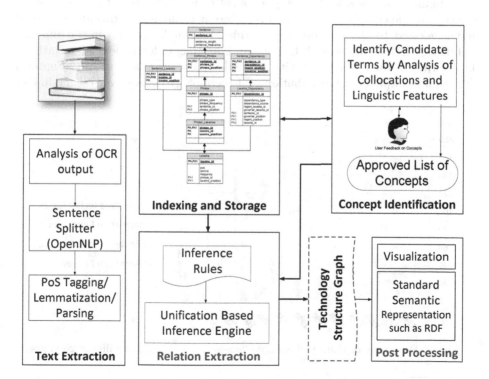

**Fig. 2.** Schematic of the Proposed Methodology

provides a suitable machine readable representation of extracted text (More information about the index scheme can be found in [18]) (3) Concept Identification marks technologies and their definitions in a semi-automatic manner (4) Parsing and Relation Extraction (RE) currently provides deep syntactic analysis of the stored sentences and extract relations between previously identified concepts by help of a unification based pattern matching over the syntactic annotations of the text (5) finally Post-processor provides a suitable representation of the extracted information e.g. a visualization for the proposed definition of Technology Structure Graph, or/and converting Technology Structure Graph to further standard representation such as RDF, and llinking the results into the Linked Open Data cloud[1].

## 5    Experimental Results

We have evaluated the proposed methodology on the C section of ACL Anthology Reference Corpus (ACL ARC)[8], which comprises of 2,435 articles from conferences in the domain of Human Language Technology. In the first step, we have been able to extract and index text from 2,003 articles. We fail to extract the text from the remaining 432 papers either because of deficiency in our heuristics for text extraction, or errors in the source XML files. The extracted text comprises of 6,168,312 tokens, 172,077 lexemes, and 230,936 sentences.

As figure 2 suggests, we then applied a set of heuristics to extract technology terms from the corpus. As a result, 147 different technology terms are extracted and suggested to the domain expert; this step finally results in 43 different technology classes where each technology class has different lexical variations. The corpus is then annotated with technology classes automatically. Figure 3 shows an example of the distribution of 4 technology classes over a time line of 25 years.

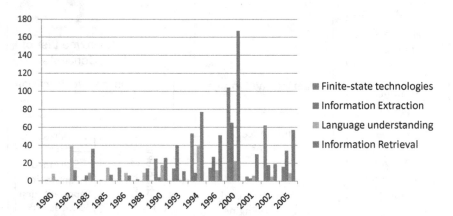

**Fig. 3.** Frequency of Four Technology Classes ordered in a time line of 25 years

---

[1] www.linkeddata.org

While we have been able to identify some of the relations automatically (Verb-Based Relations) between technology classes, the rest of the relations were extracted and tagged manually by an expert of the domain. This results in a development dataset for the proposed task of Technology Structure Mining. Details about dataset development can be found in [18]. The presented graph in Figure 1 has been generated with the help of the current post-processing module on the basis of automatically and semi-automatically extracted facts.

# 6 Conclusion and Future Work

In this paper we introduced the task of technology structure mining and proposed a formal definition for the task. Our efforts have resulted in the generation of a data set comprising of 486 sentences for training and evaluation purposes. Our future work will focus on step 4 and 5 of the method proposed in section 4. While current systems only extract verb-based relations, our experiment on the corpus of 486 sentences shows that only 10% of relations are conveyed by verbs. Therefore, extending the functionality of the relation extraction module beyond verb-based relations, e.g. relations expressed by apposition, is one of the goals of our future work.

We consider the mapping of extracted information to standard semantics and linking the information into the Linked Open Data cloud as an important step in our future research work. This comprises of mapping $\Sigma$, and $V$ from the proposed definition1 in section 3 into already published ontologies or the ontologies that are going to be developed as part of our future work.

Methodologies for the evaluation of the proposed task is the other important focus of our future research. Each step of the proposed task is subject to error and each of the proposed processes is facing accumulated errors from the previous processes. We especially would be interested to investigate the role of the quality of each of the processes in the overall result, e.g., how errors at parsing natural language sentences effects the relation extraction step, and what is the impact of this error in the overall quality of the output of the system. We consider development of the dataset as an important step towards this goal.

**Acknowledgements.** This research is supported by Science Foundation Ireland grant SFI/08/CE/I1380(Lion-2). The author wishes to express sincere gratitude to Dr Paul Buitelaar for his supervision.

# References

1. Badawy, A.M.: Technology management simply defined: A tweet plus two characters. J. Eng. Technol. Manag. 26, 219–224 (2009)
2. Maynard, D., Yankova, M., Kourakis, R., Kokossis, A.: Ontology-based information extraction for market monitoring and technology watch. In: End User Apects of the Semantic Web (2005)
3. Fry, L.W.: Technology-structure research: three critical issues. Academy of Management Journal 25, 532–552 (1982)

4. Nadeau, D., Sekine, S.: A survey of named entity recognition and classification. Linguisticae Investigationes 30, 3–26 (2007)
5. Gildea, D., Jurafsky, D.: Automatic labeling of semantic roles. Computational Linguistics 28(3), 245–288 (2002)
6. Khoo, C.S.G., Na, J.-C.: Semantic relations in information science. Annual Review of Information Science and Technology 40(1), 157–228 (2006)
7. Zelenko, D., Aone, C., Richardella, A.: Kernel methods for relation extraction. J. Mach. Learn. Res. 3, 1083–1106 (2003)
8. Bird, S., Dale, R., Dorr, B., Gibson, B., Joseph, M., Kan, M.-Y., Lee, D., Powley, B., Radev, D., Tan, Y.F.: The acl anthology reference corpus: A reference dataset for bibliographic research in computational linguistics. In: LREC 2008, Marrakech, Morocco (May 2008)
9. Hwa, R.: Learning probabilistic lexicalized grammars for natural language processing. PhD thesis, Harvard University, Cambridge, MA, USA. Adviser-Shieber, Stuart (2001)
10. Zhang, C.: Extracting chinese-english bilingual core terminology from parallel classified corpora in special domain. In: WI-IAT 2009, Washington, DC, USA, pp. 271–274. IEEE Computer Society, Los Alamitos (2009)
11. Tseng, Y.-H., Lin, C.-J., Lin, Y.-I.: Text mining techniques for patent analysis. Information Processing & Management 43(5), 1216–1247 (2007); Patent Processing
12. Oostdijk, N., Verberne, S., Koster, C.: Constructing a broad-coverage lexicon for text mining in the patent domain. In: LREC 2010, Valletta, Malta (May 2010)
13. Yoon, B., Phaal, R., Probert, D.: Structuring technological information for technology roadmapping: data mining approach. In: AIKED 2008, Stevens Point, Wisconsin, USA, pp. 417–422. World Scientific and Engineering Academy and Society, WSEAS (2008)
14. Cimiano, P., Buitelaar, P., Völker, J.: Ontology construction. In: Indurkhya, N., Damerau, F.J. (eds.) Handbook of Natural Language Processing, 2nd edn., pp. 577–605 (2010)
15. Banko, M., Cafarella, M.J., Soderland, S., Broadhead, M., Etzioni, O.: Open information extraction from the web. In: IJCAI, pp. 2670–2676 (2007)
16. Yates, A., Cafarella, M., Banko, M., Etzioni, O., Broadhead, M., Soderland, S.: Textrunner: open information extraction on the web. In: NAACL 2007, pp. 25–26. ACL, Morristown (2007)
17. Tomita, M., Kee, M., Saito, H., Mitamura, T., Tomabechi, H.: The universal parser compiler and its application to a speech translation system. In: Proceedings of the 2nd Inter. Conf. on Theoretical and Methodological Issues in Machine Translation of Natural Languages, pp. 94–114 (1988)
18. QasemiZadeh, B., Buitelaar, P., Monaghan, F.: Developing a dataset for technology structure mining. In: Proc. of IEEE International Conference on Semantic Computing (2010)

# Auto-experimentation of KDD Workflows Based on Ontological Planning

Floarea Serban

University of Zurich, Department of Informatics,
Dynamic and Distributed Information Systems Group,
Binzmühlestrasse 14, CH-8050 Zurich, Switzerland
serban@ifi.uzh.ch

**Abstract.** One of the problems of Knowledge Discovery in Databases
(KDD) is the lack of user support for solving KDD problems. Current
Data Mining (DM) systems enable the user to manually design workflows
but this becomes difficult when there are too many operators to choose
from or the workflow's size is too large. Therefore we propose to use
auto-experimentation based on ontological planning to provide the users
with automatic generated workflows as well as rankings for workflows
based on several criteria (execution time, accuracy, etc.). Moreover auto-
experimentation will help to validate the generated workflows and to
prune and reduce their number. Furthermore we will use mixed-initiative
planning to allow the users to set parameters and criteria to limit the
planning search space as well as to guide the planner towards better
workflows.

## 1   Introduction

The technology advances in the last decade facilitate the generation of large
amount of data. One of today's problems is how to process and extract patterns
from it. Knowledge Discovery in Databases (KDD) has made progress during
the last years, since the new types of data that appeared (text, image, multime-
dia) generated new algorithms to handle them. Therefore current KDD systems
incorporate more and more operators. But this creates problems for users since
they are confronted with a plethora of operators. Hence it becomes difficult to
figure out the best choice from so many options (operators).

One of the main issues of such systems is the level of user support. KDD
systems like Weka[1], RapidMiner[2], KNIME[3], EnterpriseMiner[4] or Clementine[5]
provide the user nice graphical interfaces that allow them to design KDD workflows. Users can drag and drop operators and connect them. Also explanations
about operators' functionalities and parameters are supported.

[1] http://www.cs.waikato.ac.nz/ml/weka/
[2] http://rapid-i.com/content/view/181/190/
[3] http://www.knime.org/
[4] http://www.sas.com/technologies/analytics/datamining/miner/
[5] http://www.spss.com/software/modeling/modeler-pro/

P.F. Patel-Schneider et al.(Eds.): ISWC 2010, Part II, LNCS 6497, pp. 313–320, 2010.

However taking into account the large number of operators the existing support does not help the users solve their tasks. It becomes cumbersome to select the right operator as well as to build the right workflow. One of the solutions proposed by several authors is to use AI planning techniques to automatically generate workflows [1,23,7,22]. But current implementations are limited since they support a reduced number of operators (not more than 50) as well as the generated workflows contain not more than 10 operators. The problem is that this is not enough to successfully solve Data Mining (DM) tasks.

To overcome these problems we came with the idea of using an ontology to model the DM domain that incorporates more operators [15] than the other ontologies, and use AI planning to automatically generate plans. Opposed to other approaches we use Hierarchical Task Planning (HTN) [11] since hierarchical task decomposition (from CITRUS [22]) and the knowledge available in the DM domain (CRISP-DM standard [6]) can significantly reduce the number of generated workflows. Indeed this limits the number of unwanted workflows but it is still hard for a user to decide which workflows to choose and execute. Other approaches use meta-learning to find out which operators are best for a specific data set [13,3], focusing more on classification algorithms, without a striking outcome.

Therefore we propose to use systematic auto-experimentation of generated plans to discover heuristics that prune the number of generated workflows as well as structure them. By automatic experimentation we mean running the plans retrieved from the planner in order to find the best plans or a ranking for the overall plans. Our approach will run all the plans for different data sets and try to learn from the results of the experiments. The auto-experimentation will not only improve the ranking of plans, but will also evaluate the outcome of the generated plans, acting as a validation module for the planning system. Furthermore, we propose a mixed-initiative planning approach where the user can define hints or criteria to prune the planning search process and guide the planner towards better workflows.

The remainder of the paper is structured as follows: The next section describes the current state of related research, Section 3 discusses the current state of the work and presents the future research steps. The paper closes with conclusions in Section 4.

## 2   Related Work

Several approaches try to improve user support for DM by providing intelligent assistance [21]. However all these approaches are either a proof of concept or limited to a small number of operators, therefore they are not usable for todays DM tasks. Existing IDAs do not include any support for auto-experimentation except maybe with the attempt of [1] which enable the user to execute workflows based on a ranking. Yet they don't improve the generation of workflows based on the previous executions.

*Planning and ontologies.* Considering planning and ontologies to automatically generate KDD workflows has been suggested and tried by several researchers.

They use ontologies to organize and structure the DM domain and then they use it for planning [1,7,24]. But most of the existing ontologies are rather simple, except for [23,24] which include a larger number of operators (more than 60). But even so they are not able to plan very large workflows. IDEA [1] offers an IDA that uses a prototype ontology and a DM-process planner to systematically enumerate and rank DM processes by speed and accuracy. However both the IDA and the ontology are prototypes and limited to a specific number of DM operators.

The framework presented in [24] automatically constructs DM workflows based on input and output specification of the data mining task based on a Knowledge Discovery ontology which is used for planning DM workflows. Their approach uses a Fast-Forward algorithm for planning combined with the benefits of the hierarchy from the ontology.

The approach proposed by [8] uses a KDD ontology to support KDD process design. Moreover they use a semantic matching function for the automatic composition of algorithms forming valid prototype KDD processes.

We choose to base our system on the approach of [15]. They built a DM ontology (developed within the e-Lico project[6]) which contains more than 70 operators as well as conditions and effects for each operator expressed in an extended SWRL language[7] with a set of needed built-ins. Moreover they use HTN planning to be able to reduce the number of generated workflows, but also because HTN planning proved to be better in solving real problems [20].

*Auto-experimentation of KDD workflows.* A recent discovered approach is the one of [9] which try to improve the execution of KDD workflows generated by AI planners. We can say this is the closest work to our approach. They propose a distributed architecture for automating the KDD processes as well they include a learning module which can learn from the execution of previous workflows. But they focus only on classification and regression as well they separate the execution of pre-processing actions. Their work is close to meta-learning. Our work will focus on the whole workflow and try to find different metrics to evaluate its performance. Similar to their approach we are also going to define different quality criteria which the user can set before executing the plans such that we limit the auto-experimentation. But our approach will focus on generating hypotheses based on the plans execution, which are better plans.

*Meta-learning.* Experimental databases are proposed by [2] to store Machine Learning (ML) experiments. They facilitate large-scale experimentation, guarantee repeatability of experiments and improve reusability of experiments. They also use meta-learning to determine the most appropriate ML tool for a data set. But their are focusing on a single step of the DM process, in fact on a ML tool (or algorithm).

MetaL uses the notion of meta-learning to advise users which induction algorithm to choose for a particular data-mining task [13]. One of the outcomes

---

[6] http://www.e-lico.eu/
[7] http://www.w3.org/Submission/SWRL/

of the project was a Data Mining Advisor (DMA) [12] based on meta-learning that gives users support with model selection. IDM has a knowledge module which contains meta-knowledge about the data mining methods, and it is used to determine which algorithm should be executed for a current problem.

Other work was done by the StatLog project [16] which has investigated which induction algorithms to use given particular circumstances. This approach is further explored by [3,10] which use meta-rules drawn from experimental studies, to help predict the applicability of different algorithms; the rules consider measurable characteristics of the data (e.g., number of examples, number of attributes, number of classes, kurtosis, etc.). [4] present a framework which generates a ranking of classification algorithms based on instance based learning and meta-learning on accuracy and time results. However, all these approaches are studying the execution of only one DM algorithm. Our approach will focus on entire DM workflow.

*Mixed-initiative planning.* Several approaches proposed different techniques to involve the user in the planning process, which are usually known under the name of mixed-initiative or collaborative planning. One of the approaches is the one described in [18] and known as *advisable planning* which attempts to model the behavior of the planner before starting the planning process. Our ontology already contains such mixed-initiative planning facility and based on it the user will be able to refine the planning goal and its steps. Another approach is *configurable planning* suggested by [19] which is a combination of domain-independent planning engines with higher-level abstractions like HTNs that capture and exploit domain knowledge. Since our approach is based on HTNs we are already using such domain knowledge.

## 3   Research Plan

We propose to integrate auto-experimentation of DM workflows, generated using ontological planning, in existing IDAs. The automatic experimentation approach will provide heuristics to simplify and improve the planning process as well as rank the plans according to different metrics such as accuracy, length of workflows, execution time, etc.

The main purpose of the system is to assist users in the generation and experimentation of DM workflows, as well as guiding them to configure parameters for achieving best performance. The target group consists of KDD researchers and people who are familiar with DM terminology. Later on we will try to extend it for naive users.

### 3.1   Current State

**Ontological planning.** We choose to base our approach on ontological planning [14] since the ontology offers a hierarchical structure of the DM concepts. Moreover it enables us to define conditions and effects for operators as SWRL rules which are essential for planning. The planner uses a DM ontology as a planning domain. To be able to use the ontology for planning we need to compile it in a format that the planner can understand. So far we have been involved in developing the compilation of the DM ontology (the DMWF - Data Mining Workflow ontology [14]) such that the planner can use it as a domain for generating plans. The compilation consists of compiling the TBox (terminology which does not include annotations), then the operators together with their conditions and effects (operators are classes but their conditions and effects are stored as annotations and we need to compile them separately), inputs, outputs and parameters as well as the task/method decomposition (the same goes for the task and method decomposition, the structure is saved as annotations as well). Finally we compile the ABox.

**Experiments' design.** Based on the ontology and planner we can start designing experiments. The main task was to implement an IDA-API which is able to retrieve plans starting from a goal definition and a provided data set. The IDA-API has the following features:

- The DM task can be specified in the form of main goals and optional goals.
- The meta data from the used data sets can be added as a list of facts.
- The plans can be easily retrieved.

The architecture of our system can be seen in Figure 1. We are using the IDA-API to define the DM problems and to retrieve the generated plans. The developed IDA-API uses a pre-compiled DM ontology (as described before) that is later used for planning. Then it compiles the task definition and the meta-data of the data sets used as inputs into a set of facts that can be recognized by the planning module.

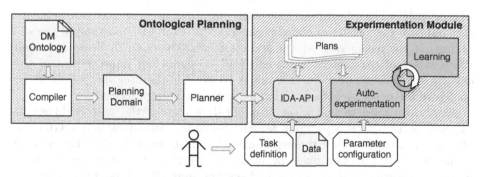

**Fig. 1.** System architecture

## 3.2 Plan for Future Research

Having laid the foundations for auto-experimentation of KDD workflows, we can start working on the future steps to achieve our contributions described in Section 1 as follows:

**Auto-experimentation to discover heuristics to prune the search plan.** Having a large set of plans to choose from is a challenge for the experimentation module. In this part of the project we develop an experimentation module which is able to automatically run experiments in an optimal way. Moreover the experiments need to be analyzed and used for learning and improving the experimentation as well as the planning process.

Firstly we need to decide which DM system to use to run the experiments. We incline towards RapidMiner [17], since it is one of the leading open source DM tools. Moreover RapidMiner is implemented in Java which makes it easy to be used with our IDA-API.

Secondly, we need to implement a module that based on the IDA-API enables us to easily define DM tasks. A simple approach would be to have that programatically but later on we need to develop a graphical interface for it.

Thirdly, we need to build a system that enables us to run DM experiments in a distributed manner. For starting, the plans will be executed in parallel, one plan in a thread. Later on we can find ways of parallelizing sub-workflows.

Also we need to study the results of the experiments and find good heuristics to reduce the number of generated plans. Then we need to evaluate the metrics and heuristics by performing several types of tasks on a large number of data sets. For that we will develop a learning module which analyzes the results of the experiments and tries to improve the auto-experimentation. As a starting point, we could use the data sets from the UCI repository[8] and later test it on larger data sets (also which are not preprocessed). Another idea would be to generate hypotheses (which would represent better plans) based on the experiments and then run them and check their accuracy. In the end, our purpose is to be able to solve one of the KDDCups[9] (for example KDDCup'98) using the system and show that the auto-experimentation module can successfully provide a reduced number of plans and qualitative rankings.

**Mixed-initiative planning.** There are many parameters that can influence and improve the results of the experiments, for example, the time of the experiments, the resources used, the accuracy, etc.. The main challenge is to find the best set of parameters which can lead to significant improvement of the experimentation module. Another one is to provide the user the possibility to configure the experiments and to influence the planning process.

Firstly, we need to find a set of qualitative metrics the user could set to improve and guide the planning search. Secondly, we will design a GUI the allows the user to set all these parameters. Then, we will allow the user to visualize and manipulate plans by integrating actions like plan step by step, pause or execute a plan, go next or go back one step. We will later try to extend this approach and allow the user to contribute not only to the formulation and development of plans, but also in the management, refinement, analysis and repair of the plans. But first we need to study and analyze all the problems raised in [5].

Finally, we will perform user tests and check if the generated system helps the users to solve their tasks better and faster than the existing DM systems.

---

[8] http://archive.ics.uci.edu/ml/
[9] http://www.sigkdd.org/kddcup/index.php

# 4    Conclusions

In this paper we introduce auto-experimentation of KDD workflows based on ontological planning. We extend upon research described in Section 2 in various ways. Firstly, we use auto-experimentation to reduce and prune the number of automatically generated workflows. Secondly, we integrate the auto-experimentation module into an IDA and allow the users to browse workflows by rankings and analyze the outcomes of their execution. Thirdly, we will provide a mixed-initiative module that allows the users to guide the planning process as well as to suggest criteria to prune the searching space.

We are optimistic and believe that the current approach will lead to different ways of ranking and structuring of the plans as well as involve the users in the planning process. The impact of our approach is the possibility to find rankings for DM workflows and heuristics to prune the planner searching space, hence reducing the time needed to generate plans and finding the best workflow for a specific DM problem through auto-experimentation.

**Acknowledgements.** This work is supported by the European Community $7^{th}$ framework ICT-2007.4.4 (No 231519) "e-Lico: An e-Laboratory for Interdisciplinary Collaborative Research in Data Mining and Data-Intensive Science".

# References

1. Bernstein, A., Provost, F., Hill, S.: Towards Intelligent Assistance for a Data Mining Process: An Ontology-based Approach for Cost-sensitive Classification. IEEE Transactions on Knowledge and Data Engineering 17(4), 503–518 (2005)
2. Blockeel, H., Vanschoren, J.: Experiment databases: Towards an improved experimental methodology in machine learning. In: Kok, J.N., Koronacki, J., Lopez de Mantaras, R., Matwin, S., Mladenič, D., Skowron, A. (eds.) PKDD 2007. LNCS (LNAI), vol. 4702, pp. 6–17. Springer, Heidelberg (2007)
3. Brazdil, P., Gama, J., Henery, B.: Characterizing the applicability of classification algorithms using meta-level learning. In: Bergadano, F., De Raedt, L. (eds.) ECML 1994. LNCS, vol. 784, pp. 83–102. Springer, Heidelberg (1994)
4. Brazdil, P., Soares, C., Da Costa, J.: Ranking learning algorithms: Using IBL and meta-learning on accuracy and time results. Machine Learning 50(3), 251–277 (2003)
5. Burstein, M., McDermott, D.: Issues in the development of human-computer mixed-initiative planning. Advances in Psychology 113, 285–303 (1996)
6. Chapman, P., Clinton, J., Kerber, R., Khabaza, T., Reinartz, T., Shearer, C., Wirth, R.: Crisp–dm 1.0: Step-by-step data mining guide. Technical report, The CRISP–DM Consortium (2000)
7. Diamantini, C., Potena, D., Storti, E.: Kddonto: An ontology for discovery and composition of kdd algorithms. In: Service-oriented Knowledge Discovery (SoKD 2009) Workshop at ECML/PKDD 2009 (2009)
8. Diamantini, C., Potena, D., Storti, E.: Supporting users in kdd processes design: a semantic similarity matching approach. In: Planning to Learn Workshop (PlanLearn 2010) at ECAI 2010, pp. 27–34 (2010)

9. Fernández, S., Súarez, R., de la Rosa, T., Ortiz, J., Fernández, F., Borrajo, D., Manzano, D.: Improving the execution of kdd workflows generated by ai planners. In: Planning to Learn Workshop (PlanLearn 2010) at ECAI 2010, pp. 19–25 (2010)

10. Gama, J., Brazdil, P.: Characterization of classification algorithms. In: Progress in Artificial Intelligence, pp. 189–200 (1995)

11. Ghallab, M., Nau, D., Traverso, P.: Automated Planning: Theory & Practice. Morgan Kaufmann, San Francisco (2004)

12. Giraud-Carrier, C.: The data mining advisor: meta-learning at the service of practitioners. In: Proceedings of Fourth International Conference on Machine Learning and Applications, p. 7 (2005)

13. Hilario, M., Kalousis, A.: Fusion of meta-knowledge and meta-data for case-based model selection. In: Siebes, A., De Raedt, L. (eds.) PKDD 2001. LNCS (LNAI), vol. 2168, pp. 180–191. Springer, Heidelberg (2001)

14. Kietz, J., Serban, F., Bernstein, A., Fischer, S.: Data mining workflow templates for intelligent discovery assistance and auto-experimentation. In: Third-Generation Data Mining: Towards Service-Oriented Knowledge Discovery SoKD 2010 (2010)

15. Kietz, J.-U., Serban, F., Bernstein, A., Fischer, S.: Towards cooperative planning of data mining workflows. In: Service-oriented Knowledge Discovery (SoKD 2009) Workshop at ECML/PKDD 2009 (2009)

16. Michie, D., Spiegelhalter, D., Taylor, C., Campbell, J.: Machine learning, neural and statistical classification (1994)

17. Mierswa, I., Wurst, M., Klinkenberg, R., Scholz, M., Euler, T.: Yale: Rapid prototyping for complex data mining tasks. In: KDD 2006: Proceedings of the 12th ACM SIGKDD International Conference on Knowledge Discovery and Data Mining, pp. 935–940. ACM, New York (2006)

18. Myers, K.: Strategic advice for hierarchical planners. In: Principles of Knowledge Representation and Reasoning-International Conference, pp. 112–123. Morgan Kaufmann Publishers, San Francisco (1996)

19. Nau, D.S.: May all your plans succeed (invited talk). In: Proceedings of the National Conference on Artificial Intelligence (AAAI) (July 2005)

20. Nau, D.S., Smith, S.J.J., Erol, K.: Control strategies in htn planning: Theory versus practice. In: IAAI Proceedings, pp. 1127–1133 (1998)

21. Serban, F., Kietz, J.-U., Bernstein, A.: An overview of intelligent data assistants for data analysis. In: Planning to Learn Workshop (PlanLearn 2010) at ECAI 2010, pp. 7–14 (2010)

22. Wirth, R., Shearer, C., Grimmer, U., Reinartz, T., Schlösser, J., Breitner, C., Engels, R., Lindner, G.: Towards process-oriented tool support for knowledge discovery in databases. In: Komorowski, J., Żytkow, J.M. (eds.) PKDD 1997. LNCS, vol. 1263, pp. 243–253. Springer, Heidelberg (1997)

23. Žáková, M., Křemen, P., Železný, F., Lavrač, N.: Planning to learn with a knowledge discovery ontology. In: Planning to Learn Workshop (PlanLearn 2008) at ICML 2008 (2008)

24. Žáková, M., Podpečan, V., Železný, F., Lavrač, N.: Advancing data mining workflow construction: A framework and cases using the orange toolkit. In: Service-oriented Knowledge Discovery (SoKD 2009) Workshop at ECML/PKDD 2009 (2009)

# Customizing the Composition of Actions, Programs, and Web Services with User Preferences*

Shirin Sohrabi

Department of Computer Science, University of Toronto
shirin@cs.toronto.edu

**Abstract.** Web service composition (WSC) – loosely, the composition of web-accessible software systems – requires a computer program to automatically select, integrate, and invoke multiple web services in order to achieve a user-defined objective. It is an example of the more general task of composing business processes or component-based software. Our doctoral research endeavours to make fundamental contributions to the knowledge representation and reasoning principles underlying the task of WSC, with a particular focus on the customization of compositions with respect to individual preferences. The setting for our work is the semantic web, where the properties and functioning of services and data are described in a computer-interpretable form. In this setting we conceive of WSC as an Artificial Intelligence planning task. This enables us to bring to bear many of the theoretical and computational advances in reasoning about action and planning to the task of WSC. However, WSC goes far beyond the reaches of classical planning, presenting a number of interesting challenges that are relevant not only to WSC but to a large body of problems related to the composition of actions, programs, business processes, and services. In what follows we identify a set of challenges facing our doctoral research and report on our progress to date in addressing these challenges.

## 1 Challenges and Background

Given a set of suitably described services, a specification of the state of the world, and a user objective, Web service composition (WSC) is the task of composing a set of services to achieve the user's objective. A popular approach to WSC is to characterize it as an Artificial Intelligence (AI) planning problem and to solve it as such (e.g., [8,9,1]). However, WSC goes far beyond the reaches of classical planning, presenting a number of interesting challenges that are relevant not only to WSC but to a large body of problems related to the composition of actions, programs, business processes, and services. For example, unlike typical classical planning tasks, knowledge of how to achieve the user's objective is often known,

---

* The doctoral research described in this paper is being developed and carried out under the supervision of Professor Sheila McIlraith.

P.F. Patel-Schneider et al.(Eds.): ISWC 2010, Part II, LNCS 6497, pp. 321–329, 2010.
© Springer-Verlag Berlin Heidelberg 2010

at least at an abstract level; plans are often plentiful, high-quality plans are not, requiring optimization of complex preferences. Another differentiating property of WSC is that it can be data intensive resulting in planning domains with tens of thousands of actions, each of which is itself a program with non-determinism and intermediate state.

Several planning-based approaches have appealed to some sort of a template or workflow structure to help define the user objective and/or to guide the composition. The agent programming language Golog (e.g., [9]), Hierarchical Task Networks (HTNs) (e.g., [14]), and Finite State Automata (FSA) (e.g., [3]) have all been used for this purpose. In the case of our work, our objective is to define a **flexible, declarative WSC template** that provides high-level guidance on how to perform a task, but can leave many of the details to run-time synthesis, should that be warranted. For many WSC problems, the task can be realized by a diversity of different services, offering comparable, but not identical services. Also unknown at the outset is the data that serves as choice points in a WSC – the availability of goods, their properties and pricing, etc. A flexible composition template streamlines the generation of a composition, while enabling the individual user to customize the composition with respect to their preferences and constraints perhaps in association with preferences and constraints of other stakeholders such as the corporation they work for, the laws of the countries in which they are doing business, etc.

The general challenge we face in our doctoral research is to investigate principled techniques for composing web services, that support user customization. This manifests itself in a number of specific research challenges. We discuss some of the major challenges below.

**Challenge 1:** The first challenge is how to customize WSC templates (the problem specification) to meet the objectives of individual users. The templates are meant to be flexible yet shared by many users. So the challenge is how to customize the templates for each individual user. In order to do so we need to address how users can specify their preferences and to that end propose a rich preference language with which to express arbitrarily complex and mutually inconsistent preferences. In addition, we would like to have a user preference language that can handle preference specifications over both the functional and the non-functional properties of services such as trust, reliability, and privacy.

**Challenge 2:** Our second challenge is how to synthesize compositions that adhere to policies and regulations. Policies and regulations are an important aspect of semantic web services. Many customers are concerned with enforcement of regulations, perhaps in the form of corporate policies and/or government regulations. Software that is developed for use by a particular corporation or jurisdiction will have the enforcement of such regulations built in. For web services that are published for use by the masses this is not the case, and the onus is often on the customer to ensure that regulations are enforced when a workflow is constructed from multiple service providers. For inter-jurisdictional or international business, different regulations may apply to different aspects of the composition.

Hence, providing a mechanism for generating compositions from templates that adhere to such regulations is the second challenge we would like to address.

**Challenge 3:** Our third challenge is how to develop heuristics to search efficiently for an optimal plan/composition. An optimal plan is a plan that has the best quality (i.e., no other plan has a better quality than this plan). Heuristic-guided search is an effective method for efficient plan generation, but the challenge is to find a heuristic that gives guidance towards optimal solutions without exhaustively searching the search space. We can use either an admissible or inadmissible set of heuristics. An admissible heuristic is a heuristic that never overestimates the cost of reaching the goal. If an admissible heuristic is used in a A*-like algorithm, then first plan found would be an optimal plan. However, finding such plan in practice may not be feasible when the search space is large. Hence, we may consider using inadmissible heuristics with a hope of finding a good-quality plan quickly instead and find a condition under which we can guarantee optimality despite the use of inadmissible heuristics.

**Challenge 4:** Our fourth challenge is how to perform and integrate online information gathering in order to gather the necessary information needed to produce high-quality compositions in the absence of complete information. Many planning-based characterizations of WSC make an assumption that there is complete information about the initial state. This assumption is often violated in many real-world settings; it is impractical or impossible to have all the information necessary to generate a composition prior to the commencement of the search for a composition. A more compelling solution is to instead gather information as it becomes necessary in the generation and optimization of the composition. In our doctoral research, we describe a means of executing web services to collect information as it is deemed necessary to inform the search for a valid, ideally optimal, composition.

**Challenge 5:** Our fifth challenge is how to optimize compositions with tens of thousands of services and with extensive data. Optimization requires considering all alternatives, at least implicitly. However, given the large volume of information available on the Web, evaluating the search space effectively is a challenging problem that has not been addressed in previous work. Consider just as an example three information sources $A$, $B$, and $C$ containing $i$, $j$, and $k$ data items, respectively. In the worst case, the system will need to access and evaluate $i * j * k$ different alternatives in order to identify the optimal combination. However, if the choice of data for $C$ can be made independently of $A$ and $B$, then the search space is worst case $k + i * j$. This decomposition and *localization* of the optimization is one of the ways that humans manage to address the complexity of optimization tasks, and it's one that we plan to incorporate into our work.

In the next section we will briefly overview a small subset of the previous work and then we will describe the progress we have made so far in addressing the above five challenges. We develop knowledge representation and reasoning techniques, describe formal properties, prove properties of our formalizations including, soundness, completeness, correctness and optimality, where relevant. We also evaluate our work experimentally to demonstrate the practical advantages of our approach.

## 2   Related Work

There is an important body of previous work that our work builds on. We note a subset of them here. In particular, a number of researchers have advocated using AI planning techniques to address the task of WSC including planners that are based on model checking (e.g., [21]) and planners that use a regression-based approach [8]. Previous work has also considered using a template or workflow to ease the task of composition including the work using Golog [10,9] and HTNs [14,13,7]. The work on the so-called Roman model is another example of a template-like approach (e.g., [3]). Also there are several proposed solutions to the information gathering problem (e.g., [9,14,6]); however, they have not examined the problem of information gathering in the context of optimizing the composition. In the absence of the need to optimize, it is often sufficient to arbitrarily select one choice among many and to ignore the rest. However, if the task is to generate a high-quality composition that optimizes for the user's preferences, then the entire space of alternatives must be considered somehow. This both alters the information gathering task and also greatly increases the search space for a composition.

While no other WSC planners can perform true preference-based planning, SHOP2 [11] and ENQUIRER [6] handle some simple user constraints. Also a notable work is the SCUP prototype planner in [7] but there are several differences to our work. In particular, their preferences are pre-processed into task networks and conflicting user preferences are detected and removed prior to invocation of their planner. Further, they do not consider handling regulations and are not able to specify preferences over the quality of services.

## 3   Progress Made to Date

Progress towards addressing the challenges presented in our doctoral research have appeared in a number of publications as summarized in Figure 1.

As mentioned above, a composition template can be represented in a variety of different ways. One way to represent a template is to use a workflow or a flowchart. This can be expressed pictorially as a schematic or alternatively in a form akin to a procedural programming language. The Algol-inspired Golog agent programming language provides one such procedural language (e.g., [12]). Indeed, the first template-based approach to WSC exploited Golog to provide a so-called generic procedure that provided a template specification of the composition [10,9]. The Golog procedures were combined with individual user constraints (e.g., *"I want to fly with a star alliance carrier"*) at run time, resulting in *dynamic binding of web services*. However, the user constraints considered were hard constraints, i.e., realizations that did not satisfy those constraints were eliminated. In [20], we make progress towards addressing our first challenge by

| # | Challenge | Approach | Progress |
|---|-----------|----------|----------|
| 1 | Customize WSC templates for each user | Use and extend preference languages LPP and PDDL3 | [20,17] [16,18] |
| 2 | Generate compositions from templates that adhere to regulations or policies | Use pruning and specify regulations as LTLs | [18] |
| 3 | Develop heuristics to search efficiently for an optimal composition | Use (in)admissible heuristic techniques developed for planning | [15,16] |
| 4 | Access & integrate online information gathering to produce a high-quality plan | Use a middle-ground execution engine | [19] |
| 5 | Optimize compositions with many services and with extensive data | Use preference decoupling | [19] |

**Fig. 1.** Challenges in our doctoral research and the progress made to address them

extending this framework to deal with *soft* user constraints (i.e., preferences). To specify user preferences, we exploit a rich qualitative preference language, proposed by Bienvenu et al. to specify users' preferences in a variant of linear temporal logic (LTL) called LPP [2]. We prove the soundness of our approach and the optimality of our compositions with respect to the user's preferences. Our system can be used to select the optimal solution from among families of solutions that achieve the user's stated objective. Our system, GologPref, was implemented in Prolog and integrated with a selection of scraped web services that are appropriate to our test domain of travel planning. Unfortunately, the implementation of the system was not optimized.

Similar to Golog, HTNs [5] provide useful control knowledge — advice on how to perform a composition. However, this how-to knowledge is specified as a *task network*. The task network provides a way of hierarchically abstracting the composition into a set of tasks that need to be performed and that decompose in various ways into leaf nodes realized by programs. While HTNs specify a family of satisfactory plans, they are, for the most part, unable to distinguish between successful plans of differing quality. In [17] we address the problem of generating preferred plans by combining the procedural control knowledge specified by HTNs with rich qualitative user preferences. Note this is work towards addressing our first challenge where templates are now specified in HTNs. The outcome of our work is a language for specifying user preferences, tailored to HTN planning, together with a provably optimal preference-based planner, **HTNPlan**, that is implemented as an extension of SHOP2 [14], a highly-optimized HTN planner for the task of WSC. To specify user preferences, we augment the preference language LPP used in [20] with HTN-specific constructs. Among the HTN-specific properties that we add to our language, is the ability to express preferences over how tasks in our HTN are decomposed into subtasks, preferences over the parameterizations of decomposed tasks, and a variety of temporal and non-temporal preferences over the task networks themselves. To compute preferred plans, we propose an approach based on forward-chaining heuristic search. Our heuristic uses an admissible evaluation function measuring the satisfaction of preferences over partial plans. We prove our approach sound and optimal with respect to the plans it generates by appealing to a situation calculus [12] semantics of our preference language and of HTN planning.

**HTNPlan** discussed briefly above is a provably optimal preference-based planner; however, with large search space, finding this optimal plan may not be feasible. As an alternative, in [16] we propose several *inadmissible* heuristics, designed specifically to guide the search quickly to a good decomposition. In particular, we designed a heuristic called the *look-ahead* heuristic that is designed specifically to address this problem. Also we decided to use the popular Plan Domain Description Language, PDDL3 [4] as our preference language instead. Similar to [17] we extended PDDL3 to support specification of preferences over HTN constructs; note this is work towards addressing our first and third challenge. To compute preferred plans, we propose a branch-and-bound algorithm, together with our set of heuristics that leverage HTN structure. The search is performed in a series of episodes, each of which returns a plan with a better quality than the last plan returned. We showed that under some condition we can guarantee optimality. The experimental evaluations of our planner shows that our HTN preference-based planner, **HTNPlan-P**, generates plans that, in all but a few cases, equal or exceed the best preference-based planners in plan quality. As such, our results shows that our approach is viable and promising to preference-based planning.

We tackle our first and second challenge in [18] by providing a mechanism for generating compositions from templates that adhere to regulations as well as extending the preference language developed earlier with preferences over service and data selection. We specify regulations as a subset of Linear Temporal Logics (LTLs), considering for the most part the *never* and *always* constructs, and use pruning to eliminate those compositions that violate such regulations. Hence, we describe a composition framework that that simultaneously optimizes, at run time, the selection of services based on functional and non-functional properties and their groundings, while enforcing stated regulations. We also use the search heuristics developed in our previous work and provide an implementation that combines HTN templates, the optimization of rich user preferences, and adherence to LTL regulations all within one system. Experimental evaluation on our system, **HTNWSC-P**, shows that our approach can be scaled as we increase the number of preferences and the number of services.

## 4   Current and Future Research

Much of the AI-related work on WSC that relates it to an AI planning problem performs composition offline prior to execution. Recent research on WSC has argued convincingly for the importance of optimizing quality of service and user preferences. While some of this optimization can be done offline, many interesting and useful optimizations are data-dependent, and must be done following execution of at least some information-providing services. In our recent work [19], we examine this class of WSC problems, attempting to bridge the gap between offline composition and online information gathering with a view to producing high-quality compositions without excessive data gathering. Our investigation is performed in the context of our preference-based HTN web service composition system [18]. We propose a way to address the critical information-gathering component of preference-based WSC as well as optimization. This need to actually execute services to gather data, as well as the potential size and nature of the resultant optimization problem truly distinguishes our WSC task from previous work on preference-based planning. To this end we propose a notion of middle-ground execution that enables information gathering during generation of a WSC. We further propose a notion of *localized data optimization* in which the optimization task can be decomposed into smaller, local optimization problems, while preserving global optimality. We showed that our approach to data optimization can greatly improve both the quality of compositions and the speed with which they are generated.

In future work, we plan to further improve our approach, ideally in addressing all of our challenges. In particular, in addressing our second challenge, we specified regulations in a subset of LTLs and we plan to further improve our work by considering the full expressive power of LTLs instead. Furthermore, we plan to improve our online information gathering procedure. Currently, the HTN structure embodies the place where information gathering is necessary. In future, we like to improve our procedure to remove this restriction by possibly having a pre-processing step influenced by query optimization techniques to remove or push forward the "hard" information gathering steps. Finally, we would like to design a user-friendly interface that possibly not only takes the user's preferences, objectives, policies, but also interacts with the user in a mixed-initiative manner during the composition construction time.

In conclusion, the need to compose actions, programs and web service is pervasive not only on the Web but in general software engineering or manufacturing settings where we like to describe and possibly reuse software with the desire to customize. We hope to make principled theoretical and practical contributions to these fundamental problems.

## Acknowledgements

We gratefully acknowledge funding from the Natural Sciences and Engineering Research Council of Canada (NSERC) and the Ontario Ministry of Innovations Early Researcher Award (ERA). Note some of the materials presented here has appeared in parts in previous publications as noted in Figure 1.

## References

1. Bertoli, P., Kazhamiakin, R., Paolucci, M., Pistore, M., Raik, H., Wagner, M.: Continuous orchestration of Web services via planning. In: Proceedings of the 19th Int'l Conference on Automated Planning and Scheduling, ICAPS 2009, pp. 18–25 (2009)

2. Bienvenu, M., Fritz, C., McIlraith, S.: Planning with qualitative temporal preferences. In: Proceedings of the 10th Int'l Conference on Knowledge Representation
   and Reasoning, KR 2006, pp. 134–144 (2006)
3. Calvanese, D., Giacomo, G.D., Lenzerini, M., Mecella, M., Patrizi, F.: Automatic
   service composition and synthesis: the Roman Model. IEEE Data Eng. Bull. 31(3),
   18–22 (2008)
4. Gerevini, A., Long, D.: Plan constraints and preferences for PDDL3. Tech. Rep.
   2005-08-07. Department of Electronics for Automation, University of Brescia, Brescia, Italy (2005)
5. Ghallab, M., Nau, D., Traverso, P.: Hierarchical Task Network Planning. Automated Planning: Theory and Practice. Morgan Kaufmann, San Francisco (2004)
6. Kuter, U., Sirin, E., Nau, D.S., Parsia, B., Hendler, J.A.: Information gathering
   during planning for Web service composition. In: McIlraith, S.A., Plexousakis,
   D., van Harmelen, F. (eds.) ISWC 2004. LNCS, vol. 3298, pp. 335–349. Springer,
   Heidelberg (2004)
7. Lin, N., Kuter, U., Sirin, E.: Web service composition with user preferences. In:
   Bechhofer, S., Hauswirth, M., Hoffmann, J., Koubarakis, M. (eds.) ESWC 2008.
   LNCS, vol. 5021, pp. 629–643. Springer, Heidelberg (2008)
8. McDermott, D.V.: Estimated-regression planning for interactions with Web services. In: Proceedings of the 6th Int'l Conference on Artificial Intelligence Planning
   and Scheduling, AIPS 2002, pp. 204–211 (2002)
9. McIlraith, S., Son, T.: Adapting Golog for composition of semantic Web services.
   In: Proceedings of the 8th Int'l Conference on Knowledge Representation and Reasoning, KR 2002, pp. 482–493 (2002)
10. McIlraith, S., Son, T., Zeng, H.: Semantic Web services. IEEE Intelligent Systems.
    Special Issue on the Semantic Web 16(2), 46–53 (2001)
11. Nau, D.S., Au, T.C., Ilghami, O., Kuter, U., Murdock, J.W., Wu, D., Yaman, F.:
    SHOP2: An HTN planning system. Journal of Artificial Intelligence Research 20,
    379–404 (2003)
12. Reiter, R.: Knowledge in Action: Logical Foundations for Specifying and Imple-
    menting Dynamical Systems. MIT Press, Cambridge (2001)
13. Sirin, E., Parsia, B., Hendler, J.: Template-based composition of semantic Web
    services. In: AAAI 2005 Fall Symposium on Agents and the Semantic Web (2005)
14. Sirin, E., Parsia, B., Wu, D., Hendler, J., Nau, D.: HTN planning for Web service
    composition using SHOP2. Journal of Web Semantics 1(4), 377–396 (2005)
15. Sohrabi, S., Baier, J., McIlraith, S.A.: HTN planning with quantitative preferences
    via heuristic search. In: 8th International Conference on Automated Planning and
    Scheduling (ICAPS) Workshop on Oversubscribed Planning and Scheduling (2008)
16. Sohrabi, S., Baier, J.A., McIlraith, S.A.: HTN planning with preferences. In: Pro-
    ceedings of the 21st Int'l Joint Conference on Artificial Intelligence, IJCAI 2009,
    pp. 1790–1797 (2009)
17. Sohrabi, S., McIlraith, S.A.: On planning with preferences in HTN. In: 12th Inter-
    national Workshop on Non-Monotonic Reasoning (NMR 2008), pp. 241–248 (2008)
18. Sohrabi, S., McIlraith, S.A.: Optimizing Web service composition while enforcing
    regulations. In: Bernstein, A., Karger, D.R., Heath, T., Feigenbaum, L., Maynard,
    D., Motta, E., Thirunarayan, K. (eds.) ISWC 2009. LNCS, vol. 5823, pp. 601–617.
    Springer, Heidelberg (2009)

19. Sohrabi, S., McIlraith, S.A.: Preference-based Web service composition: A middle ground between execution and search. In: Patel-Schneider, P.F., et al. (eds.) ISWC 2010, Part II. LNCS, vol. 6497, pp. 321–329. Springer, Heidelberg (2010)
20. Sohrabi, S., Prokoshyna, N., McIlraith, S.A.: Web service composition via generic procedures and customizing user preferences. In: Cruz, I., Decker, S., Allemang, D., Preist, C., Schwabe, D., Mika, P., Uschold, M., Aroyo, L.M. (eds.) ISWC 2006. LNCS, vol. 4273, pp. 597–611. Springer, Heidelberg (2006)
21. Traverso, P., Pistore, M.: Automatic composition of semantic Web services into executable processes. In: McIlraith, S.A., Plexousakis, D., van Harmelen, F. (eds.) ISWC 2004. LNCS, vol. 3298, pp. 380–394. Springer, Heidelberg (2004)

# Adding Integrity Constraints to the Semantic Web for Instance Data Evaluation

Jiao Tao

Department of Computer Science, Rensselaer Polytechnic Institute,
Troy, NY 12180, USA

**Abstract.** This paper presents our work on supporting evaluation of integrity constraint issues in semantic web instance data. We propose an alternative semantics for the ontology language, i.e., OWL, a decision procedure for constraint evaluation by query answering, and an approach of explaining and repairing integrity constraint violations by utilizing the justifications of conjunctive query answers.

**Keywords:** Instance Data, Evaluation, Integrity Constraint, OWL.

## 1 Introduction: Motivation and Challenges

With the rapid growth of semantic web [1] technologies, a lot of semantic web applications such as Twine, Freebase, TrueKnowledge, Hakia, etc, are emerging on the web. To consume the data generated by these applications, it is critical to evaluate the data and ensure it meets the needs of users first. The data on the semantic web includes the ontologies that describe the schema of the domain, and the instance data that refers to the ground level data particular to the applications. There has been a lot of research aimed at ontology evaluation [2] [3] [4], however there is little, if any, research focusing on instance data evaluation, even though instance data usually accounts for orders of magnitude more than ontologies on the web.

We have identified three categories of issues [5] that may occur in instance data which are syntax errors, logical inconsistencies, and integrity constraint issues: syntax errors are the issues indicating that the syntax representation of instance data does not conform to the corresponding syntax specifications such as RDF/XML, N3, N-Triple, Turtle, etc; logical inconsistencies are the issues showing that the underlying logical theory of the instance data includes contradictory axioms; integrity constraint issues are the issues caused by the failure of the instance data to follow the restrictions that are imposed by the integrity constraints. While the evaluation of first two categories of issues are well studied [6] [7] [8] [9] and are being supported by existing tools such as W3C RDF validation service[1], the DL reasoner Pellet[2], Chimaera [10], the BBN validator[3], and ODEval [11], evaluation support for integrity constraint issues in semantic web instance data still remains an open research problem.

---

[1] http://www.w3.org/RDF/Validator/
[2] http://clarkparsia.com/pellet/
[3] http://projects.semwebcentral.org/projects/vowlidator/

P.F. Patel-Schneider et al.(Eds.): ISWC 2010, Part II, LNCS 6497, pp. 330–337, 2010.

The concept of integrity constraint (IC) was invented in the field of relational databases where ICs are used to ensure data consistency [12]. As an important functionality, almost all database systems support IC checking. The main approach for checking ICs in databases is to represent ICs as part of the database schema and translate the constraints to corresponding queries. Whenever there is an update to the data, the queries are executed first to see if the ICs are violated, therefore preventing potential constraint violations. One might wonder if similar approach can be used for the task of evaluating IC issues on the semantic web: modeling ICs as part of the domain knowledge using knowledge representation languages on the semantic web such as OWL [13] then translating IC axioms to queries and validating ICs by corresponding query answering. The standard semantics of OWL is based on Description Logics (DL) which has the following two characteristics:

- Open World Assumption (OWA): i.e., a statement cannot be inferred to be false on the basis of failures to prove it.
- Absence of the Unique Name Assumption (UNA): i.e., two different names may refer to the same object.

Due to the above characteristics, what triggers constraint violations in closed world systems, such as databases, leads to new inferences in standard OWL systems. Therefore, it is difficult to use OWL for IC evaluation. In this paper, we present our work on supporting evaluatin of integrity constraint issues in semantic web instance data by enabling OWL as an IC language.

## 2   Related Work

Several existing proposals on enabling OWL as an IC language combine OWL with different formalisms such as rules, epistemic queries, or epistemic logics. The rule-based approach [14] [15] expresses ICs as rules with a special predicate and check ICs by examining if the special predicate is entailed by the hybrid DL-rule knowledge base (KB). With this approach the developers need to be familiar with rules. The epistemic query-based approach [16] expresses ICs as epistemic queries and evaluates ICs by checking the epistemic query answers. However the complexity results of this approach in expressive DLs are still unknown. The epistemic DL-based approach extends DLs with epistemic logics and expresses ICs with epistemic DL axioms. With this approach, IC evaluation is to determine if the IC axioms are epistemic-entailed by the DL KB. However, this approach has two limitations: first, it focuses on less expressive DLs; second, it adopts the strict UNA which is not compatible with OWL since two different OWL individual names might refer to the same object.

In this paper, we focus on approaches that reuse OWL as an IC language. Our closest related work is a minimal Herbrand model-based approach [17]. With this approach, an OWL IC axiom is satisfied if all minimal Herbrand models of the KB satisfy it. This approach may result in counterintuitive results or modeling burden: first, existentially quantified individuals can satisfy ICs, which is not desirable for IC evaluation; second, with this approach, if an IC needs to be satisfied only by individual names, then a special concept $O$ has to be added into the original IC axiom, and every individual name should be asserted as an instance of $O$. This adds a significant maintenance burden on ontology developers while still not capturing the intuition behind ICs; third, the disjunctions and ICs may interact in unexpected ways.

# 3    Research Objectives and Plan

This work is aimed to support evaluating IC issues in semantic web instance data. For this purpose, we identify the following research objectives and plan:

- Enabling OWL as an IC language. Aside from being an ontology language, OWL is also an IC language that can be used to represent constraints that the instance data has to satisfy.
- Providing decision procedures for IC evaluation. Given a set of instance data and IC axioms, the decision procedures decide if the ICs are violated by the instance data.
- Providing explanation services for IC violations which explain why certain ICs are violated and recommend how to repair the violations.
- Implementation and Evaluation. Implementing a prototype and evaluating IC issues in instance data.

# 4    Research Progress

Until now we have completed the review of the state of the art, and compared the different approaches of enabling OWL as an IC language. Due to the various issues of these approaches that we have discussed in Section 2, we decided that designing an alternative semantics for OWL which supports ICs and correctly captures the intuitions behind ICs would be a promising solution. The work that we have done so far includes an IC semantics for OWL, a decision procedure for IC evaluation, an approach to explain and repair IC violations, and a partial implementation.

## 4.1    IC Semantics for OWL

In this section, we will describe an IC semantics [18] that we have proposed for OWL [19] which is based on DL $\mathcal{SROIQ}$[20].

In the IC semantics, we adopt a weak form of UNA[4]: two individual names with different identifiers are assumed to be different by default unless their equality is required to satisfy the axioms in the KB. We formalize this notion of weak UNA by Minimal Equality (ME) models. Given a $\mathcal{SROIQ}$ KB $\mathcal{K}$, $\mathcal{I}$ and $\mathcal{J}$ are two $\mathcal{SROIQ}$ interpretations of $\mathcal{K}$, we say $\mathcal{J} \prec_= \mathcal{I}$ if: (1)$\forall C \in N_C$, $\mathcal{J} \models C(a)$ iff $\mathcal{I} \models C(a)$; (2) $\forall R \in N_R$, $\mathcal{J} \models R(a,b)$ iff $\mathcal{I} \models R(a,b)$; (3) $E_\mathcal{J} \subset E_\mathcal{I}$ where

---

[4] With UNA, two different names always refer to different entities.

$E_{\mathcal{F}} = \{\langle a, b\rangle \mid a, b \in N_I \text{ s.t. } \mathcal{F} \models a = b\}$. The Minimal Equality (ME) models, i.e., $Mod_{ME}(\mathcal{K})$, are the models of $\mathcal{K}$ with minimal equality between individual names. Formally, we define

$$Mod_{ME}(\mathcal{K}) = \{\mathcal{I} \in Mod(\mathcal{K}) \mid \nexists \mathcal{J}, \mathcal{J} \in Mod(\mathcal{K}), \mathcal{J} \prec_= \mathcal{I}\}$$

First, we define an *IC-interpretation* $\mathcal{I}, \mathcal{U} = (\Delta^{\mathcal{I}}, \cdot^{\mathcal{I}, \mathcal{U}})$ where $\mathcal{I} = (\Delta^{\mathcal{I}}, \cdot^{\mathcal{I}})$ is a $\mathcal{SROIQ}$ interpretation and $\mathcal{U}$ is a set of $\mathcal{SROIQ}$ interpretations. The IC-interpretation function $\cdot^{\mathcal{I}, \mathcal{U}}$ maps concepts, roles, and individuals as follows:

$$C_a{}^{\mathcal{I}, \mathcal{U}} = \{x^{\mathcal{I}} \mid x \in N_I \text{ s.t. } \forall \mathcal{J} \in \mathcal{U}, x^{\mathcal{J}} \in C_a{}^{\mathcal{J}}\},$$

$$R^{\mathcal{I}, \mathcal{U}} = \{\langle x^{\mathcal{I}}, y^{\mathcal{I}}\rangle \mid x, y \in N_I \text{ s.t. } \forall \mathcal{J} \in \mathcal{U}, \langle x^{\mathcal{J}}, y^{\mathcal{J}}\rangle \in R^{\mathcal{J}}\},$$

$$a^{\mathcal{I}, \mathcal{U}} = a^{\mathcal{I}}, \quad (R^-)^{\mathcal{I}, \mathcal{U}} = \{\langle x^{\mathcal{I}}, y^{\mathcal{I}}\rangle \mid \langle y^{\mathcal{I}}, x^{\mathcal{I}}\rangle \in R^{\mathcal{I}, \mathcal{U}}\},$$

$$(C \sqcap D)^{\mathcal{I}, \mathcal{U}} = C^{\mathcal{I}, \mathcal{U}} \cap D^{\mathcal{I}, \mathcal{U}}, \quad (\neg C)^{\mathcal{I}, \mathcal{U}} = (N_I)^{\mathcal{I}} \setminus C^{\mathcal{I}, \mathcal{U}},$$

$$(\geq nR.C)^{\mathcal{I}, \mathcal{U}} = \{x^{\mathcal{I}} \mid x \in N_I \text{ s.t. } \#\{y^{\mathcal{I}} \mid \langle x^{\mathcal{I}}, y^{\mathcal{I}}\rangle \in R^{\mathcal{I}, \mathcal{U}} \text{ and } y^{\mathcal{I}} \in C^{\mathcal{I}, \mathcal{U}}\} \geq n\},$$

$$(\exists R.\text{Self})^{\mathcal{I}, \mathcal{U}} = \{x^{\mathcal{I}} \mid x \in N_I \text{ s.t. } \langle x^{\mathcal{I}}, x^{\mathcal{I}}\rangle \in R^{\mathcal{I}, \mathcal{U}}\}, \quad \{a\}^{\mathcal{I}, \mathcal{U}} = \{a^{\mathcal{I}}\}.$$

where $(N_I)^{\mathcal{I}} = \{x^{\mathcal{I}} \mid x \in N_I\}$, $C_a \in N_C$ (atomic concepts), $R \in N_R$ (atomic roles), $a \in N_I$ (individual names), $C, D$ are concepts.

Then, the satisfaction of axiom $\alpha$ in an IC-interpretation $\mathcal{I}, \mathcal{U}$, denoted as $\mathcal{I}, \mathcal{U} \models \alpha$, is defined in Table 1. Note that, there are also four kinds of ABox axioms ($C(a)$, $R(a, b)$, $a = b$, $a \neq b$). Their semantics is given by encoding them as TBox axioms ($\{a\} \sqsubseteq C$, $\{a\} \sqsubseteq \exists R.\{b\}$, $\{a\} \sqsubseteq \{b\}$, $\{a\} \sqsubseteq \neg\{b\}$, resp.).

Given a $\mathcal{SROIQ}$ KB $\mathcal{K}$ and a $\mathcal{SROIQ}$ axiom $\alpha$, we say $\mathcal{K}$ IC-satisfies $\alpha$, i.e., $\mathcal{K} \models_{IC} \alpha$, iff $\forall \mathcal{I} \in \mathcal{U}, \mathcal{I}, \mathcal{U} \models \alpha$, where $\mathcal{U} = Mod_{ME}(\mathcal{K})$. We define an *extended KB* as a pair $\langle \mathcal{K}, \mathcal{C}\rangle$ where $\mathcal{K}$ is a $\mathcal{SROIQ}$ KB interpreted with the standard semantics and $\mathcal{C}$ is a set of $\mathcal{SROIQ}$ axioms interpreted with the IC semantics. We say that $\langle \mathcal{K}, \mathcal{C}\rangle$ is valid if $\forall \alpha \in \mathcal{C}, \mathcal{K} \models_{IC} \alpha$, otherwise there is an IC violation. Note that, the IC-satisfaction has a closed world flavor: given an atomic concept C (R resp.), if $\mathcal{K} \not\models_{IC} C(a)(R(a, b)$ resp.) then we conclude $\mathcal{K} \models_{IC} \neg C(a)$ ($\neg R(a, b)$ resp.). We have verified [18] that this CWA[5] and the weak UNA addresses the issues caused by the OWA and absence of UNA of OWL standard semantics, therefore enabling OWL as an IC language.

**Table 1.** Axiom satisfactions in IC-interpretation $\mathcal{I}, \mathcal{U}$

| Type | Axiom | Condition on $\mathcal{I}, \mathcal{U}$ |
|---|---|---|
| TBox | $C \sqsubseteq D$ | $C^{\mathcal{I}, \mathcal{U}} \subseteq D^{\mathcal{I}, \mathcal{U}}$ |
| RBox | $R_1 \sqsubseteq R_2$ | $R_1^{\mathcal{I}, \mathcal{U}} \subseteq R_2^{\mathcal{I}, \mathcal{U}}$ |
| | $R_1 \ldots R_n \sqsubseteq R$ | $R_1^{\mathcal{I}, \mathcal{U}} \circ \ldots \circ R_n^{\mathcal{I}, \mathcal{U}} \subseteq R^{\mathcal{I}, \mathcal{U}}$ |
| | $\text{Ref}(R)$ | $\forall x \in N_I : \langle x^{\mathcal{I}, \mathcal{U}}, x^{\mathcal{I}, \mathcal{U}}\rangle \in R^{\mathcal{I}, \mathcal{U}}$ |
| | $\text{Irr}(R)$ | $\forall x \in N_I : \langle x^{\mathcal{I}, \mathcal{U}}, x^{\mathcal{I}, \mathcal{U}}\rangle \notin R^{\mathcal{I}, \mathcal{U}}$ |
| | $\text{Dis}(R_1, R_2)$ | $R_1^{\mathcal{I}, \mathcal{U}} \cap R_2^{\mathcal{I}, \mathcal{U}} = \emptyset$ |

---

[5] With CWA, a statement is inferred to be false if it is not known to be true, which is the opposite of OWA.

## 4.2   IC Evaluation

In this section, we describe a decision procedure for IC evaluation. That is, deciding if a KB IC-satisfies an IC axiom. First, we present the translation rules from IC axioms to DCQ$^{\mathbf{not}}$ queries. Then we show that IC evaluation can be reduced to corresponding DCQ$^{\mathbf{not}}$ query answering. Due to space limitation we do not introduce DCQ$^{\mathbf{not}}$ here. Please refer to [18] for more details.

The translation rules are similar in the spirit to the Lloyd-Topor transformation [21] but instead of rules we generate DCQ$^{\mathbf{not}}$ queries. The idea behind the translation is to translate an IC axiom into a query such that when the IC is violated the query is true. In other words, whenever the answer set of the query is not empty, we can conclude that the IC is violated. The translation contains two operators: $\mathcal{T}_c$ for translating concepts and $\mathcal{T}$ for translating axioms:

$$\mathcal{T}_c(C_a, x) := C_a(x)$$
$$\mathcal{T}_c(\neg C, x) := \mathbf{not}\, \mathcal{T}_c(C, x)$$
$$\mathcal{T}_c(C_1 \sqcap C_2, x) := \mathcal{T}_c(C_1, x) \wedge \mathcal{T}_c(C_2, x)$$
$$\mathcal{T}_c(\geq nR.C, x) := \bigwedge_{1 \leq i \leq n} (R(x, y_i) \wedge \mathcal{T}_c(C, y_i)) \bigwedge_{1 \leq i < j \leq n} \mathbf{not}\, (y_i = y_j)$$
$$\mathcal{T}_c(\exists R.\mathsf{Self}, x) := R(x, x)$$
$$\mathcal{T}_c(\{a\}, x) := (x = a)$$
$$\mathcal{T}(C_1 \sqsubseteq C_2) := \mathcal{T}_c(C_1, x) \wedge \mathbf{not}\, \mathcal{T}_c(C_2, x)$$
$$\mathcal{T}(R_1 \sqsubseteq R_2) := R_1(x, y) \wedge \mathbf{not}\, R_2(x, y)$$
$$\mathcal{T}(R_1 \ldots R_n \sqsubseteq R) := R_1(x, y_1) \wedge \ldots R_n(y_{n-1}, y_n) \wedge \mathbf{not}\, R(x, y_n)$$
$$\mathcal{T}(\mathtt{Ref}(R)) := \mathbf{not}\, R(x, x)$$
$$\mathcal{T}(\mathtt{Irr}(R)) := R(x, x)$$
$$\mathcal{T}(\mathtt{Dis}(R_1, R_2)) := R_1(x, y) \wedge R_2(x, y)$$

where $C_a$ is an atomic concept, $C_{(i)}$ is a concept, $R_{(i)}$ is a role, $a$ is an individual name, $x$ and $y_{(i)}$ are variables.

**Example 1.** *Suppose* $\alpha : \mathtt{Product} \sqsubseteq \exists\mathtt{hasProducer}.\mathtt{Producer}$, *then we have:*

$$\mathcal{T}(\mathtt{Product} \sqsubseteq \exists\mathtt{hasProducer}.\mathtt{Producer})$$
$$:= \mathcal{T}_c(\mathtt{Product}, x) \wedge \mathbf{not}\, \mathcal{T}_c(\exists\mathtt{hasProducer}.\mathtt{Producer}, x)$$
$$:= \mathtt{Product}(x) \wedge \mathbf{not}\, (\mathtt{hasProducer}(x, y) \wedge \mathcal{T}_c(\mathtt{Producer}, y))$$
$$:= \mathtt{Product}(x) \wedge \mathbf{not}\, (\mathtt{hasProducer}(x, y) \wedge \mathtt{Producer}(y))$$

We now obtain the main decision procedure for IC evaluation:

**Theorem 1.** *Given an extended KB* $\langle \mathcal{K}, \mathcal{C} \rangle$ *with expressivity* $\langle \mathcal{SRI}, \mathcal{SROIQ} \rangle$ *(*$\langle \mathcal{SROIQ}, \mathcal{SROI} \rangle$ *resp.),* $\mathcal{K} \models_{IC} \alpha$ *iff the query answers* $\mathbf{A}(\mathcal{T}(\alpha), \mathcal{K})$ *are empty, i.e.,* $\mathcal{K} \not\models \mathcal{T}(\alpha)$, *where* $\alpha \in \mathcal{C}$.

We require $\langle \mathcal{K}, \mathcal{C} \rangle$ to be less expressive than $\langle \mathcal{SRI}, \mathcal{SROIQ} \rangle$ or $\langle \mathcal{SROIQ}, \mathcal{SROI} \rangle$ because, otherwise, the disjunctive individual (in)equivalence axioms in $\mathcal{K}$ and the cardinality restrictions in $\mathcal{C}$ will cause some problematic interactions such that the IC axioms are satisfied in different ways at different interpretations of $\mathcal{K}$ and IC evaluation can not be reduced to query answering. Please refer to [22] for more details.

## 4.3   IC Violation Explanation and Repair

By Theorem 1, $\mathcal{K}$ violates an IC axiom $\alpha$ if $\mathcal{K} \models \mathcal{T}(\alpha)$. To explain the violations of $\alpha$, we just need to justify why the query entailment $\mathcal{K} \models \mathcal{T}(\alpha)$ holds. In this section, we present our recent work on justification of conjunctive query answers [23] and show how to use the justifications to explain and repair IC violations.

Given a DCQ$^{uol}$ entailment $\mathcal{K} \models^\sigma Q$ where $\mathcal{K}$ is a $\mathcal{SROIQ}$ DL KB, $Q$ is a DCQ$^{not}$, $\sigma$ is an assignment mapping query variables to individuals, a justification for $\mathcal{K} \models^\sigma Q$ is $\mathcal{J} = \langle \mathcal{J}_+, \mathcal{J}_- \rangle$ where $\mathcal{J}_+$ and $\mathcal{J}_-$ are the positive and negative justifications respectively such that: (1) $\mathcal{J}_+ \subseteq \mathcal{K}$, $\mathcal{J}_+ \models^\sigma Q$; $\forall S \subseteq \mathcal{K}$, $\mathcal{J}_+ \cup S \models^\sigma Q$; $\forall \mathcal{J}' \subset \mathcal{J}_+$, $\mathcal{J}' \not\models^\sigma Q$. (2) $\mathcal{K} \cup \mathcal{J}_-$ is consistent; $\mathcal{K} \cup \mathcal{J}_- \not\models^\sigma Q$; $\forall T \supseteq \mathcal{J}_-$, $\mathcal{K} \cup T \not\models^\sigma Q$; $\forall \mathcal{J}' \subset \mathcal{J}_-$, $\mathcal{K} \cup \mathcal{J}' \models^\sigma Q$. That is, the existence of $\mathcal{J}_+$ in $\mathcal{K}$ and the absence of $\mathcal{J}_-$ from $\mathcal{K}$ are sufficient for $\mathcal{K} \models^\sigma Q$ to hold.

**Example 2.** *Suppose we have the following KB $\mathcal{K}$ and IC axiom $\alpha$*
$$\mathcal{K} = \{\text{Toy} \sqsubseteq \text{Product}, \text{ToyProducer} \sqsubseteq \text{Producer}, \text{Toy}(p_1), \text{Product}(p_1),$$
$$\text{hasProducer}(p_1, s_1), \text{hasProducer}(p_1, s_2)\}$$
$$\alpha : \text{Product} \sqsubseteq \exists \text{hasProducer.Producer}.$$

Then $\mathcal{T}(\alpha)$ is $Q \leftarrow \text{Product}(x) \wedge \mathbf{not}\,(\text{hasProducer}(x, y) \wedge \text{Producer}(y))$ and query answer $\mathbf{A}(Q, \mathcal{K}) = \{\sigma : x \rightarrow p_1, y \rightarrow s_1, \sigma' : x \rightarrow p_1, y \rightarrow s_2, \sigma'' : x \rightarrow p_1, y \rightarrow p_1\}$ is not empty indicating $\alpha$ is violated.

To explain the violation of $\alpha$, we first compute the justifications for $\mathcal{K} \models Q$. We have proposed algorithms for justification computation [23]. According to the algorithms, the justifications for $\mathcal{K} \models Q$ w.r.t. $\sigma$ are:

$$\mathcal{J}_1 = \langle \mathcal{J}_{1+}, \mathcal{J}_{1-} \rangle = \langle \{\text{Product}(p_1)\}, \{\text{Producer}(s_1)\} \rangle,$$
$$\mathcal{J}_2 = \langle \mathcal{J}_{2+}, \mathcal{J}_{2-} \rangle = \langle \{\text{Toy}(p_1), \text{Toy} \sqsubseteq \text{Product}\}, \{\text{Producer}(s_1)\} \rangle.$$

Since the entailment $\mathcal{K} \models^\sigma Q$ contributes to the non-emptiness of query answers $\mathbf{A}(Q, \mathcal{K})$, the violation of $\alpha$ can be explained by $\mathcal{J}_1$ and $\mathcal{J}_2$. That is, $\alpha$ is violated because $p_1$ is a product (by $\mathcal{J}_{1+}/\mathcal{J}_{2+}$) and $p_1$ does not have a known producer (by $\mathcal{J}_{1-}/\mathcal{J}_{2-}$).

To repair the above violations, we need to invalidate $\mathcal{K} \models Q$. According to the definition of query entailment justification, $\mathcal{K} \models Q$ holds because the existence of positive justifications and the absence of negative justifications. Therefore, we can either remove a minimal hitting set (mhs)[6] of positive justifications from $\mathcal{K}$

---

[6] Given a collection of sets, a set which intersects all sets in the collection in at least one element is called a hitting set. The minimal hitting set is the hitting set of smallest size.

such that no positive justification exists in $\mathcal{K}$, or add a negative justification to $\mathcal{K}$. For instance, the positive justifications for $\mathcal{K} \models^{\sigma} Q$ and its mhs are:

$$\mathcal{J}_{+}{}^{All} = \{\mathcal{J}_{1+}, \mathcal{J}_{2+}\} = \{\{\texttt{Product}(p_1)\}, \{\texttt{Toy}(p_1), \texttt{Toy} \sqsubseteq \texttt{Product}\}\},$$
$$mhs(\mathcal{J}_{+}{}^{All}) = \{H_1, H_2\} = \{\{\texttt{Product}(p_1), \texttt{Toy} \sqsubseteq \texttt{Product}\}, \{\texttt{Product}(p_1), \texttt{Toy}(p_1)\}\}.$$

So we can update $\mathcal{K}$ to one of the three KBs: (1) $\mathcal{K}_1' = \mathcal{K} \setminus H_1$; (2) $\mathcal{K}_2' = \mathcal{K} \setminus H_2$; (3) $\mathcal{K}_3' = \mathcal{K} \cup \mathcal{J}_{1-}$. Similarly we can invalidate $\mathcal{K} \models Q$ w.r.t. $\sigma'$ and $\sigma''$. It is easy to verify that after the updates to $\mathcal{K}$ constraint $\alpha$ is satisfied.

## 5   Further Research Plan

The research that remains to be done includes:

- Proposing a decision procedure for IC evaluation for the fully expressive $\mathcal{SROIQ}$ KBs. The existing query answering-based decision procedure only works when KBs are less expressive than $\mathcal{SROIQ}$ KBs. To address this issue, we need to explore other approaches such as Tableau-based approaches.
- Optimizing the algorithms for query answer justifications. The computation complexity of the algorithms is a linear function in most cases except that the complexity follows the power law when the queries include negation of conjunction. We need to utilize some optimizations to solve this problem.
- Finishing the implementation and evaluating the integrity issues in semantic web instance data by employing the IC modeling, evaluation, and explanation approaches that we have proposed. The evaluation should cover functionality, performance, and scalability aspects.

## 6   Conclusions

Our work addresses the issues of supporting IC evaluation for instance data on the semantic web. We propose an alternative semantics for OWL that adopts CWA and a weakened UNA thus enabling OWL to serve as an IC language. Further, we describe a decision procedure for IC evaluation and an approach for IC violation explanation and repair. With the contributions of this work, users can easily discover and fix the issues in the data. This enables them to obtain data that is checked and appropriate for their uses, thus improving the process of interactions between various parties on the web.

## References

1. Berners-Lee, T., Hendler, J., Lassila, O.: The Semantic Web. Scientific American 284(5), 34–43 (2001)
2. Gmez-Prez, A.: Some Ideas and Examples to Evaluate Ontologies. In: AIA 1995, p. 299 (1995)
3. Gmez-Prez, A.: Evaluation of Ontologies. Int. J. Intell. Syst. 16(3), 391–409 (2001)

4. Baclawski, K., Matheus, C.J., Kokar, M.M., Letkowski, J., Kogut, P.A.: Towards a Symptom Ontology for Semantic Web Applications. In: McIlraith, S.A., Plexousakis, D., van Harmelen, F. (eds.) ISWC 2004. LNCS, vol. 3298, pp. 650–667. Springer, Heidelberg (2004)
5. Tao, J., Ding, L., Bao, J., McGuiness, D.: Instance Data Evaluation for Semantic Web-Based Knowledge Management Systems. In: HICSS 42, pp. 1–10 (1942)
6. Bechhofer, S., Volz, R.: Patching Syntax in OWL Ontologies. In: McIlraith, S.A., Plexousakis, D., van Harmelen, F. (eds.) ISWC 2004. LNCS, vol. 3298, pp. 668–682. Springer, Heidelberg (2004)
7. Parsia, B., Sirin, E., Kalyanpur, A.: Debugging OWL ontologies. In: WWW 2005, pp. 633–640 (2005)
8. Wang, H., Horridge, M., Rector, A., Drummond, N., Seidenberg, J.: Debugging OWL-DL Ontologies: A Heuristic Approach. In: Gil, Y., Motta, E., Benjamins, V.R., Musen, M.A. (eds.) ISWC 2005. LNCS, vol. 3729, pp. 745–757. Springer, Heidelberg (2005)
9. Plessers, P., Troyer, O.D.: Resolving Inconsistencies in Evolving Ontologies. In: Sure, Y., Domingue, J. (eds.) ESWC 2006. LNCS, vol. 4011, pp. 200–214. Springer, Heidelberg (2006)
10. McGuinness, D.L., Fikes, R., Rice, J., Wilder, S.: An Environment for Merging and Testing Large Ontologies. In: KR 2000, pp. 483–493 (2000)
11. Corcho, Ó., Gómez-Pérez, A., González-cabero, R., Suárez-figueroa, C.: ODEval: A Tool for Evaluating RDF(S), DAML+OIL, and OWL Concept Taxonomies. In: AIAI 2004, pp. 369–382 (2004)
12. Codd, E.F.: A Relational Model of Data for Large Shared Data Banks. ACM Commun. 13(6), 377–387 (1970)
13. Smith, M.K., Welty, C., McGuiness, D.: OWL Web Ontology Language Guide (2004)
14. Eiter, T., Ianni, G., Lukasiewicz, T., Schindlauer, R., Tompits, H.: Combining Answer Set Programming with Description Logics for the Semantic Web. AI 172(12-13), 1495–1539
15. Motik, B.: A Faithful Integration of Description Logics with Logic Programming. In: IJCAI 2007, pp. 477–482 (2007)
16. Calvanese, D., Giacomo, G.D., Lembo, D., Lenzerini, M., Rosati, R.: EQL-Lite: Effective First-Order Query Processing in Description Logics. In: IJCAI 2007, pp. 274–279 (2007)
17. Motik, B., Horrocks, I., Sattler, U.: Bridging the Gap between OWL and Relational Databases. In: WWW 2007, pp. 807–816 (2007)
18. Tao, J., Sirin, E., Bao, J., McGuiness, D.: Integrity Constraints in OWL. In: AAAI 2010 (2010)
19. Motik, B., Patel-Schneider, P.F., Grau, B.C.: OWL 2 Web Ontology Language Direct Semantics (2009)
20. Horrocks, I., Kutz, O., Sattler, U.: The Even More Irresistible SROIQ. In: KR 2006, pp. 57–67 (2006)
21. Lloyd, J.W.: Foundations of Logic Programming (1987)
22. Tao, J., Sirin, E., Bao, J., McGuiness, D.: Integrity Consrtiants in OWL. Technical report, Rensselaer Polytechnic Institute
23. Tao, J., Sirin, E., McGuiness, D.: Towards Justification of Conjunctive Query Answers in Description Logics. Technical report, Rensselaer Polytechnic Institute

# Abstract: The Open Graph Protocol Design Decisions

Austin Haugen

Facebook, US

The Open Graph protocol enables any web page to become a rich object in a social graph. It was created by Facebook but designed to be generally useful to anyone. While many different technologies and schemas exist and could be combined together, there is not a single technology which provides enough information to richly represent any web page within the social graph. The Open Graph protocol builds on these existing technologies and gives developers one thing to implement. Developer simplicity is a key goal of the Open Graph protocol which has informed many of the technical design decisions. This talk will explore the motivation of the Open Graph protocol and the design decisions which went into creating it.

P.F. Patel-Schneider et al.(Eds.): ISWC 2010, Part II, LNCS 6497, p. 338, 2010.

# Evaluating Search Engines by Clickthrough Data

Jing He and Xiaoming Li

Computer Network and Distributed System Laboratory, Peking University, China
{hejing,lxm}@pku.edu.cn

**Abstract.** It is no doubt that search is critical to the web. And it will be
of similar importance to the semantic web. Once searching from billions
of objects, it will be impossible to always give a single right result, no
matter how intelligent the search engine is. Instead, a set of possible
results will be provided for the user to choose from. Moreover, if we
consider the trade-off between the system costs of generating a single
right result and a set of possible results, we may choose the latter. This
will naturally lead to the question of how to decide on and present the
set to the user and how to evaluate the outcome.

In this paper, we introduce some new methodology in evaluation of
web search technologies and systems. Historically, the dominant method
for evaluating search engines is the Cranfield paradigm, which employs a
test collection to qualify the systems' performance. However, the modern
search engines are much different from the IR systems when the Cranfield
paradigm was proposed: 1) Most modern search engines have much more
features, such as snippets and query suggestions, and the quality of such
features can affect the users' utility; 2) The document collections used in
search engines are much larger than ever, so the complete test collection
that contains all query-document judgments is not available. As response
to the above differences and difficulties, the evaluation based on implicit
feedback is a promising alternative employed in IR evaluation. With this
approach, no extra human effort is required to judge the query-document
relevance. Instead, such judgment information can be automatically predicted from real users' implicit feedback data. There are three key issues
in this methodology: 1) How to estimate the query-document relevance
and other useful features that useful to qualify the search engine performance; 2) If the complete "judgments" are not available, how can
we efficiently collect the most critical information from which the system performance can be derived; 3) Because query-document relevance
is not only feature that can affect the performance, how can we integrate
others to be a good metric to predict the system performance. We will
show a set of technologies dealing with these issues.

## 1   Introduction

Search engine evaluation is critical for improving search techniques. So far, the
dominant method for IR evaluation has been the Cranfield evaluation method.
However, it also has some disadvantages. First, it is extremely labor intensive
to creating relevance judgments. As a result, we often have a limited number of

P.F. Patel-Schneider et al.(Eds.): ISWC 2010, Part II, LNCS 6497, pp. 339–354, 2010.

queries to experiment with. Second, the Cranfield paradigm was proposed for evaluating traditional information retrieval (IR) systems, but it cannot reflect some new features in modern search engines. The modern search engines usually provide more than a ranked document list, such as snippet and related query suggestion. The quality of such new features can affect search users' utility.

As a very promising alternative, automatic evaluation of retrieval systems based on the implicit feedback of users has recently been proposed and studied. One important direction is to leverage the large amount of clickthrough data from users to evaluate retrieval systems. Since clickthroughs are naturally available when users use a search engine system, we can potentially use this strategy to evaluate ranking methods without extra human effort. On the other hand, the clickthrough data can not only reflect the quality of the retrieved documents, but also some other features of search engines. Both factors make it an attractive alternative to the traditional Cranfield evaluation method.

In this paper, we introduce two categories of methods of evaluating search engines based on clickthrough data. The methods of first category are for comparing two search engines. The basic idea of this method is to interleave the retrieved results from two search engines, and the search engine which gets more click on its results wins. The methods of second category infer the document relevance first and utilize the relevance information to evaluate the search engines. The document relevance is usually estimated from a probabilistic click model, and we can reorder the retrieved documents to more efficiently collect the relevance information for evaluation. Finally, we propose a new metric that is able to embed snippet generation quality in the evaluation.

## 2    Rule Based Methods: Interleaving and Comparing

The basic idea of methods in this category is to interleave the search results returned by different systems for the *same* query in a somewhat random manner and present a merged list of results to the user. The users would then interact with these results in exactly the same way as they would with normal search engine results, i.e., they would browse through the list and click on some promising documents to view. The clickthroughs of users would then be recorded and a system that returned more clicked documents would be judged as a better systems.

We can easily see that any method in this category consists of two functions: (1) an *interleaving function* which determines how to combine the ranked lists of results returned from two systems, and (2) a *comparison function* which decides which ranked list is preferred by the user based on the collected user clickthroughs. In general, the interleaving function and the comparison function are "synchronized" to work together to support relative comparison of retrieval systems. There have been two major methods proposed to instantiate these two functions: the balanced interleaving method [1] and the team-draft interleaving method [2].

## 2.1   Methods

The balanced interleaving method was proposed in [1]. It merges two lists by taking a document from each list alternatively so as to ensure that the numbers of documents taken from both lists differ by at most one [1]. Specifically, its interleaving function works as follows. It starts with randomly selecting a list (either $A$ or $B$) as the current list $L$. It then iteratively pops the top document $d$ from the current list $L$ and appends $d$ to the merged list $M$ if $d$ is not already in $M$. After each iteration, it would update the current list so that it would point to a different list. This process is repeated until the two lists are empty. To tell which system (list) performs better, the balanced interleaving method looks at which documents are clicked by the user and assumes that list $A$ performs better than $B$ if $A$ has contributed more clicked documents in the stage of constructing the merged list than $B$.

**Table 1.** Example for Interleaving Strategies

| interleaving | $A$ | $(a, b, c, d)$ |
|---|---|---|
| lists | $B$ | $(b, c, a, d)$ |
| balanced | A-B | $(a, b, c, d)$ |
| | B-A | $(b, a, c, d)$ |
| | A-B-A-B | $(a^{(A)}, b^{(B)}, c^{(A)}, d^{(B)})$ |
| team | A-B-B-A | $(a^{(A)}, b^{(B)}, c^{(B)}, d^{(A)})$ |
| draft | B-A-B-A | $(b^{(B)}, a^{(A)}, c^{(B)}, d^{(A)})$ |
| | B-A-A-B | $(b^{(B)}, a^{(A)}, c^{(A)}, d^{(B)})$ |

The team-draft interleaving method was proposed in [2] to prevent some bias in the balanced interleaving method. In each round, it start with randomly selecting a list $L$(either $A$ or $B$), and appends $M$ with $L$'s most preferred document that has not been in $M$. And then it turns to the other list and does the same thing. The rounds continue until all the documents in $A$ and $B$ are in $M$. To predict which system(list) performs better, the team-draft method counts the number of clicked documents selected from list $A$ and $B$ respectively. It assumes that $A$ performs better than $B$ if there are more clicked documents selected from $A$. Formally, it scores system $A$ by $score(A) = \sum_{d_c} \delta(d_c \in T_A)$, where $d_c$ is a clicked document and $\delta$ is a binary indicator function.

In Table 1, we show examples of merging lists $A$ and $B$ using balanced and team-draft function.

The common drawback of both methods is that they are not sensitive to the positions of the clicked documents in the ranked lists. We propose an improvement to the balanced method to overcome this limitation. Specifically, in this new method (called preference-based balanced interleaving), we would interleave the ranked lists in the same way as the balanced method, but make prediction about which system is better based on a new preference-based comparison function (ppref).

A preference relation between two documents indicates that one document is more relevant than the other (with respect to a query), which we denote by $d_i >_p$

$d_j$. It has been shown that some preference relations extracted from clickthrough data are reliable [3]. We would first try to convert the collected clickthroughs into a set of preference relations based one two rules: (1) a clicked document are preferred to the skipped documents above it; (2) a clicked document is more relevant than the next unclicked one. Both rules have been justified in some previous work [3]. Now, our key idea is to design a preference-based measure to score each ranked list by treating these inferred *incomplete* preference relations between documents as our golden standard. In this study we use precision of preference(ppref)[4].

## 2.2   Evaluation of Interleaving Strategies

In this section, we propose a simulation-based approach for evaluating and comparing different interleaving methods systematically. Our basic idea is to first systematically generate many test cases, each being a pair of ranked lists of documents, and then apply each interleaving method to these test cases and evaluate its performance. Thus our overall approach consists of two components: 1) metrics to be used for measuring the performance of an interleaving method or comparing two methods, and 2) simulation strategies to be used to generate the test cases.

**Metrics.** Intuitively, we would expect an ideal method to merge the two lists in such a way that we can differentiate the ranking accuracy of the two systems accurately based on the collected clickthroughs. Thus our first criterion is the *effectiveness of differentiation* (EoD) of a method. Moreover, it is also important that the utility of the merged list from a user's perspective is high. Thus a second criterion that we consider is the *utility to users*(UtU) of a method.

To quantify the utility for users of an interleaving method, in general, we may apply any existing IR relevance evaluation measures (MAP is this paper) to the merged result list generated by the interleaving method. The effectiveness of differentiation of a method can be measured based on the accuracy of the method in predicting which system is better. Again, since we use simulation to generate test cases, we will have available the utilities of the two candidate lists (by MAP is this paper) and decide which is better. By comparing the prediction result of an interleaving method with this ground truth, we may measure the accuracy of the prediction. Since there are three types of results when comparing system $A$ and $B$, i.e., (1) System $A$ wins; (2) System $A$ loses; or (3) the two systems tie. In general, we can associate different costs with different errors. However, it is not immediately clear how exactly we should set these costs. Leaving this important question for future research, in this paper, we simply assume that all correct predictions have zero cost, all "tie errors" have a cost of 1, and all opposite predictions have a cost of 2. With these costs, given two candidate ranked lists and the prediction of an interleaving method, we will be able to measure the accuracy of the method with the cost of the prediction; a lower cost would indicate a higher accuracy. Since the cost value is presumably not sensitive to specific queries, the absolute value of cost is meaningful, but we mainly use it to compare different interleaving methods in this paper.

**Generation of Test Cases.** A test case for our purpose consists of the following elements: (1) two ranked lists of documents of length $n$: $A$ and $B$, (2) a set of $n_r$ documents assumed to be relevant, $R$, and (3) the number of documents a user is assumed to view, $K$ ($K \leq n$). We assume that once a merged list of results are presented to the user, the user would view sequentially (starting from the top) $K$ results and click on any relevant document encountered. Thus, given a merged list of results, the clickthroughs can be uniquely determined based on $R$ and $K$. Thus in general, we can use a straightforward strategy to generate a large random sample to compare different interleaving methods. However, such a blind test provides limited information on how an interleaving method performs in different scenarios.

It would be much more informative and useful to compare different interleaving methods in various scenarios such as known item search vs. high-recall search (modeled by relevant document number $n_r$), comparing two systems that have similar retrieval results vs. very different retrieval results (modeled by kendall's $\tau$, comparing two similar-performance or different-performance retrieval systems (modeled by relative average precision $RAP$) or obtaining clickthroughs from a patient user vs. an impatient user (modeled by viewing number $K$). Thus each such possible scenario should be simulated separately to understand relative strengths and weaknesses of two methods in these different scenarios. It is possible that we may find out that some method tends to perform better in some scenarios, while others perform better for other scenarios. We can stop the sampling process when we have sufficient test cases to obtain a relatively reliable estimate of the UtU and EoD of the interleaving methods being evaluated.

**Results.** In our experiments, we control the first parameter $n$ by setting it to 10 and vary all the other parameters to simulate different evaluation scenarios. The ground truth about which system is better is decided based on the average precision of the top 10 documents for each system. Variations of other parameters and the scenarios simulated are summarized in Table 2.

We show the results of these three methods in all the different scenarios in Table 3. Because the UtU results for these three methods are very similar, it is

**Table 2.** Summary of Evaluation Scenarios

| Scenario | Variation | Parameter Setting |
|---|---|---|
| Topic | Known-Item Search | $n_r = 1$ |
| | Easy Topic | Prec@10Doc = 0.6 |
| | Difficult Topic | Prec@10Doc = 0.3 |
| Result Similarity | High | $0.5 < \tau < 1.0$ |
| | Low | $-1 < \tau < -0.5$ |
| Precision Similarity | High | $0 < RAP < 0.2$ |
| | Low | $0.3 < RAP$ |
| User Patience | High | $k = 8$ |
| | Medium | $k = 5$ |
| | Low | $k = 3$ |

**Table 3.** MAP and Cost for Specific Scenarios

| Topic | Result Sim. | Prec. Sim. | K | Cost of Prediction | | |
|---|---|---|---|---|---|---|
| | | | | Balanced | Team | Preference |
| Known-item | Low | Low | 3 | **0.61**(0.24) | 0.63(0.25) | **0.61**(0.24), (0.00, 0.03$^-$) |
| | | | 5 | **0.38**(0.23) | 0.41(0.27) | **0.38**(0.23), (0.00, 0.07$^-$) |
| | | | 8 | **0.15**(0.13) | 0.18(0.19) | **0.15**(0.13), (0.00, 0.17$^-$) |
| | Low | High | 3 | **1.00**(0.00) | **1.00**(0.00) | **1.00**(0.00), (0.00, 0.00) |
| | | | 5 | **0.93**(0.06) | 0.95(0.08) | **0.93**(0.06), (0.00, 0.02$^-$) |
| | | | 8 | **0.60**(0.24) | 0.70(0.32) | **0.60**(0.24), (0.00, 0.14$^-$) |
| | High | High | 3 | **1.00**(0.00) | **1.00**(0.00) | **1.00**(0.00), (0.00, 0.00) |
| | | | 5 | **0.80**(0.23) | 0.87(0.29) | **0.80**(0.23), (0.00, 0.08$^-$) |
| | | | 8 | **0.26**(0.19) | 0.48(0.47) | **0.26**(0.19), (0.00, 0.46$^-$) |
| Hard | Low | Low | 3 | **0.31**(0.23) | 0.31(0.26) | **0.31**(0.21), (0.00, 0.00) |
| | | | 5 | 0.26(0.22) | 0.27(0.22) | **0.11**(0.11), (0.58$^-$, 0.59$^-$) |
| | | | 8 | 0.35(0.33) | 0.27(0.25) | **0.13**(0.20), (0, 63$^-$, 0.52$^-$) |
| | Low | High | 3 | 0.85(0.27) | 0.86(0.34) | **0.81**(0.40), (0.05$^-$, 0.06$^-$) |
| | | | 5 | 0.77(0.36) | 0.79(0.42) | **0.68**(0.55), (0.12$^-$, 0.14$^-$) |
| | | | 8 | 0.73(0.44) | 0.72(0.47) | **0.57**(0.64), (0.22$^-$, 0.21$^-$) |
| | High | High | 3 | 0.78(0.24) | 0.84(0.31) | **0.75**(0.47), (0.04$^-$, 0.11$^-$) |
| | | | 5 | 0.68(0.37) | 0.73(0.56) | **0.56**(0.35), (0.18$^-$, 0.23$^-$) |
| | | | 8 | 0.69(0.48) | 0.66(0.57) | **0.39**(0.36), (0.43$^-$, 0.41$^-$) |
| Easy | Low | Low | 3 | 0.29(0.22) | 0.32(0.24) | **0.06**(0.08), (0.79$^-$, 0.81$^-$) |
| | | | 5 | 0.17(0.15) | 0.19(0.15) | **0.01**(0.02), (0.94$^-$, 0.95$^-$) |
| | | | 8 | 0.15(0.16) | 0.07(0.07) | **0.00**(0.00), (1.00$^-$, 1.00$^-$) |
| | Low | High | 3 | 0.71(0.48) | 0.72(0.55) | **0.54**(0.63), (0.24$^-$, 0.25$^-$) |
| | | | 5 | 0.66(0.46) | 0.66(0.53) | **0.41**(0.50), (0.28$^-$, 0.28$^-$) |
| | | | 8 | 0.65(0.42) | 0.56(0.43) | **0.34**(0.48), (0.48$^-$, 0.39$^-$) |
| | High | High | 3 | 0.69(0.31) | 0.74(0.56) | **0.59**(0.36), (0.14$^-$, 0.20$^-$) |
| | | | 5 | 0.77(0.35) | 0.67(0.60) | **0.45**(0.35), (0.41$^-$, 0.33$^-$) |
| | | | 8 | 0.85(0.38) | 0.58(0.57) | **0.37**(0.35), (0.56$^-$, 0.36$^-$) |

omitted here due to the space constraint. In general, the preference method performs better than the other two methods and the balanced method is preferred to the team-draft method. For known-item search, it's always better to use the balanced or preference method. For searches with more relevant documents, if users are expected to view very few top documents, the balanced and preference method is also preferred, but when the user is patient and willing to view more documents, team-draft and preference method may be more appropriate.

## 3    Evaluation Based on Click Models

The interleaving method can compare the performance of two ranked list for a specific query. However, it has two problems. First, we cannot get the confidence level about the comparison result. Second, it is a little expensive to compare the search engine pairly to get the relative performance for a large number of search engines.

To address the these problems, various unsupervised graphical click models have been recently proposed [5,6,7,8]. Click models connect the document relevance and users' behaviors with probability graphical models. They provide a principled approach to infer relevance of the retrieved documents. In general, an examined document is inferred to be more relevant if it is clicked. The click models can not only give the document relevance estimation, but also give how reliable it is.

## 3.1 Click Models

Click models usually model two types of search user behaviors: *examination* and *click*. Once a user submitted a query, the search engine returns a ranked list of snippets, each of which corresponds to a retrieved document (the document list is denoted as $D = \{d_1, \ldots, d_M\}$). Then, the user usually *examine* these snippets and *click* on those which seem to be relevant to his information need. Usually, the examination and click events on a document $d_i$ are modeled as two binary variable (denoted as $E_i$ and $C_i$ respectively). Then the documents relevance are connected with these events by some hypotheses.

One category of hypotheses define the click behavior. The most commonly used hypothesis in this category is *examination hypothesis* [9]. For a document $d_i$, it defines that the click behavior ($C_i$) depends on both examination ($E_i$) and document relevance ($r_{d_i}$). When a document is not examined, it cannot be clicked. And once it is examined, the probability of click is proportional to its relevance.

Another category of hypotheses defines the examination behavior. Most existing work has a common assumption called *cascade hypothesis*, stating that users examine the document in a top-down manner, i.e., if one document at rank $i$ is not examined, all documents below it would not be examined. Besides the *cascade hypothesis*, click models define different functions determining the examination probability. The *cascade model* [10] assumes that the users would continue examine until the first click, and they stop examining after first click. The *dependent click model* [8] assumes that users continue examining until a click, and the probability of keeping examining after a click depends on the position. The *click chain model* [5] assumes that users may stop even when they does not click, and the probability of keeping examination after click depends on clicked document's relevance. While the *user browsing model* [6] does not use the *cascade hypothesis* and assumes that the examination probability is determined by its absolute rank and its distance to the previous clicked document.

All examination and click events can be modeled as nodes in the graphical click model, and the hypotheses on relationship between these events are modeled as edges in the graph. In this graphical click model, only the click variables are observed. The task is to estimate the parameters such as document relevance, and it usually can be done in EM algorithm (or other more efficient method for some specific model). With these models, we can estimate the how these document relevances distribute, so we can not only get the expected relevance value, but also know how reliable the estimation is. In general, the relevance estimation for a document is more reliable if it is examined more frequently. It has shown that *click chain model* (denoted as CCM) can predict user behavior accurately with the document relevance it estimates. In the later experiments, we use CCM to predict the document relevance.

## 3.2 Efficiently Collecting Relevance Information

With the relevance information collected by a click model, we can use them to evaluate a search engine with some graded relevance based IR metrics (such as *DCG* or *RBP*). One challenge of using this method is that the evaluation results may be questionable for some tail queries, due to the limited number of query submissions. Intuitively, the retrieved documents contribute differently for evaluating IR systems. The main idea here is to measure the benefit of collecting a document's relevance, thus it can guide us collect the relevance of documents which can bring more benefit.

The first intuition is that the benefit of examining a document is affected by its relevance uncertainty reduction. For example, if we have been very confident about the relevance of a document, it provides little information by further examining this document. Therefore, it is a good choice to move up the document with larger relevance uncertainty reduction. Naturally, the relevance uncertainty of a document can be measured by the variance of its inferred relevance, i.e., the larger variance is, we are more uncertain about the relevance. We can formulate the inferred relevance variance reduction from examining a document $d_i$ as follows:

$$
\Delta V(r_i) = P(c_i = 1|\hat{r}_i)\Delta_1 V(r_i) \\
+ P(c_i = 0|\hat{r}_i)\Delta_0 V(r_i) \tag{1}
$$

Where $r_i$ and $\hat{r}_i$ are the inferred and actual relevance of $d_i$ respectively; $P(c_i = 1|\hat{r}_i)$ and $P(c_i = 0|\hat{r}_i)$ are clicking and skipping probability given the actual relevance level $\hat{r}_i$ respectively. According to *examination hypothesis*, $P(c_i = 1|\hat{r}_i = r) = r$. Unfortunately, we don't know the exact value of $\hat{r}_i$ in reality, so we approximate by replacing it with inferred relevance $r_i$. $\Delta_1 V(r_i)$ and $\Delta_0 V(r_i)$ are variance reduction for clicking and skipping cases respectively.

The second intuition is that we should encourage the users to examine deeper, because it helps to collect more documents' relevance information. The users would generally stop examining when their information need has been satisfied. To encourage the users to examine deeper, we can delay their satisfaction by moving the relevant documents down. Though this strategy obviously sacrifices the user utility, the effect may not be very serious because only a very limited percent of queries are employed for evaluation purpose in real search engine. Assuming the ranked list is $(d_1, \ldots, d_n)$, the *list benefit function* $b(d_1, \ldots, d_n)$ can be defined as:

$$
b(d_1, \ldots, d_n) = \sum_i P(e_i = 1)b_1(d_i) \tag{2}
$$

Generally, the users examine the top document, but the probability of examining deeper documents depends on the documents ranked above. The above document relevance factor can be plugged in the list benefit function:

$$b(d_1, \ldots, d_n) = b_1(d_1) + P(e_2 = 1|\hat{r}_1)b(d_2, \ldots, d_n)$$

Where $P(e_2|\hat{r}_1)$ is the probability of continuing examining document $d_2$ given actual relevance $\hat{r}_1$.

Obviously, an optimal presented document list is to maximize the benefit function. Unfortunately, this problem is intractable. We approximate it by a greedy algorithm: at each step, we select the document that leads to maximal benefit and append it to the end of the presented list. We approximate the benefit of examining below by the maximal document benefit of the unselected documents. Thus, the weight of a document $d_i$ in an unselected document set $D$ can be formulated as:

$$w(d_i) = b_1(d_i) + P(e_{next}|r_i) \max_{\substack{d_j \in D \\ i \neq j}} b_1(d_j) \tag{3}$$

The third intuition is that the highly ranked documents contribute more to the overall performance score. Most IR metrics model this in their formulas. For example, in $DCG$, the document at rank $k$ can weighted as $\frac{1}{log_2(k+1)}$; Thus it generally requires to infer relevance of highly ranked document more accurately. Click models can infer a distribution of document relevance, so the mean and variance of $DCG$ value distribution can be expressed as:

$$E(DCG(A)) = \sum_{d_i \in A} \frac{E(r_i)}{\log_2(A_i + 1)}$$

$$V(DCG(A)) = \sum_{d_i \in A} \frac{V(r_i)}{\log_2^2(A_i + 1)}$$

In evaluating one search engine, the purpose is to reduce the uncertainty of the metrics score, so the contribution of each document does not only depend on its variance reduction but also its original rank. The weight of document at rank $k$ is $1/\log_2^2(k + 1)$, so the benefit of examining document $d_i$ is:

$$b_2(d_i) = \frac{\Delta V(r_i)}{\log_2^2(A_i + 1)} \tag{4}$$

For list benefit function(Equation 2) and document weight function(Equation 3), we can replace the document benefit function $b_1(d_i)$ by $b_2(d_i)$.

Finally, in the context of comparing two systems, retrieved documents have different effect on distinguishing two systems. Therefore, in addressing the comparison problem, it benefits from moving up the documents that are ranked differently. The mean and variance of $DCG$ score difference from two ranked lists $A$ and $B$ can be expressed as:

$$E(\Delta DCG_{A,B}) = \sum_{d_i \in A \cup B} (W_A(d_i) - W_B(d_i))V(r_i)$$

$$V(\Delta DCG_{A,B}) = \sum_{d_i \in A \cup B} (W_A(d_i) - W_B(d_i))^2 V(r_i)$$

$$W_S(i) = \begin{cases} \frac{1}{\log_2(S_i+1)} & \text{if } d_i \in S. \\ 0 & \text{otherwise} \end{cases}$$

We expect to reduce the uncertainty of $\Delta DCG$ value, so one document's contribution to the overall performance difference can be formulated as:

$$b_3(d_i) = (W_A(d_i) - W_B(d_i))^2 \Delta V(r_i) \tag{5}$$

The corresponding list benefit and document weight can be expressed by replacing $b_1(d_i)$ to $b_3(d_i)$ in Equation 2 and 3 respectively.

### 3.3    Experiments and Results

The first experiment is a user study, which is designed to resemble the common search engine usage. We recruited 50 college students to answer ten questions with the help of our system. These questions contain both open and close questions and vary in difficulty and topic. For each question, we designed a query and collected 10 results from search engines $A$ and $B$ respectively. A user was presented with a question following ten ranked snippet results in a page. The user can answer the question by examining and click the results.

We implemented five reordering functions in our system: three of them presented the results $A$ from one search engine, and two of them interleaved results $A$ and $B$ from two search engines. For one-system results reordering, the baseline function ($base1$) presents the results $A$ unchanged. Two other reordering functions ($fun1$ and $fun2$) use benefit function in Equation 1 and Equation 4 respectively. For two-system results reordering, the baseline function($base2$) presents the results $A$ and $B$ alternatively, i.e., balanced interleaving method used in [1]. The another reordering function($fun3$) determines the presented list using benefit function in Equation 5.

As a golden standard, we asked three assessors to provide the relevance judgments for the results. The relevance level is then normalized into a value between 0 and 1. The engine's performance on a query can be measured by DCG based on the actual relevance judgments(denoted as $D\hat{C}G$). As mentioned, $DCG$ can also be calculated based on the inferred relevance(denoted as $DCG$).

For one-system evaluation task, it is measured by the relative difference (denoted as $rerr$) between $D\hat{C}G$ and $DCG$. For two-system comparison task, it is measured by the ratio that predicted comparison result is incorrect.

The results from the user study is presented in Table 4. In the one-system evaluation results we can find that the both $fun1$ and $fun2$ perform much

**Table 4.** User Study Results

| ID | Evaluation | | | Comparison | |
|----|------|------|------|-------|------|
| | base1 | fun1 | fun2 | base2 | fun3 |
| 1 | 0.77 | 0.27 | **0.26** | C | C |
| 2 | 1.33 | 1.47 | **1.32** | E | C |
| 3 | **1.64** | 1.97 | 2.07 | C | C |
| 4 | 1.32 | **0.45** | 0.69 | C | C |
| 5 | 0.43 | **0.09** | 0.32 | E | C |
| 6 | 1.61 | **0.48** | 0.58 | C | C |
| 7 | **0.43** | 0.47 | 1.24 | C | C |
| 8 | 0.70 | 0.24 | **0.02** | C | C |
| 9 | **0.01** | 0.02 | 0.17 | C | C |
| 10 | 0.47 | 0.44 | **0.26** | E | C |
| All | 0.87 | **0.59** | 0.09 | 0.30 | **0.00** |

better than *base*1 on most questions. In the two-system comparison results, C denotes correct and E denotes error. We find that *fun*3 can distinguish two search engines accurately on all questions but *base*3 can do only 7 out of the 10 questions.

We further conduct a simulation study. The simulation experiment is deployed in the similar manner as that of Section 2.2. But the user behavior is synthesized from a click model instead of clicking all relevant documents. For one-system evaluation problem, we conduct the experiment for *base*1, *fun*1 and *fun*2. For two-system evaluation problem, besides *base*2 and *fun*3, we also test *fun*1 and *base*3.

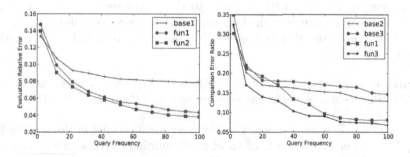

**Fig. 1.** One-system Evaluation (Left) and Two-system Comparison (Right)

In *base*3, it uses the balanced strategy[1] to merge the documents from the two ranked lists into the presented list. We present the one-system evaluation and two-system comparison results in Figure 1.

The experimental results suggest that at the first few query submissions, *base*1 and *base*3 are good choices for one-system evaluation and two-system comparison. When the query is submitted a little more, *fun*2 and *fun*3 are generally good choices for one-system evaluation and two-system comparison problems respectively.

# 4    Beyond Document Relevance

All methods use document relevance to evaluate the quality of information retrieval component. There are some research to investigate the correlation between the Cranfield paradigm experiment and user study. Some recent evaluation work [11,12] reported the positive correlation between the Cranfield evaluation based on document relevance and real user study. However, in the use of a real search engine, the effectiveness and the satisfaction of a user is also affected by some factors other than retrieval system performance. Our problem is: how to embed the quality of other components in search engine for the evaluation.

The document snippet is a very important feature for modern search engine. Good snippets can guide the users to find out the relative documents from the retrieved results, or even contain relative information itself. On the contrary, users may miss relevant documents or waste time clicking and examining irrelevant documents due to bad snippets. Turpin et al. [13] investigated this problem and showed that it makes difference by including snippet quality in search engine evaluation. In this paper, we interpret traditional IR metric *precision* as *effective time ratio* of the real user, i.e., the ratio between the time used in reading relevant information and the total search time, and extend it in the scenario when the search engines provide document snippets.

## 4.1    Effective Time Ratio

Intuitively, for a search engine with snippets, the users' satisfaction is affected by both retrieval system and snippet generation qualities. *Precision* is one of most important metrics in IR. In order to derive the metric that can be used to evaluate search engines with document snippet, we first interpret *precision* as *effective time ratio* (denoted as $ETR$) when using retrieval system without snippet presentation.

**Definition 1.** *(Effective Time Ratio): Effective Time Ratio is the ratio between effective time used in get relevant information and the total search time.*

We assume an IR system presents to the user a ranked list of documents for a query. We further assume the time spent on examining each document is identical (denoted as $T$). Thus a user needs $T \cdot N$ time to examine top $N$ documents, but the *effective time*, which is used to examine relevant documents, is $T \cdot \sum_{i=1}^{N} R(d_i)$, where $d_i$ is the $i$-th ranked document in the result list, and $R$ is a binary function, indicating relevance of a document. With very simple derivation, we can find that $ETR@N$ is identical to $P@N$, so we can interpret *precision* as *effective time ratio* when using the retrieval system without providing document snippets.

However, the modern search engines usually present the users a list of snippets instead of documents. In this scenario, a user examines $i$-th snippet, which is generated from $i$-th ranked document ($i$ is initially assigned 1). If he finds the snippet is relevant, he would click and examine the corresponding document; otherwise, he would examine the next snippet. After examining a document, the users may quit search or continue to examine $i + 1$-th snippet. As in most click models, this process uses examination hypothesis and cascade hypothesis.

According to the definition of *effective time ratio*, it is easy to include snippet quality. The effective time used in getting really relevant information include the time spent on reading the snippet and the original text of the relevant documents. The total time is composed by two parts: 1) the time spent on reading all snippets of the (top) retrieved documents, and 2) the time spent on reading the text of the clicked documents, whose snippets seem to be relevant. We further assume that the time spent on reading a snippet and a document are two static values($T_1$ and $T_2$). In the top $N$ documents, only the relevant documents with relevant snippets can provide relevant information, so the *effective time* is $\sum_{i=1}^{N} R(s_i) \cdot R(d_i) \cdot (T_1 + T_2)$(where $d_i$ the $i$-th ranked document, $s_i$ is its snippet, and $R$ is a function indicating whether a document or a snippet is relevant or not). Total time spent on reading snippets is $N \cdot T_1$. The user would read all documents with relevant snippet, so the total time spent on reading documents is $\sum_{i=1}^{N} T_2 \cdot R(s_i)$. Thus the *effective time ratio* can be formulated as Equation 6. We can rewrite it by the time rate between reading document and reading snippet ($c = T_2/T_1$) in Equation 7.

$$ETR@N = \frac{(T_1 + T_2) \cdot \sum_{i=1}^{N} R(d_i)R(s_i)}{T_1 \cdot N + T_2 \cdot \sum_{i=1}^{N} R(s_i)} \tag{6}$$

$$= \frac{(1 + c) \sum_{i=1}^{N} R(d_i)R(s_i)}{N + c \cdot \sum_{i=1}^{N} R(s_i)} \tag{7}$$

To further understand the effect of including snippet quality, we would compare versions of *effective time ratio* implementation with and without snippet quality factor. A snippet is good if it can indicates the document relevance accurately. Otherwise it may mislead users to miss a relevant document or waste time reading an irrelevant document. Here we define that a snippet is relevant iff it indicates the corresponding document is relevant. Therefore the quality of snippet can be qualified by possibility two types of errors.

**Definition 2.** *(First Type Error) Error of generating a relevant snippet for an irrelevant document*

**Definition 3.** *(Second Type Error) Error of generating an irrelevant snippet for a relevant document*

The first type of errors would lead the users to waste time clicking and examining irrelevant documents, and the users would miss relevant documents because of the second type of errors. Given a snippet generation algorithm, we define $p_1 = Pr(R(s) = 1|R(d) = 0)$ as the conditional possibility of making *first type error* and $p_2 = Pr(R(s) = 0|R(d) = 1)$ as the conditional possibility of making *second type error*. The ratio between expected expected effective time and expected total time (denoted as $EETR$) can be expressed as in Equation 8. Because $P@N$ is also an implementation of *effective time ratio* without considering snippet factor, this equation describes the relations of *effective time ratio* with and without including snippet quality.

$$EETR@N = \frac{(1+c)(1-p_2)}{c(1-p_1-p_2) + \dfrac{1+cp_1}{P@N}} \tag{8}$$

We can derive three properties of *expected effective time ratio* from this equation. Proposition 1 shows that a search engine with an error-free snippet generation algorithm has higher *expected effective time ratio* than a retrieval system without snippet.

**Proposition 1.** $p_1 = 0, p_2 = 0 \Rightarrow EETR@N > P@N$

Proposition 2 validates that the $EETR$ values can reflect the underlying retrieval performance. Similarly, proposition 3 validates that the $EETR$ values can reflect snippet generation quality.

**Proposition 2.** $p_1(A) = p_1(B), p_2(A) = p_2(B), P@N(A) > P@N(B) \Rightarrow$ $EETR@N(A) > EETR@N(B)$

**Proposition 3.** $p_1(A) > p_1(B), p_2(A) > p_2(B), P@N(A) = P@N(B) \Rightarrow$ $EETR@N(A) < EETR@N(B)$

## 4.2   Experiments and Results

To validate the *effective time ratio* metric, a user study is designed to resemble the common search engine usage. 10 college students are employed to collect information for 50 questions with the help of a search engine. The questions are uniformly distributed in open and close categories, topic categories and difficulty degrees. We collect 100 results for each question and present 10 on one page. Once ending a question, the user was asked to answer the question and to report his satisfaction, whose values ranging from 1 to 4 (the higher the better). The satisfaction values are then compared with scores of various IR metrics including the proposed *effective time ratio*.

The first group of IR metrics use the document relevance only, including P@N, DCG, RR and cumulated precision. The reason for using unnormalized version of the metrics is that the total number of relevant documents is unknown. The second group of metrics have the same forms as those in the first group, but one document is considered to be relevant iff the document and its snippet are both relevant [13]. The third groups of metrics are *effective time ratio* and its extensions. As extending precision to cumulated precision, we can also define the *cumulated effective time ratio* as the sum of *effective time ratio* at the cutoffs where both the document and the snippet are relevant.

There are two parameters in the metric *effective time ratio*: cutoff $N$ and document/snippet reading time rate $c = T_2/T_1$. For $N$, we can tune it in the experiment. For $c$, we can estimate it from the user study log and a one-month commercial search engine log. The estimated $c$ value is 8.25 for the former log and is 10.36 for the later one. We also find that most $c$ values are near 10, so we use 10 as $c$ value in ETR.

A good evaluation metric is supposed to reflect the users' satisfaction in using a search engine to find out relevant information for a need. In this paper, we follow Huffman and Hochster's work [12] to use correlation between metric score and user reported satisfaction. If the correlation is larger, the metric can reflect the users' satisfaction better.

**Table 5.** Correlations for Metrics

| D | RR | P3 | P5 | P10 | DCG3 | DCG5 | DCG10 | CP3 | CP5 | CP10 |
|---|---|---|---|---|---|---|---|---|---|---|
| **Rel** | **0.497** | 0.396 | 0.407 | 0.286 | 0.412 | 0.431 | 0.366 | 0.330 | 0.345 | 0.286 |
| **S-D** | RR | P3 | P5 | P10 | DCG3 | DCG5 | DCG10 | CP3 | CP5 | CP10 |
| **Rel** | **0.467** | 0.344 | 0.343 | 0.253 | 0.366 | 0.375 | 0.320 | 0.283 | 0.287 | 0.222 |
| **ETR** | ETR3 | ETR5 | ETR10 | ETR20 | CETR3 | CETR5 | CETR10 | CETR20 | | |
| | 0.469 | **0.537** | 0.513 | 0.383 | 0.312 | 0.314 | 0.239 | 0.122 | | |

Table 5 presents the correlation results for three groups of metrics. The highest score in each group is in bold. It shows that the *effective time ratio* has overall highest correlation with the users' satisfaction, and *RR* also has relative high correlation. Surprisingly, though *average precision* is the most commonly used metric in IR, *cumulated precision* and *cumulated effective time ratio* work not so well when compared with the real users' satisfaction. Another finding is that metrics at cutoff 5 can reflect users' satisfaction better. It may be because the user can see about 5 snippets without scrolling the mouse at the search engine result page.

## 5   Conclusion

In this paper, we introduce two categories of methods for evaluating search engines based on clickthrough data. Both methods model the noise in click data. The interleaving and comparing method simply combines results from two search engine and counts the clicks for system comparison. We show that the comparison can be more accurate by considering rank information. On the other hand, the click model based methods formulate the user behavior as a graphical click model and can get both value and confidence of document relevance estimation. Thus we can get reliable evaluation results of search engines. Besides, it reminds us to develop an algorithm to reorder the retrieved result to collect relevance information more efficiently. Moreover, we observe that the document relevance alone is handicapped for search engine evaluation, and we propose a new metric called *effective time ratio*. We show that this metric can reflect the users' utility better than the existing metrics employing document relevance only.

## References

1. Joachims, T.: Unbiased evaluation of retrieval quality using clickthrough data. In: SIGIR Workshop on Mathematical/Formal Methods in Information Retrieval (2002)

2. Radlinski, F., Kurup, M., Joachims, T.: How does clickthrough data reflect retrieval quality? In: CIKM 2008: Proceeding of the 17th ACM Conference on Information and Knowledge Mining, pp. 43–52. ACM, New York (2008)
3. Joachims, T., Granka, L., Pan, B., Hembrooke, H., Radlinski, F., Gay, G.: Evaluating the accuracy of implicit feedback from clicks and query reformulations in web search. ACM Transactions on Information Systems (TOIS) 25 (2007)
4. Carterette, B., Bennett, P.N., Chickering, D.M., Dumais, S.T.: Here or there. In: Macdonald, C., Ounis, I., Plachouras, V., Ruthven, I., White, R.W. (eds.) ECIR 2008. LNCS, vol. 4956, pp. 16–27. Springer, Heidelberg (2008)
5. Guo, F., Liu, C., Kannan, A., Minka, T., Taylor, M., Wang, Y.M., Faloutsos, C.: Click chain model in web search. In: WWW 2009: Proceedings of the 18th International Conference on World Wide Web, pp. 11–20. ACM, New York (2009)
6. Chapelle, O., Zhang, Y.: A dynamic bayesian network click model for web search ranking. In: WWW 2009: Proceedings of the 18th International Conference on World Wide Web, pp. 1–10. ACM, New York (2009)
7. Dupret, G.E., Piwowarski, B.: A user browsing model to predict search engine click data from past observations. In: SIGIR 2008: Proceedings of the 31st Annual International ACM SIGIR Conference on Research and Development in Information Retrieval, pp. 331–338. ACM, New York (2008)
8. Guo, F., Liu, C., Wang, Y.M.: Efficient multiple-click models in web search. In: Proceedings of the Second International Conference on Web Search and Web Data Mining, WSDM 2009, Barcelona, Spain, February 9-11, pp. 124–131 (2009)
9. Richardson, M., Dominowska, E., Ragno, R.: Predicting clicks: estimating the click-through rate for new ads. In: WWW 2007: Proceedings of the 16th International Conference on World Wide Web, pp. 521–530. ACM, New York (2007)
10. Craswell, N., Zoeter, O., Taylor, M., Ramsey, B.: An experimental comparison of click position-bias models. In: WSDM 2008: Proceedings of the International Conference on Web Search and Web Data Mining, pp. 87–94. ACM, New York (2008)
11. Allan, J., Carterette, B., Lewis, J.: When will information retrieval be "good enough"? In: SIGIR 2005: Proceedings of the 28th Annual International ACM SIGIR Conference on Research and Development in Information Retrieval, pp. 433–440. ACM, New York (2005)
12. Huffman, S.B., Hochster, M.: How well does result relevance predict session satisfaction? In: SIGIR 2007: Proceedings of the 30th Annual International ACM SIGIR Conference on Research and Development in Information Retrieval, pp. 567–574. ACM, New York (2007)
13. Turpin, A., Scholer, F., Jarvelin, K., Wu, M., Culpepper, J.S.: Including summaries in system evaluation. In: SIGIR 2009: Proceedings of the 32nd International ACM SIGIR Conference on Research and Development in Information Retrieval, pp. 508–515. ACM, New York (2009)

# Abstract: Semantic Technology at The New York Times: Lessons Learned and Future Directions

Evan Sandhaus

New York Times, US

At last year's International Semantic Web Conference, The New York Times Company announced the release of our Linked Open Data Platform available at http://data.nytimes.com. In the subsequent year, we have continued our efforts in this space and learned many valuable lessons. In our remarks, we will review these lessons; demonstrate innovative prototypes built on our linked data; explore the future of RDF and RDFa in the News Industry and announce an exciting new milestone in our Linked Data efforts.

P.F. Patel-Schneider et al.(Eds.): ISWC 2010, Part II, LNCS 6497, p. 355, 2010.

# What Does It Look Like, Really? Imagining How Citizens Might Effectively, Usefully and Easily Find, Explore, Query and Re-present Open/Linked Data

mc schraefel

IAM Group, Electronics and Computer Science
University of Southampton
Southampton, UK, SO17 1BJ
mc+iswc@ecs.soton.ac.uk

**Abstract.** Are we in the semantic web/linked data community effectively attempting to make possible a new literacy - one of data rather than document analysis? By opening up data beyond the now familiar hand crafted Web 2 mash up of data about X plus geography, what are we trying to do, really? Is the goal at least in part to enable net citizens rather than only geeks the ability to pick up, explore, blend, interogate and represent data sources so that we may draw our own statistically informed conclusions about information, and thereby build new knowledge in ways not readily possible before without access to these data seas? If we want citizens rather than just scientists or statisticians or journalists for that matter to be able to pour over data and ask statistically sophisticated questions of comparison and contrast betewen times, places and people, does that mission re-order our research priorities at all? If the goal is to enpower citizens to be able to make use of data, what do we need to make this vision real beyond attending to Tim Berners-Lee's call to "free your data"? The purpose of this talk therefore will be to look at key ineraction issues around defining and delivering a useful, usable *data explorotron* for citizens. In particular, we'll consider who is a "citizen user" and what access to and tools for linked data sense making means in this case. From that perspective, we'll consider research issues around discovery, exploration, interrogation and representation of data for not only a single wild data source but especially for multiple wild heterogeneous data sources. I hope this talk may help frame some stepping stones towards useful and usable interaction with linked data, and look forward to input from the community to refine such a new literacy agenda further.

**Keywords:** interaction, design, user experience, linked data.

## 1 Introduction

What does interacting with the Semantic Web or Linked Data actually look like? And if we understand that, what are the challenges in making those interactions possible? And for whom do we imagine we design these interactions to support? Whose

P.F. Patel-Schneider et al.(Eds.): ISWC 2010, Part II, LNCS 6497, pp. 356–369, 2010.

problems do we solve with any of the tools we create? When are these imagined users of our work actually real people? How do we know?

Some of my colleagues and I have been thinking about these semantic web and user interaction questions since the first eponymous meeting at the WWW 2004 conference launching the First International Semantic Web User Interaction (SWUI) Workshop. The series kicked off with a memorable head to head session between Jim Hendler and Ben Shneiderman. Violent agreement rocked the sessions.

Since then, we have had a workshop on the themes of identifying interaction challenges in a semantic web context annually at venues from CHI, the ACM's annual Conference on Human Factors, to a virtual workshop between MIT and Zurich, and more frequently, here at ISWC.

One of the high notes of this annual series was the 2006 SWUI in Atlanta where Tim Berners-Lee revealed during his talk the revised Semantic Web Layer Cake that included, at last, a user interaction layer on top no less.

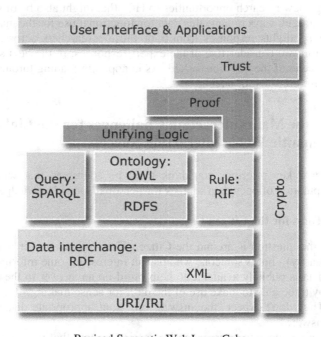

Revised Semantic Web Layer Cake

6 years on from that first outing, I've been asked to give an invited talk about Human Computer Interaction here at ISWC, for which I am honored. We still struggle to get good HCI oriented papers submitted to ISWC. Some of us who do submit them still struggle to get our them accepted at ISWC.

Indeed, I'd like to take this opportunity to acknowledge some of our colleagues who have been an ongoing steering committee for SWUI: Duane Degler, Lloyd Rutledge, Avi Bernstein, David Karger, Jennifer Goldbeck. I'd also like to acknowledge the Web Science Foundation as a constant sponsor for the workshops we've held, more often than not, the founder of the post workshop feast (or at least coffee and nibbles during sessions). In particular, Wendy Hall, Nigel Shadbolt and Tim Berners-Lee.

While I am delighted to have the opportunity to talk for half an hour or more on my favorite concerns about the intersections of interaction and linked data research, I would rather take this space to give some time to other researchers in the field whose thinking in this space is already well grounded in practice. Indeed, David Karger and I recently enjoined HCI and Human Computer Information Retrieval (HCIR) researchers who deal with large data sets of various domains to help us frame an agenda to excite other HCI/HCIR researchers to consider the opportunities for new research in this area of massive open data.

These conversations have encouraged me to reach out to these experts again, specifically to hear their formative thoughts on what they see as key challenges to make open data/linked data to in particular useful and usable by regular citizens. In particular we asked that they consider what may be new or special about this kind of data that brings new research opportunities to HCI that might also be of interest to the researchers who seek ways to tame this data for functional use for the machine.

This paper spotlights responses from 5 of these researchers across industry and academia. Before we consider five of the expert responses, in the next section, allow me to set the scene of the questions asked. As is apparent reading through the replies, a few themes for consideration recur.

## 2 Eliciting the Main Interaction Challenges for the Linked Data/Semantic Web's Interaction Success

What are 1 or 2 key priorities you think must be addressed that will aid citizen-focused manipulation of open data sources for personal/social knowledge building?

### 2.1 Focus: Tools for the Citizen User

The focus of the question is around the Citizen User: a citizen user is not a domain expert (necessarily) - but is someone who has an interest in some information, and the (structured) data is publicly available to help build up an answer to the question, and they are happy to be able to make use of the data for sense making - for building new knowledge. They don't expect "the answer" but want appropriate data to build up a sense of an answer.

So, we are not expecting to create an interaction system that provides *The Answer*, but rather facilitates:

- discovery,
- interrogation,
- manipulation,
- annotation,
- representation of heterogeneous open data sources.

What, therefore are key challenges that in your view we MUST address/prioritize to support citizen based exploration of the freed data of sites like data.gov, data.gov.uk and related sources?

**Example Scenarios:**

> o *Where Should i Live?*
> - where data exists on pollution, hospital waiting times, transportation, political representation in a region, crime stats,
> o *for whom should i vote?*
> - where there may be data on a party's voting record and individual members' voting records, regional crime stats, etc
> o *Is this a good school?*
> - where data may come from league tables, student reports on their views of instructors from all over, house prices, grocery locations, transport
> o *What about drug reactions?*
> what other drugs have people taken with my condition, and what's been the success rate whether self-reported or by other measures?

To make such citizen-based exploration possible, what should be our research agenda? Our concerns from the back end to serve the front end? How do we move from the current high geek expert tools to citizen tools? How do we help people used to thinking about issues with the data to think about issues *for the person using the data* as the best path into solving problems of serving the data. Not all of you may think that that IS the best way, which is fine. Alternatives requested, too.

Some examples of issues we encounter regularly within data:

- geographical boundaries in different data sets don't match up  (hospital trusts don't map to crime regions)
- not always clear what information in the data is - meaningless labels
- data is incomplete or messy

## 2.2 Thesis and Background

My rationale in posing these particular questions is the following: with the emphasis on "freeing data" it seems we are *de facto* potentially establishing or requiring a new literacy - a literacy about data rather than documents; that we've moved from the page if you will to the cell. And that requires new kinds of knowledge - what to do with the data.

In the pre-printing press era literacy was the purview of a select few - the religious cast - who had access to manuscripts. With the press (and the middle class) literacy of documents becomes more wide spread.  Is the era of linked data going to be the same now, where data and what to do with it has been the purview of statisticians or those trained in statistics - have access to the data, and produce the results for the rest of us?

If the goal is to believe that access to data is a Public Good, what does that mean for interaction? For a basic data literacy? Does this understanding of data in the 21$^{st}$ Century start with mash ups for all? Where do we as technologists / researchers /designers begin? Similarly, in order to apply that knowledge of data manipulations to the Interface, we also need services to enable normal web-literate citizens to engage the data - find it, explore it, manipulate it, and re-present it where that "it" may be sourced from many heterogeneous sources.

### 2.3  Audience: At the Coal Face of Digging the Semantic Web

Most of the researchers in the Semantic Web community work in terms of dealing with representing the data for the machine efficiently rather than thinking primarily about people accessing and manipulating the data directly.

The goal of these interviews is to help people working on the problems of machines processing data to find it meaningful to connect potentially instead with what citizens need to be able to do with the data, where those citizens are not any more geeks than are the current users of the web.

## 3  Expert Responses

The following responses are direct reports in their own words of responses to the above framing.

### 3.1  Daniel Tunkelang, Technical Lead, Google

Here's some of my admittedly US-centric thinking about patent and census data:

**Patents.** Despite the availability of patent data through public (e.g., USPTO, WIPO) and private (e.g. Google) repositories and the regular appearance of patents in the news, the average citizen (at least in the US) seems to have little ability to either understand or influence how patents work. Some things that we could do to make this data more accessible:

**Exposing the links between related resources** (e.g., patent applications and prosecution histories). Even people familiar with patents may not be aware of resources like PAIR, the USPTO portal that offers the full history of a patent or pending patent application. And the interfaces make it inefficient and painful to navigate among resources.

**Relationships between patents and entities.** People invent patents; patents are assigned to companies; people work for companies; companies acquire other companies or their assets.

**Connections between similar patents.** Even the simple classification system used by the US patent system is not well exposed in interfaces. But I'm thinking of far more than that: connecting patents using link analysis of the citation and entity graphs and computing content-based similarity using information extraction.

Patent law itself is pretty complex, and there's more required here than exposing the raw data in a nicer interface. For example, technical terms should be linked to glossary entries where possible. Links to non-patent-art should also be connected to published documents where possible. And ultimately the value of all of these efforts would require policy changes that would make it easier for citizens to participate. But there's a chicken-and-egg problem: today's citizens are ill-equipped to participate, so there is little motivation for policy change.

**Census Data.** It should be straightforward for the average citizen to access public demographic information, whether at a national level, a neighborhood level, or anything in between. But I'm not aware of any interface that makes it easy to do so. The best tools are designed for professionals who invest time in developing reusable queries for generating reports. But it's not just that the tools are complicated.

There's a vocabulary problem--Census data is classified using codes that are familiar for government agencies but not necessarily for citizen users.

Another challenge is that data is collected a varying geographical and temporal granularities, so users need to be able to explore to discover the data that best matches their information needs (i.e., they might not find it at precisely the granularity they had in mind).

### 3.2 David Huynh, Research Scientist, Google

I don't think I can tell you the *main* interaction challenges, because every interaction challenge seems roughly equally important. But anyway, here are a few thoughts.

**(1) URIs are for machines to unambiguously identify entities to operate on, but URIs are practically useless for humans to perform the same task.** For humans, images and identifying details (race, gender, birth year, profession for a person; industry, size, location for a company; etc.) are what help us unambiguously identify entities. Sure, there's clearly a realization that raw URIs shouldn't be shown to users, but there is not yet a realization that we need something else to do for humans what URIs do for machines--unambiguously identifying entities. The Freebase Suggest (search) widget provides an example of how to do this:

Freebase Interaction image

Note that each suggested entity is labeled ("Actor", "US President", etc.) and described with brief but identifying details on hover. Semantic UIs should strive to communicate to their users that things on the screen are representations of entities rather than mere text, and help users unambiguously identify entities.

The iPhone interface has nuances that make it so pleasant and fun to use, nuances that somehow let the user "feel" the interface. In the same way, semantic UIs must somehow get users to "feel" the semantic entities. After using my first iPhone for a few months, one day while reading a paperback book, upon reaching the end of a page, I instinctively placed my thumb at the bottom of the page and pushed it upward, only to realize that it's not an iPhone. That's what we want here with semantic UIs. After using semantic UIs for a while, when a user sees a plain piece of text like "ford", the reflex should be, "it's ambiguous / raw / bare".

And yes, it's about details, details, details.

**(2) Direct manipulation techniques for en-masse data editing seem to be a game changer for a class of users--folks who can handle Excel but are not familiar with scripting or don't have time and patience for scripting.** My work on Gridworks has already started to address some of the design challenges.

Just because data is open doesn't mean it's clean or it's formatted in a way that you can use. This inconvenience seems to be swept under the rug sometimes. The semantic web community has focused so much on semantics that perhaps not enough effort has been spared for addressing syntax. But obviously, without syntax, there is no semantic.

It's also quite important to make sure that these tools are generic, rather than RDF-specific. Our goal here isn't to shoehorn everything into RDF. The goal is to increase awareness, desire, and demand for structured data, potentially linked. Let each user decide which format might serve their own purpose at this time. As people use these tools more and more, gradually, structured data, linked, will become natural to them.

**(3) Entity Reconciliation.** One surprising thing I've learned from Gridworks is how ready people are to want entity reconciliation. And not just with Freebase but with their own databases. This persuaded me to generalize the reconciliation support in Gridworks, as per the Reconciliation Service API (http://code.google.com/p/freebase-gridworks/wiki/ReconciliationServiceApi)

I could almost claim that, by making reconciliation easy, Gridworks makes it obvious why one would want reconciliation. It's almost as if the tool makes people think in a certain way.

So, don't start with "RDF". You would have already lost. Start with what users ask for. Then if possible, let the tools nudge their thinking toward RDF or whatever that's ideal in the long term.

**(4) Help build the upcoming structured data web.** Or do research on the semantic web. Pick one (since you can't do both).

### 3.3   Ed Chi, Principal Scientist, Augmented Social Cognition, Xerox Parc

The issues you raised was precisely the inspiration for my Ph.D. Thesis work on creating a visualization spreadsheet. The idea was that if people can easily use spreadsheets, then they ought to be able to take that model further and start creating visualizations using them, and the thesis was an exploration to find out how to design such systems. I think of ManyEyes, and Jeff Heer's later works to be in the same direction.

We have since learned a lot about user-contributed content on systems like Wikipedia, Delicious, Twitter, and they show a very interesting participation architecture that consists of readers, contributors, and leaders. Not all users want to be leaders, and not all users want to contribute. We have sometimes use the derogatory term of "lurkers" to describe "readers", which I think is a bit unfair. Ronald Burt's work have shown that a lot of us would like to be brokers of information among social groups, but there are also need for an audience, or followers, who might become brokers later, but not everyone all at once.

I believe that data manipulation of open data sources to follow the same curve. Yes, some cancer patients will want to read all they can about their condition, and do the analytical work, and others (not necessarily because of tool limitations) would prefer to take a backseat, and let others curate the information for them. What's interesting is that they might want very simple interactions that enable for basic sorting of data, or maybe even services that interpret the data for them (e.g. doctor experts), but they would prefer someone else does the bulk of the work (even if it becomes very easy due to tool development).

Consider a typical usage scenario: I am reading several medical journal articles. Data all in tables, in PDF format. need to extract the data from the tables and plot them. Ahem! Good luck. Let's go and type them all in by hand.

So, given that, what can we do?

First, it's quite clear that much of the hard work remains in data import and cleaning. To democratize data analytics and manipulation, the bulk of the difficulty is dealing with data acquisition. Unfortunately, most of this is engineering and not sexy research, so there aren't really innovative work in this area.

By not exciting I mean, I don't know of a single tool that enables me to grab tables out of PDFs. Worse still, if I have browse around on the net, and I find the data I need in web pages, often they're in HTML tables that are very hard to cut and paste into my excel spreadsheet. What tool is really out there for my information extraction tasks? Tables are just one example. Other problems include things that are locked in databases, but barely visible to end-users:

- say I want to analyze all of the flights from US to Europe over the last month. How do I get the data? Do I perform lots of searches on travel websites to extract that? Do I go to airlines one by one and examine their schedule (in PDF or HTML format), and get the data that way?
- say I want to plot the price of harddrives by dollar per MB in the last decade. Again, where do I get the data? How do I clean it, so that I can plot them?

Some information extraction (AI-style algorithms, and some machine learning techniques) are making some inroad in this area. Some recent work on entity extraction with human in the loop seems pretty good. So if I have a document, and want to find all interesting entities in them, and make a cross-index of related entities, there now seems to be some good research tools that do that. I also believe that mixed-initiative research for data import is sorely needed. We're doing a bit of this work in my lab at the moment. That is: human in the loop. The machine does some extraction, then human says, ah, that's not quite right, fix it this way, and machine do more, and then human fix again.

Second, there is the issue of data literacy. What kind of visualization works with what kind of data? What analytic technique is appropriate? Early work by Jock Mackinlay pointed to the possibility of automating some of these design choices, and we haven't made a huge amount of progress in this area.

Some tree Viz seems pretty automate-able, as are stacked graphs, population analyses, tables. I tend to favor simple visualizations that are understandable to lay people these days. Visual literacy is a huge problem that will take decades to overcome, so I favor simple vis these days.

Wizards, try-visualization-refine loops have all been tried in research. We need to stop inventing new visualizations, but actual usable tools for people here. By going to vertical domains, we will learn how to solve this problem. We need curriculum in visuzliation that is part of basic education.

That said, Vis researchers need to work on real scenarios more often. Go into medicine, and you see a lot of data analytics problems that are huge, and often not about visualizing generic trees. Often, it's visualizing protein interaction networks, or seeing evolving relationships. Go into another field, say, transportation, and you realize you need to combine infovis with geo-viz. It's often not about new visualizations, but about how to put vis components together.

### 3.4  Lloyd Rutledge, Computer Science, Open University, The Netherlands

In summary: less emphasis on grand new interfaces. More on familiar interfaces, but under user control and independent from the data. The user doesn't notice anything, thus no new literacy. The user only stops noticing that information access doesn't work the way it obviously should. Can we thus take large-scale data from multiple civil sources and have users access it in a way so unified and quick that there is nothing remarkable about it (finally!)? A "new literacy"? Computers as devices require(d) a new literacy. The Web didn't: users of the Web feel it acts they way they always knew it should (even though they actually couldn't imagine it beforehand). We do new things with the Web in new ways, but they feel familiar once you start.

To me, it seems that the same will be true of the end-user front-end applications of the type of use of the Semantic Web for which we dream and strive. The users will not notice the difference. They won't really notice anything. What will happen is:

- They want some data and they ask for it in a reasonable commonsense way, probably in ways they already (think to) ask for information
- Appropriate and correct data comes back.
- It comes back in the form of a presentation that makes perfect, common sense. The form of presentation itself is not remarkable apart from the data it presents.

This all happens with interfaces users have long been familiar with. The end user won't notice. At least on the per interaction basis. Perhaps over a longer period the user will day "It seems getting information used to be buggier and clunkier". The end users have no new literacy to learn. Their current literacy just works better.

This was just about data access. Could we argue the same way for data input and sharing? I think so. Users add data using means they already know. They and other users get this data back in ways that make sense. This data gets combined with other data, but that makes sense, of course? Only Semantic Web researchers know how remarkable this last step is. No one else will notice a thing. Researchers strive for a grand new SW interface that will put SW in the mainstream like a magic bullet application. But why make a new interface "paradigm"? A more appropriate challenge is to get tried-and-true interfaces to work the way they should with data placed on and accessed from the Semantic web the way it should.

There is no new literacy. There is new technology, good practice and science on the side of Semantic Web developers to make existing literacy work with data that is the way it should be.

*One problem is that familiar interfaces are often controlled by non-user parties who also own and isolate the data.* Challenges are thus having user control the form of interface and unrestricted access to public data, and have these two control issues be separate from each other. And to have the users not notice anything: they just have their interface, and they just ask for and browse data. And they just simply get it. No barriers based on who is providing the interface or who is providing the data or if one needs to be linked on the other.

A *second challenge*: allowing seamless combination of public, institutional and private data, all in the same interface, but with the corresponding security and data sharing/blocking. An example is combining civil databases on medications and medical services with your own medical records.

I [don't] mean to poopoo the work of [previous SWUI presented work]. But I think we need to encourage other challenges now. For one, there are too many submissions for new types of interfaces to the Semantic Web that aren't research and don't work, as we've seen in various journal submissions.

Making a new interface is often an attractive project for programmers, but most fall short of burden of proof, and there are only so many new interfaces possible. Not only unproven, many proposed new interfaces just don't work (such as big fat graphs).

But even the successful new interfaces aren't that new and aren't necessarily attached to the Semantic Web. All the new SW interfaces work with any amount of data of any origin. Their newness is more about what computers make possible for data access than was the SW makes possible.

These successful new SWUI's also don't require new literacy. They are natural extensions of familiar interfaces, some of which go back to paper. Like hypertext, they have a "retrospective obviousness", despite being hard to imagine beforehand (and hard to develop the first time). Thus no new literacy needed. SWUI that do require new literacy, like large RDF graphs (BFD's) and queries, even assisted, and even visually assisted (mostly), tend IMHO not to catch on.

So what is new is the type and scale of data that gets to these interfaces. In the civil service data example, what we need is multiple civil service branches to have their data in familiar, even in 21st century SWUI's like facet browsers and autocompletion, but seamlessly. When another institute doesn't have their data on the SW/LW, end users should find it strange. Not "Why do they have any RDF files?" but "Why can't I get at this information? Why do I have to use their website to see it? Why do I have to jump back and forth from their website to my (semantic, but they don't know it) browser?"

### 3.5 Abraham Bernstein, Dynamic and Distributed Information Systems Group Univesrity of Zurich

There is plenty of data out there, but as you point out the linking is abysmal. I am not sure where I read it but even the connections between the LOD datasets is only very brittle ... I am not even talking of data repositories such as data.gov or even worse department of statistic excel sheets from different geographical regions etc...

So, the single most important question is how to integrate a multitude of sources. *ASIDE:* Yes, it is true, we are very far away from actually understanding what the "best" way to interact with linked data is (assuming there is such a thing) and approaches such as faceted browsing, David and your stuff, NLP, etc. are a only a first step. Lost of work needed here - mostly of a good UI nature. The crux of the Semantic Web is that it adds heterogeneous (even previously unknown) data sources to the mix. Most of the UI approaches so far assume that the data already has been integrated "nicely" into one data-set. Exhibit, e.g., is great, but the most difficult work has already been done: the data integration. So if we really ask ourselves what the Semantic Web brings new to the picture in contrast to "just" interacting with Graph-based data then it is the data-integration problem.

So if we want to bring the Semantic Web to fruition we need to think how we can help our citizen user to combine heterogeneous data sources. My hunch is that it will need a combination of (possibly novel) UI metaphors, a sprinkle of good AI, some social computation, good software engineering. How can I substantiate this hunch?

- I think the first point is clear: We need to find out what UI metaphor is best used to integrate information. Personally, I have no clue if anybody has systematically explored citizen user data integration. I am aware of many projects doing it for pros, but not a lot of work on casual users.
- A sprinkle of AI is needed, as I believe that mixed-initiative might help to ease the bruden of data integration. To that end some statistical processing (e.g., for finding candidates for joins), maybe some rules (e.g., to encode otherwise collected background knowledge), and guided interaction (e.g., using planing techniques) might be helpful.
- social computation will probably help the enterprise by enabling the exchange of integration recipes.
- Good software engineering is needed to build some robust prototypes to test these ideas.

So finding the right interaction metaphor for integrating data seems to be the single, biggest challenge.

### 3.6 Others in the Discourse

The above commentaries represent specific contributions requested for this presentation of voices. In related conversations, a few more relevant points emerged that are germane to this discussion. Ben Shneiderman, Computer Science, University of Maryland, maps the process articulated above of discovery, exploration, interrogation, and re-presentation with parallel discussion going on in the visual analytics (VA) world, where a similar process (discovery, exploration, interrogation, presentation) is central. He recommends Thomas & Cook's online book "Illuminating the Path" (nvac.pnl.gov). There is also a 16-step process model in the *Readings in Information Visualization* (1999) that is also useful for construing stages within data engagment to be mapped. Since then, he notes, a variety of process models (e.g. Systematic Yet Flexible) have been described, tied to different data types. Shneiderman continues,

> The current term for this [data processing for sense making] in the VA world is "data wrangling" to describe the rough & tumble effort to get, clean, merge, filter, convert, extract, present, and share. Also part of this process will be discovering what is missing in the data or when the meaning has changed for an attribute or attribute value. In many cases, natural language processing methods are needed to clean messy text data, network analysis helps (as in DDupe for entity resolution http://www.cs.umd.edu/projects/linqs/ddupe/), and increasingly Mechanical Turk workers are being engaged.

Steve Drucker of Microsoft Research poses the question about how to sustain general UI's versus application-specific approaches in this citizen-user context: can general or data agnostic interfaces be adapted to specific conditions? By "condition" we might consider different capabilities or specific devices. "What would be the logical workflow to enable this to be a convenient and compelling usage condition?" An example case may be adapting Huynh's Exhibit to work within Excel.

Natasha Noy of Stanford University's Biomedical Information Research queries the representation challenges of making visible distinctions that may need to be made about data access in terms of provenance and trust: how does exposure of provenance and trust get represented across these borders? For example there is data that may be linked but which most citizens would not get near: health record data, even, intriguingly, if anonymized may remain protected. Thus there may be data known to exist, but not accessible. How might these cases be incorporated into tools that would expose sources for possible querying?

# 4 If You Build It...

Several strands emerge across these responses to the challenge of what is the interaction for the data web to be like?

David Huynh is keen to foreground entitities free from URLS, as URI's are codes for machines, not concepts meaningful for people. Likewise under the for people heading, making practices easy for people to perform makes new processes' value almost "obvious." Make alignment operations easy fast and intuitive, Huyhn argues; use familiar interaction approaches like direct manipulation, and people want it.

Lloyd Rutledge also talks about naturalizing what seems to be new now – having raw data sources from organizations available – into practices that simply illuminate a gap if they are not there. Like Huyhn, Rutledge suggests that the machine readable remain machine readable: one should not wonder "where's the RDF" – but "where's the information I can use for this problem." While Huyhn evolves facetted browsing into more spaces, Rutledge plays down the need for new interaction or new interaction paradigms. For Rutledge, simply getting the anticipated right data back from an interaction is a big win. He postulates this experience not as a "new literacy" but as the "current literacy just working better."

Ed Chi turns the focus away from manipulating extant data to supporting capture of new data, whether personal or public. He likewise owns that not all citizens will want to do the raw data manipulation work anymore than all readers of Wikipedia contribute content to it. Hence services that enable data basics like simple sorting may be invaluable for light touch exploration. Sometimes, just getting data out of a fixed source may be the win: simple tools to remove and convert tabular data in PDF's to new metadata encoded data may also be a boon for personal use, and the ability to share/contribute new open data or raw data or linkable data – rich data – to the world quickly. This table cutter may be applied manually, but Chi also makes the case for more AI type scrapers to go and make the data that has not yet been formally freed. Then being able to wrap visualizers around the data semi-automatically at least via wizards may also help make the information accessible now as information, not just data.

Daniel Tunkelang laments the fact that in his experience there may be copious amounts of data already freed, such as patent or census data, but that its availability offers little or no opportunity for the citizen either to explore it or influence the process. Tunkelang wants obvious relationships between data better excavated, exposed and presented, such as patents and prosecution histories. Services already exist for histories of patents; these have yet to be linked to the patents themselves. Perhaps even more relevant to the searcher, similar patents are not obvious. Prior art could also be linked automatically.

Finally Abraham Bernstein echoes Tunkelang's and Chi's sentiments to say there is already a lot of data available that screams out for almost native linking of well expressed interactions (semantic zooming, eg nation to city to neighborhood to street) to AI to blend with mixed initiative to both find associated data (patents to their prior art components or drawings), to human computing/mixed initiatives to help enrich data where there are gaps.

Drucker wishes to see the use case that will show how general UIs may help work in specific contexts and Shneiderman shows where there are existing paradigms in visual analytics that may be useful to frame the practices to be represented (discovery, exploration, interrogation, re-presentation), and Noy suggests we consider representation issues for boundary conditions of the fully open to the partially exposed.

Intriguingly, there are few examples here of particular interaction designs. The closest we get is Huynh's approach to facets. Another opportunity for research in the SW/UI space may be to taxonomize the approaches that may be useful for the types of exchanges rich meta/data affords beyond facet browsing. What, as well, do mixed explorations look like that blend documents and data?

## 5 Concludium

The above exchanges are background or subtext to the formal conference presentation to be presented at ISWC 2010, and I heartily thank the participants who agreed to share their voices in this context.

From these, readers can see that those of us who are investigating how to support rich data sources exploration are intrigued by a variety of different properties in the space. All of us, however, seem to come from a core starting point: what are desirable and sensible processes for people? If one is looking at patents, what data is associated with patents and why not bring those sources together? Similarly, if one is looking at a census, why not make it possible easily to add new annotations to that data or connect related sources or represent relationships? If there is a hole in a data set, similarly why not find ways to automate citizen-directed scraping to enrich such sparse data sets?

Most of these questions have been acknowledged at least in conversations within the semantic web community, and many of them predate the semantic web, going back to hypertext. So we may wish to ask ourselves: where are the great semantic web applications that are meaningful to citizen users by doing these apparently simple, obvious, things?

This is the 9th Semantic Web conference. If we do not have these kinds of apparently simple and sensible interactions by now, is it time for us to look at our program and ask if there's something we should be doing differently? And if not why not? And if not, how else do we get to a useful and usable semantic web of linked data for citizen users? Which of these challenges will we own to say that by the tenth anniversary conference, we will prove not that we can manage a even more triples in less time, but that we can delight a citizen by solving one of her data-related problems?

May year ten be the year of the Semantic Web Citizen (?).

# Author Index